Hugh Robert Mill

The Realm of Nature : an outline of Physiography

Hugh Robert Mill

The Realm of Nature : an outline of Physiography

ISBN/EAN: 9783337024987

Printed in Europe, USA, Canada, Australia, Japan

Cover: Foto ©berggeist007 / pixelio.de

More available books at **www.hansebooks.com**

THE
REALM OF NATURE

AN OUTLINE OF PHYSIOGRAPHY

BY

HUGH ROBERT MILL

D. Sc. Edin.

FELLOW OF THE ROYAL SOCIETY OF EDINBURGH; OXFORD UNIVERSITY
EXTENSION LECTURER

WITH 19 COLOURED MAPS AND 68 ILLUSTRATIONS

LONDON
JOHN MURRAY, ALBEMARLE STREET
1892

AUTHOR'S PREFACE

It is the aim of this volume to illustrate the principles of science by applying them to the world we live in, and to explain the methods by which our knowledge of Nature has been acquired and is being daily enlarged. An attempt is made to define the place of physical science in the sphere of human knowledge, and to show the interrelations of the various special sciences. The greater part of the book is occupied by an outline of the more important facts regarding the structure of the Universe, the form, material, and processes of the Earth, and the relations which they bear to Life in its varied phases. Such descriptions must necessarily be brief, and they are consequently apt to appear more dogmatic than the discoveries of science warrant; but care has been taken to minimise this result. References to original memoirs are given in cases where the facts or theories are not yet fully accepted, and the student is urged whenever it is practicable to read and study these works.

The Fahrenheit scale of temperature and the British system of weights and measures are used throughout, as these are most familiar to the class of readers expected.

The division into numbered paragraphs is intended to facilitate the frequent cross-references, which are necessary in order to bring out the interdependence of the various departments of Nature. The illustrations are meant to elucidate the text rather than to serve as pictures. With the exception of those marked in the list with an asterisk (*), which are adapted from other sources, they were designed and drawn by the author and Mrs. H. R. Mill for this book. The maps have been specially compiled by Mr. J. G. Bartholomew, who has spared no pains to make them accurate and complete.

The book has been planned and written with the constant advice and assistance of Professor Knight, editor of the series, to whom the author desires to record hearty thanks. Mr. Murray, the publisher of the series, has also made valuable suggestions, and the title of the book, *The Realm of Nature*, is due to him.

Thanks are given to many of the author's teachers and friends who have kindly revised the proofs of chapters referring to the departments in which they are authorities, especially to Professor P. G. Tait, Professor R. Copeland (the Astronomer-Royal for Scotland), Dr. A. Buchan, Dr. John Murray of the *Challenger*, Professor James Geikie, Mr. H. M. Cadell, Mr. J. Arthur Thomson, and Mr. A. J. Ramsay.

H. R. M.

HERIOT-WATT COLLEGE,
    EDINBURGH, *August* 1891.

CONTENTS

CHAP.
- I. THE STUDY OF NATURE
  Definition and scope of Physiography . §§ 1-23, p. 1
- II. THE SUBSTANCE OF NATURE
  Properties of matter and measurement of space §§ 24-48, p. 15
- III. ENERGY, THE POWER OF NATURE
  Work —Wave-motion—Light—Heat—Electricity—Magnetism . . . . . . §§ 49-80, p. 30
- IV. THE EARTH A SPINNING BALL
  Figure of the Earth — Results of rotation: polarity, direction, latitude, longitude, time, terrestrial magnetism . . . . . §§ 81-99, p. 49
- V. THE EARTH A PLANET
  The Moon—Tides—Earth's orbit—The Sun—The Earth's share of sun-heat . . . §§ 100-125, p. 65
- VI. THE SOLAR SYSTEM AND UNIVERSE
  Planets—Comets—Meteors—Stars—Nebular and Meteoritic hypotheses . . . §§ 126-144, p. 84
- VII. THE ATMOSPHERE
  Air, composition and properties . §§ 145-162, p. 98
- VIII. ATMOSPHERIC PHENOMENA
  Warmth in air—Dew, mist, clouds, rain, snow and hail—Lightning—Circulation of atmosphere—Permanent and seasonal winds . . . §§ 163-185, p. 111
- IX. CLIMATES OF THE WORLD
  Configuration and climate—Isotherms—Isobars—Warmth and winds of January and July—Climate of British Islands—Storms—Weather forecasts §§ 186-213, p. 131

Contents

CHAP.

X. THE HYDROSPHERE
   Land and Water—Oceans and Seas—Tides—River and sea-water—Temperature of water—Oceanic currents
   §§ 214-250, p. 157

XI. THE BED OF THE OCEANS
   Divisions of the Lithosphere—Mean sphere level—Abysmal and Transitional Areas—Beach-formation—Marine deposits—Coral islands . . §§ 251-282, p. 188

XII. THE CRUST OF THE EARTH
   Rocks—Temperature of the Crust—Volcanoes—Earthquakes—Origin of Mountains . §§ 283-304, p. 214

XIII. ACTION OF WATER ON THE LAND
   Weathering of Rocks—Springs—Rivers—Mountains of circumdenudation—Lakes—Glaciers §§ 305-340, p. 234

XIV. THE RECORD OF THE ROCKS
   Fossils—Classification of rocks—Evolution of continents
   §§ 341-353, p. 262

XV. THE CONTINENTAL AREA
   Form of the continents, their mountain and river systems—Configuration of the British Islands §§ 354-392, p. 274

✓ XVI. LIFE AND LIVING CREATURES
   Classification and functions of plants and animals—Floral zones and Faunal realms . . §§ 393-417, p. 307

✓ XVII. MAN IN NATURE
   Civilisation and environment—Races of Mankind—Geography—Man's power in Nature. §§ 418-436, p. 326

APPENDICES

I. *Some Important Instruments*
   Weights and measures—Mariner's compass—Barometers—Thermometers—Hygrometers—Anemometers—Deep-sea soundings . . . . §§ 437-443, p. 343

II. *Curves and Maps*
   Graphic representations — Map-projections — Contour-lines . . . . . . §§ 444-446, p. 350

III. *Derivations of Scientific Terms* 355

INDEX 361

LIST OF MAPS

(Compiled by J. G. BARTHOLOMEW, F.R.G.S.)

		PAGE
1.	Magnetic Conditions of the Earth, after Admiralty Chart, 1885	62
2.	Earthquake Regions and Volcanoes	92
3.	Isotherms for January, after A. Buchan	104
4.	Isotherms for July, after A. Buchan	112
5.	Isobars and Winds for January, after A. Buchan	120
6.	Isobars and Winds for July, after A. Buchan	128
7.	Permanent Winds, Calms, and Storms	136
8.	Mean Annual Rainfall of Land and Salinity of Ocean, after Loomis, J. Y. Buchanan, and others	144
9.	British Islands, Isotherms for January, after A. Buchan	152
10.	British Islands, Isotherms for July, after A. Buchan	152
11.	Configuration of the Globe	192
12.	Equidistant Coastal Lines	208
13.	Drainage Areas of Continents and Co-tidal Lines of Oceans, after J. Murray and others	262
14.	Evolution of Continents, after J. Geikie	272
15.	Ocean Surface Isotherms, Coral Reefs, Rising and Sinking Coasts, after A. Buchan, H. B. Guppy, and others	288
16.	British Islands, Physical Configuration	304
17.	British Islands, Mean Annual Rainfall, and Co-tidal Lines, after A. Buchan and Charts	304
18.	Vegetation Zones of Continents and Oceanic Currents, after Engler and others	320
19.	Faunal Realms, after Sclater and A. R. Wallace	336

LIST OF ILLUSTRATIONS

FIG.		SECT.	PAGE
1.	Interrelation of the Sciences	21	12
2.	Four right angles	31	18
3.	Angular measurement of distance	33	20
4.	*Inverse Squares	36	22
5.	Swing of a pendulum	54	33
6.	Wave-motion	57	35
7.	Prismatic Refraction	62	37
8.	Diagram of the Solar Spectrum	63	39
9.	Curvature of the Earth	81	49
10.	Direction of rotation round North Pole	88	53
11.	Direction of rotation round South Pole	88	54
12.	*Diagrammatic section of the Earth	93	58
13.	Revolution of a non-rotating body	102	67
14.	Revolution of a body rotating in the same time	102	67
15.	Problem of the Earth and Sun	106	70
16.	*Ellipse representing Earth's Orbit	109	72
17.	*Cause of the Seasons	121	80
18.	Angle of Light Rays	124	82
19.	*Zones of Climate	125	83

List of Illustrations

FIG.		SECT.	PAGE
20. Mercurial Barometer		146	99
21. Atmospheric Refraction		150	101
22. *Theoretical Circulation of the Atmosphere		177	122
23. *Daily Range of Atmospheric Temperature		182	126
24. *Daily Range of Atmospheric Pressure		183	127
25. Sea-breeze		184	128
26. Land-breeze		184	128
27. Distribution of Atmospheric Temperature in latitude		187	132
28. Curves of monthly mean temperature		191	136
29. *Isobars of Anticyclone		205	147
30. *Isobars of Cyclone		207	149
31. *Proportion of land and sea in different latitudes		214	157
32. Curves of temperature in river entrances		232	171
33. *Curves of temperature in the Ocean		235	173
34. Temperature section of Red Sea and Indian Ocean		236	175
35. Circulation of Water by Wind		240	178
36. Section across Atlantic Ocean 20° N.		258	194
37. Steep slopes		260	195
38. *Slopes of the Gulf of Guinea		263	197
39. *Formation of a Beach		265	199
40. *Darwin's Theory of Coral Islands		281	211
41. Murray's Theory of Coral Islands		282	212
42. Illustration of Rock structures		290	218
43. *Ideal Section of a Volcano		295	224
44. *Earthquake Wave		300	227
45. Anticline and Syncline		302	230
46. *Production of thrust-planes in rocks		302	231

List of Illustrations

FIG.		SECT.	PAGE
47.*Section of the Alps		303	232
48. Mellard Reade's Theory of Mountain Origin		304	233
49.*The origin of Springs		314	240
50.*Artesian Wells		314	240
51. Origin of River Windings		323	247
52. Embankment of a River		324	248
53.*Ideal Section of Falls of Niagara		330	252
54. Map of a Glacier		337	258
55. Section of Loch Goil		339	260
56. Typical Section of a Continent		356	277
57. Section across South America in 18° S.		359	280
58. Section across North America in 36° N.		363	282
59. Section along North America in 90° W.		367	285
60. Section across Australia in 26° S.		370	287
61. Section across Africa on the Equator		374	290
62. Section across Asia in 90° E.		380	295
63.*Climate and Vegetation in latitude and altitude		405	316
64. Photographic Barograph		439	345
65. Mercurial Thermometer		440	346
66. Cylindrical Projection		445	352
67. Conical Projection		445	353
68. Contour-lines		446	353

CHAPTER I

THE STUDY OF NATURE

1. **Physiography** means literally the description of Nature. In order to describe anything we must know something about it, and in order to know something about anything we must study it. Knowledge obtained by the best method of study is science, and it differs from knowledge otherwise obtained in being so clear and definite that every step leading to the final result may be recalled and tested, if any doubt should arise as to its trustworthiness. Hence description based on science is clear and full, and this is the kind of description required in Physiography.

2. **Nature** means all creation; not only all created things but also all the changes they undergo. The scope of Physiography is thus immense but not unlimited. It includes everything of which we can gain knowledge in the Earth and beyond it, and every change now happening or of which a record has been left, together with the causes of all these changes. It is, however, customary to exclude the First Cause of all from consideration in connection with the account of facts and immediate causes. Theology—the study of the Creator—is in itself an immense field of science, and although it accounts for the origin of Nature, it may be readily separated from the study of natural facts and phenomena. The chief reason for separating Theology from Physiography is that authorities are greatly divided as to the right means of studying the former science, while every one is agreed as to the right method of studying Nature.

A description of the steam-engine which did not refer to Watt and other inventors and give something of their biography would not be held satisfactory unless some explanation of the omission were given, such as the desire to avoid controversy. For the same reason in a general description of Nature it is necessary to insist on the relation of Nature to God, and explain why this relation is not more fully dealt with.

3. **Science** is organised and accurate knowledge, and consequently its range has no limits; it is equally necessary in order to understand Nature and the supernatural. Science deals with everything, and its first duty is to classify by observing resemblances and differences.

4. **Comparison and Description.**—Suppose that we were comparing the tastes of different kinds of fruit in a garden. It is not enough to pluck bunches of red currants, black currants, gooseberries, and raspberries off the bushes and eat them. Each bunch must be classified into berries and leaves or stems; the former are to be tasted, the latter to be thrown away and thought no more of. Without this precaution the taste of gooseberries might be compared with that of black currant leaves, and different tasters would give irreconcilable reports. When we compare the various things around us, a preliminary classification is quite as much required to ensure that we compare things that are comparable. If we were to take into account *mountains, pain, rivers, happiness, air, beauty,* and *motion,* the description would be very confused and puzzling. When several people who have had the same opportunity of seeing, describe an event, the descriptions are almost sure to differ among themselves. This is because a different impression is produced on each mind, and the various subjective sensations of interest, or fear, or pleasure, or surprise, are confused to a greater or less degree with the objective facts. A scientific description should be as free as possible from all subjective colouring; a mountain must not be described as impressive in its grandeur or beautiful in its colouring, but as of so many feet in height and composed of such and such materials. Nature presents us with so many pheno-

mena to observe, and these are to all rightly constituted minds so full of wonder, beauty, and charm that we are apt to be dazzled and distracted, and even if our attention is roused it is too often satisfied by the first superficial impressions. It is only by putting aside these and looking at bare facts and abstract principles that we can truly understand our natural surroundings and so fully appreciate "all the wonder and wealth" of the Universe in its deepest meaning.

5. **Real Things.**—The first classification of things is into (*a*) Things that exist only in our own minds ; (*b*) things that exist outside of us and independent of us. Emotions, feelings, tastes, and beliefs belong to the former class and are termed subjective things. Facts and phenomena which exist whether we know of them and understand them or not, are termed objective or *real* things. The real things of Nature are the objects of physical science, and they alone fall to be considered here. The one test of reality in Nature is permanence. Only those things are real which can neither be created nor put out of existence by human power. Subjective things, such as pain, happiness, beauty, may be very readily produced and destroyed, hence however vivid the impression of them may be they are not real in our sense of the word and form no part of Physiography.

6. **Definition of Physiography.**—Physiography is an account of physical science as a whole. It describes the substance, form, arrangement, and changes of all the real things of Nature in their relations to each other, giving prominence to comprehensive principles rather than to isolated facts. This definition of the *term* Physiography is simply a definite statement of the meaning of the *word* Physiography (§§ 1, 12).

7. **Use of the Senses.**—Our senses are windows of knowledge through which alone information enters the mind, and through which alone we are able to study the things outside us. Instruments and apparatus of various kinds are of value only in making the evidence of the senses more precise or more appropriate to the object of study. All the senses—sight, hearing, touch, and the less used

smell and taste—are limited in their scope, and liable to get out of order through disease or neglect. But even when in full health and within their own range they are not fully trustworthy. If an object present different appearances when looked at through different windows, we are justified in supposing that the windows are not equally trustworthy. A few simple experiments show us that this is the case with the windows of knowlédge. Every one is familiar with optical illusions proving the imperfection of the sense of sight. A coin spinning quickly looks like a hazy sphere, but we know it to be a flat disc. Strobic circles which seem to whirl rapidly when the card on which they are printed is moved slightly, and designs appearing in their complementary colours on looking at a blank wall have been made familiar by their use as advertisements. Mountains always look steeper than they really are; in a slight haze on a wide moor a rabbit close at hand may be mistaken for a distant deer, and the most familiar object is often passed unrecognised if in an unusual place. One well-known experiment shows that touch is as fallacious as sight. When a pea or small ball is rolled on a table by the middle finger crossed over the forefinger of the same hand, so that both fingers touch the object, the impression produced is that there are two peas, not one. Similarly if one hand has been held in hot water and the other in cold water, and then both are plunged into a mixture of hot and cold, the mixture will be pronounced cold by the heated hand and hot by the chilled one. The deceitfulness of the senses may impose upon the most acute and practised mind if taken unawares. When Sir Humphry Davy discovered potassium he showed a piece of it to Dr. Wollaston, one of the most accurate observers who ever lived. Wollaston saw the silvery lustre of the new metal, weighed it in his hand and said, "How ponderous it is!" Davy in reply threw the metal into a basin of water, where it floated lightly on the surface. Wollaston's illusion is the more striking because at that time he was the only man who was in the habit of handling platinum, a metal which, bulk for bulk, is twenty-five times heavier than potassium.

The Study of Nature

8. Use of Reason.—In spite of such cases of deception, we trust our senses and are rarely deceived by them. Reason, man's supreme gift, examines, weighs, extends, and judges the evidence of the senses. It requires a course of reasoning to let us know that a tall man far off on a straight road is not a dwarf close at hand, or that the Moon rising behind a wood is not a yellow plate hung in the trees. Long practice has made the operation of reason so swift and smooth that we are seldom conscious of an interval between seeing and understanding. Reason makes the senses satisfactory means for acquiring knowledge, although reason alone can give no information about natural things. Just as the senses may be greatly aided by instruments and apparatus, reason may be greatly aided by mathematics. And as accurate measurements, on which the value of all scientific observations depend, can only be made by means of suitable apparatus, sometimes of a very elaborate nature, so accurate reasoning, which is essential in all scientific discussions, can only be fully carried out by mathematical processes which are sometimes difficult and complicated.

9. Common Sense is the name which practical people give to the best and easiest way of doing their work, and the simplest and completest way of gaining knowledge or explaining any difficulty. Common sense consists of reasoning on the evidence of the senses, but without keeping account of the process. When this common-sense method is made precise and accurate, it becomes the Scientific Method of gaining knowledge. The two guardians of thought in science are *Accuracy* and *Definiteness*. The scientific man deals with phenomena as the banker does with money, counting and recording everything with scrupulous exactness. The student should remember that for the practical purposes of life the knowledge of what are called scientific facts is unimportant compared with the power of using the scientific method. It is really more scientific to repeat a quotation from a political speech correctly, or to pass on a story undistorted, than it is to know of the rings of Saturn or the striation of diatoms.

10. Accuracy in observation usually takes the form of

correct measurement of mass, space, or time, by means of suitable instruments. Accuracy is always to be striven for, but it can never be attained. This fact is only fully realised by scientific workers. The banker can be accurate because he only counts or weighs masses of metal which he assumes to be exactly equal. The Master of the Mint knows that two coins are never exactly equal in weight, although he strives by improving machinery and processes to make the differences as small as possible. When the utmost care is taken the finest balances which have been constructed can weigh 1 lb. of a metal with an uncertainty less than the hundredth part of a grain. In other words, the weight is not accurate but the inaccuracy is very small, and its greatest possible amount is known to less than $\frac{1}{700000}$ part of the mass weighed. In weighing out tea or sugar a grocer is content if the inaccuracy is not more than about $\frac{1}{500}$ of the mass. No person is so stupid as not to feel sure that the height of a man he sees is between 3 ft. and 9 ft.; some are able by the eye to estimate the height as between 5 ft. 6 in. and 5 ft. 8 in.; measurement may show it to be between 5 ft. 6¾ in. and 5 ft. 7 in., but to go closer than that requires many precautions. Training in observation and the use of delicate instruments thus narrow the limits of approximation. Similarly with regard to space and time, there are instruments with which one-millionth of an inch, or of a second, can be measured, but even this approximation, although far closer than is ever practically necessary, is not accuracy. In the statement of measurements there is no meaning in more than six significant figures, and only the most careful observations can be trusted so far. The height of Mount Everest is given as 29,002 ft.; but here the fifth figure is meaningless, the height of that mountain not being known so accurately that two feet more or less would be detected. Similarly the radius of the Earth is sometimes given as 3963·295833 miles, whereas no observation can get nearer the truth than 3963·30 miles.

11. **Definiteness** in thought and description does not require perfect accuracy in observation. We must always

be definite in order to be clear. If he wishes to be definite in thought the student must never rest content with the dubious "I think" or the vague "about," but endeavour after the clear "I know" and the precise "with a probable error of." Vagueness and indecision are utterly foreign to the scientific method. It often happens that there is no definite knowledge concerning some fact; then all that the scientific method of description permits is to say, "There is no information," and to wait until the scientific method of observation has found out something. The difficulty is not overcome by guessing, or by calling the unknown unknowable. There is a place for speculation and imagination in the scientific method (§ 18), but it is a place apart, which must be shut off, for if speculations are not kept in strict quarantine they are certain to infect our conceptions of facts with their own fatal vagueness.

12. **Scientific Terms.**—Definite words are necessary for the expression of definite ideas, hence scientific terms have to be employed. A term has one definite meaning which does not change with time. The rush of affairs drifts words from their original meanings, as ships drag their anchors in a gale, but terms sheltered from common use hold to their moorings for ever. The word *let*, for example, has drifted in 200 years from meaning *hinder* until now it means *permit*; but the term *bisect* has remained unaltered in significance for twenty centuries. Many scientific terms are derived from the Greek and have an unfamiliar appearance; a list of all those employed in this book, together with their derivation, is given in Appendix III.

13. **Classification** of the facts and processes of Nature is necessary before we can form definite ideas concerning them; but the definiteness of classification is an artificial restriction. In Nature one thing merges into another by imperceptible degrees, and although, for example, we can readily class typical metals and non-metals, typical igneous and sedimentary rocks, typical plants and animals, there are in each of these pairs of classes many cases which cannot be referred with certainty to either side of the dividing line. Nature is discrete only within certain limits, and its

classes are never so definite as to isolate one from another, the unity of Nature being as marked as its diversity.

14. **Natural Law** is the order in which things have been observed to happen. The fact that there is order and not chance in the way things happen is one of the chief discoveries of science. It is the discovery on which all science depends, because knowledge could never be definite and accurate if it were not based on orderly phenomena. It is impossible that there can be any exception to a law of Nature, or any contradiction of it. Much has been written as to the impossibility of miracles because they would be breaches of the laws of Nature. If there is evidence, however, that a miracle did happen, the law of Nature it appears to contravene must be restated so as to take account of the new phenomenon. It is because the law expands to admit apparent exceptions that we say there can be no exceptions. We have, strictly speaking, no right to assume that things will continue to happen in the order in which they have happened hitherto. Nothing in time past has been more regular and uniform in its recurrence than the appearance of the Sun rising and setting. This regular order is a natural law, yet we cannot say certainly that the Sun will rise to-morrow; merely that its rising is very highly probable. The law of gravitation, the laws of heat, light, sound, and of all other observed facts, are similarly the summary of observations in the past; and although each new verification increases the probability that the laws will continue to hold good, that probability never becomes certainty.

15. **Probability.**—The probability of 7,000,000 to 1 is so great that all but very cautious people think of it as certainty. It represented the chance of a passenger arriving alive at the end of a railway journey in the United Kingdom in the year 1890. The probability that the Sun will rise to-morrow is far greater than this, because no failure has ever been recorded in the past. The laws of Nature, although only expressions of very high probability as regards the future, may be assumed as quite certain for all the practical purposes of life.

16. **Cause and Effect.**—The relation of Cause and Effect is the fundamental law of Nature. There is no recorded instance of an effect appearing without a previous cause, or of a cause acting without producing its full effect. Every change in Nature is the effect of some previous change and the cause of some change to follow; just as the movement of each carriage near the middle of a long train is a result of the movement of the one in front and a precursor of the movement of the one behind. Facts or effects are to be seen everywhere, but causes have usually to be sought for. It is the function of science or organised knowledge to observe all effects, or phenomena, and to seek for their causes. This twofold purpose gives richness and dignity to science. The observation and classifying of facts soon becomes wearisome to all but the specialist actually engaged in the work. But when reasons are assigned, and classification explained, when the number of causes is reduced and the effects begin to crystallise into essential and clearly related parts of one whole, every intelligent student finds interest, and many, more fortunate, even fascination in the study.

17. **Inductive and Deductive Reasoning.**— Reason may be applied to the study of facts in two different ways. *Inductive Reasoning* is the arduous process of finding the meaning of phenomena by collecting and classifying facts and thinking out their causes. *Deductive Reasoning* is the shorter operation of finding what effects must result from the operation of a known cause. It is often supposed that since we can observe facts alone the inductive method of reasoning is the only one which can be employed in studying Nature, but the number of facts even in one small department is so great that life is not long enough for the labour of collecting, classing, and discussing them all.

18. **The Scientific Method** of discovering the causes of phenomena involves the use of both inductive and deductive reasoning linked together by *imagination*, a mental power which is as essential to the scientific discoverer as it is to the poet. After observing a considerable number of facts the investigator imagines a possible cause or explanation,

and this possible explanation is termed an *hypothesis*. Then he reasons deductively from the assumed explanation, usually employing mathematics for the purpose, and so arrives at a number of additional facts which must exist if the hypothesis be true. These predicted facts may not be familiar or may not occur naturally at all. In the latter case it is necessary to seek them by making *experiments*, and so important is this aid in some cases that the expression Experimental Science is often used in the sense of physical science. If the facts predicted to exist in certain circumstances by hypothesis are not found, and if others which the hypothesis could not account for appear, the hypothesis is proved to be erroneous, or, at least, incomplete. Renewed inductive reasoning from the wider basis of ascertained facts must then furnish material for a fresh effort of imagination and a new hypothesis to be similarly tested, and, if necessary, rejected in turn. Should the facts agree with those deduced from the hypothesis there is a probability of its being true, but a great many tests must be thought of, applied, and found realised before the hypothesis is accepted as a true and complete explanation. An explanation of facts found, tested, and proved to be true and complete in this way is called a *theory*, and when a theory is confirmed by a great number of observations it is accepted as a Law of Nature.

19. **Proof of a Theory.**—The process of testing an hypothesis requires great caution in order to prevent mistakes. A long time and the labour of many observers are often necessary to perfect a theory or demolish an incorrect hypothesis. When Newton imagined the hypothesis of universal gravitation, according to which the force that causes a stone to fall to the ground also controls the motion of the Moon round the Earth and of the Earth round the Sun, he deduced from the hypothesis that the Moon in its orbit should fall toward the Earth 15 feet in a minute. Careful observation of the Moon's motion showed that it was only bent toward the Earth 13 feet in a minute, and therefore Newton abandoned his hypothesis as untrue. Thirteen years later a new measurement of the size of the

Earth, and consequently of the distance of the Moon, gave him more accurate data, and applying these to his hypothesis and to the observations, he found that the discrepancy vanished. This assured him of the truth of his hypothesis, which has ever since taken rank as a theory and a law of Nature.

20. **Test of a Law of Nature.**—A law of Nature has no exceptions (§ 14); the only test by which a theory can be accepted as of this rank is the successful prediction of future effects. The theory of gravitation enables astronomers to calculate the relative position of the Sun, Moon, planets, and stars as seen from all parts of the Earth's surface. This is regularly done by a government office in London, and the positions for stated times each day are published three years in advance in the *Nautical Almanac*. From the tables of this work the captains of ocean-going vessels are able to work out their exact place on the ocean by observations of the positions of the heavenly bodies (§ 92). The smallest deviation from truth in the expression of the law of gravitation would throw the results into confusion and lead to almost certain shipwreck. No such confusion has ever occurred, and every successful sea-voyage is one proof more that the law of gravitation was fully understood in the past, and holds in the present. The appointments made for the appearance of the Sun, Moon, and planets amongst special groups of stars at definite times in the *Nautical Almanac* are analogous to the appointments for the arrival of trains at stations made in official railway time-tables. Observation of the fulfilment of time-table predictions very soon demonstrates that the hypothesis in accordance with which they are framed is not exact, and cannot be depended upon for timing watches or determining our position on the Earth.

21. **Magnitude of Nature.**—The Scientific Method is applicable to the acquisition of knowledge of any kind, but it has been most used in the study of Nature. It is necessary that each scientific investigator should confine himself to one department of Nature in which he finds the facts and tries to reason out the theories connecting them. Thus we

are apt to form the impression that Physics, Chemistry, Astronomy, Geology, Geography, Meteorology, Biology are definite sciences, distinct from each other, dealing with different orders of facts which are accounted for by independent theories. These sciences do not completely cover the

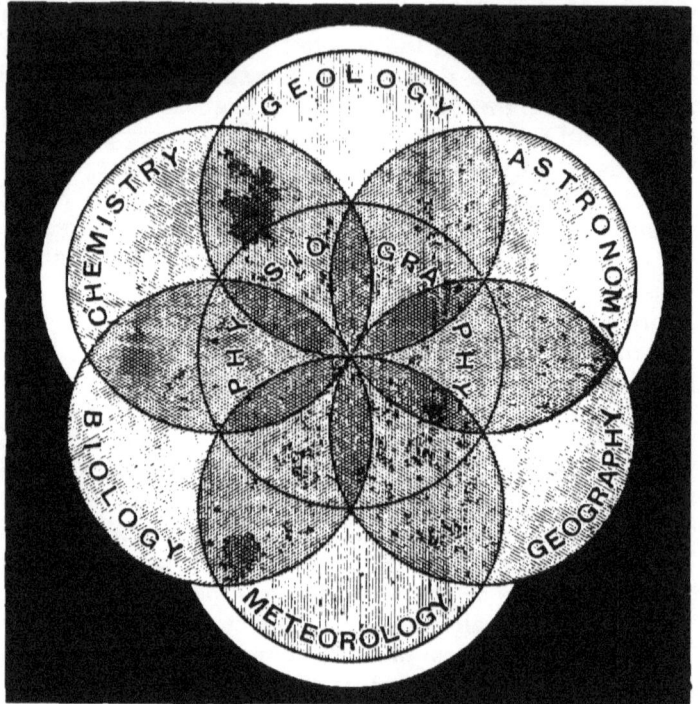

Fig. 1.—Inter-relation of the Sciences.

field of Nature as the coloured blocks of counties do the map of England. They traverse the field rather like the railway lines which radiate from London. The main line of each science is distinct and easily followed, but the branches interlace with one another in a very complex manner, and though the network is very comprehensive, a mere fraction of the vast surface is after all covered with the lines of definite knowledge. The inter-relations of the sciences are shown in

Fig. 1 by representing each as a circle cutting all the others, for on the outskirts of every science there are regions in which another science shares the explanation of phenomena. Chemistry, for example, is called in to aid astronomy in interpreting the spectra of the stars, to aid geology in explaining the composition of rocks, to aid biology in determining the changes of substance in living creatures. Physics or Natural Philosophy in a sense includes every other branch of physical science, although portions of Biology and Geography extend beyond its limits.

22. **Physiography and the Special Sciences.**—By division of labour the various parts of a watch are constructed by different workmen, and by the specialisation of science the different realms of Nature are explored by different investigators. In order to have a watch, however, the results of divided labour must be combined, and in order to have a just conception of the Universe the results of specialised research must be fitted harmoniously together. This is the function of Physiography, which has consequently a unique value in mental training, being at once an introduction to all the sciences and a summing up of their results. It enables a beginner to obtain a quicker insight into any of the special sciences and a fuller grasp of it, while at the same time a student versed in any one special science is enabled to appreciate far more fully than an unversed one its relation to all others and to the system of the Universe.

23. **Physiography and Nature.**—The natural Universe may be compared to a gorgeous carpet of rich design. In order to understand such a web we might follow out the pattern thread by thread. Selecting first a red thread of the weft, we notice how it passes above and below the threads of the warp, across the fabric and back again, to and fro until the end. Next a blue thread may be followed in the same way, and so with all the separate colours. The course of each thread has explained something, but the results of all must be brought together to give a complete explanation. In some such way each special science unravels one of the threads of the Universe, but that thread is

so interwoven with the clues of other sciences that a general knowledge of them all is necessary to understand the effect produced by one. Pursuing the simile a step farther, we may note how one observer sees in the rich world-carpet nothing but a number of coloured threads intricately interwoven; the taste of another is so much gratified by the colour and design that he enjoys the beauty without thinking of the parts or the process; while a third loses sight of material and beauty alike in admiration of the genius of the designer and the skill of the craftsman. Thus the typical man of science, poet, and theologian look differently on the multiform unity of Nature, which has a true though different meaning for each.

BOOKS OF REFERENCE

T. H. Huxley, *Science Primers—Introductory.* Macmillan and Co.

W. S. Jevons, *Principles of Science.* Macmillan and Co.

CHAPTER II

THE SUBSTANCE OF NATURE

24. **Matter.**—Diverse and innumerable as the things around us seem to be, the number of kinds is reduced greatly when they are tested by trying to destroy them. Only what cannot be annihilated is real, according to our definition (§ 5). Tested thus, air, wood, marble, vinegar, to take a few random examples, appear unreal, for they can be produced and destroyed. Closer study shows that though burning destroys both wood and air it produces at the same time other things—ashes, water, carbonic acid, nitrogen— exactly equal in amount though different in properties. Vinegar and marble are both destroyed by mixing them, but other things—calcium acetate, carbonic acid, water— appear in exactly the same amount. So with all the things we see or feel, their properties and appearance can be completely changed, but the amount of substance that exists in them cannot be increased or diminished by any power which man has learned to wield. Substance is thus a real thing, of which air, wood, marble, vinegar and the rest are kinds. The term *Matter* is applied to everything, however diverse in appearance, which we see and touch, as Man is the term used to include every human being in the world. The difference between some kinds of matter is as slight and superficial as that between soldiers and chimney-sweeps; between other kinds it may be compared to that which separates Europeans from Negroes.

25. **Energy.**—There is another real thing which does

not appeal so directly to our senses as matter does; fifty years ago it was unknown and a long course of reasoning was necessary to convince investigators of its existence and reality. Nothing appears more readily produced or destroyed than motion, heat, or light. Motion is destroyed in a railway train by applying the brake, in a bullet by contact with the target. Heat can be destroyed by using it up in a steam-engine; the visible motion of an engine can be destroyed in turning a dynamo-electric machine; electric currents can be destroyed in an incandescent lamp; light can be destroyed by allowing it to fall on a black surface. Hence none of these things is real in itself. But when motion is stopped in a train heat is invariably produced, the wheels sometimes becoming red-hot. When heat is destroyed in a steam-engine, visible motion is produced; when motion is destroyed in a dynamo-electric machine, electricity is produced; when electricity is destroyed in a lamp, light is produced; and when light is destroyed by falling on a black surface, heat is produced. More than this, the amount of heat, motion, electricity, light produced is the precise equivalent of what is destroyed in producing it. All are capable of doing work of some kind, and this power of doing work can neither be created nor destroyed, its amount can neither be increased nor diminished. *Energy* is the name given to this real thing.

26. **Matter and Energy in Nature.**—Besides matter and energy nothing has been proved to have an independent existence. The whole of Nature consists of the two grand parts, that which works and that which is worked on. The two are quite inseparable, for work of every kind has been proved to necessarily involve motion through a large or a small space in straight or curved lines, and motion is incomprehensible except as some piece of matter moving. It is only through matter that we recognise energy, and only through energy that we recognise matter. It has been proved in some cases, and is possibly true in all, that the properties which distinguish different kinds of matter from each other are due to the different amounts of energy with which they are associated.

27. Matter is that which occupies space. This definition is in many ways the most satisfactory; but although attempts to say what matter is have been made by philosophers in all ages, no really sufficient definition has ever been arrived at. Matter is often defined as that which can be perceived by the senses.

28. Mass is the term used to denote quantity of matter. Thus when the mass of the Sun is spoken of as being 300,000 times that of the Earth, it is meant that the Sun contains 300,000 times as much matter as the Earth contains. Mass is usually measured out by the balance, and it is common to speak of the mass of any portion of matter as its *weight* (§ 38), although on the same principle we might speak of a man's health as his appetite. The unit of mass in British Possessions and the United States is the pound; in almost all other civilised nations it is the kilogram. (See § 437.)

29. Volume and Density.—*Volume* is the amount of space occupied by a body, and if matter were of one kind and always in the same state, the same mass would always fill the same volume. But matter exists in many forms, and if, for example, we compare together charcoal, lithium, coal, granite, arsenic, lead, and platinum, we find that the same volume contains very different quantities of matter. Indeed, the mass of a cubic inch of platinum is twice that of a cubic inch of lead, four times that of arsenic, eight times that of granite, sixteen times that of coal, thirty-two times that of lithium, and sixty-four times that of charcoal. So that these parcels of matter are packed with different degrees of tightness, as much as is present in 64 cubic inches of charcoal being packed within the limits of 1 cubic inch of platinum. The amount of matter in a unit of volume is called its *density*; thus in the list given above the density of each substance mentioned is twice that of the preceding. The unit of density universally employed is that of water, and calling this 1 the densities given above run:—

Charcoal.	Lithium.	Coal.	Granite.	Arsenic.	Lead.	Platinum.
0·34	0·59	1·33	2·70	5·96	11·36	21·53

The density of each kind of matter is very distinctive; that of quartz, for example, is 2·6, that of the diamond 3·5; and by means of this difference diamond buyers at once detect any attempts at fraud. The term *specific gravity* is often used to express the ratio of the density of substances to that of water.

30. **Form.**—The form which different kinds of matter assume varies greatly, and can be easily changed. Pure kinds of matter, *i.e.* elements and compounds (§§ 42, 45), when allowed to solidify or separate out of solution frequently assume a shape of beautiful symmetry—metallic bismuth, alum, or quartz, for example—and these definite forms are spoken of as *crystals*. Mixtures, and sometimes pure kinds of matter, have no special form naturally, but occur as they were moulded in the cavity or vessel containing them, or as they were broken off from larger pieces. These are often spoken of as *amorphous* or formless. The forms of crystals are so characteristic that the minutest trace of some compounds may be recognised by their appearance under the microscope.

31. **Angular Measurement.**—In considering the form and position of bodies regard must be had to the properties of space, and especially to the nature and use of angles. An angle is the inclination of two lines which meet at a point and may be measured by a certain definite amount of turning done by a line; the angle APB in Fig. 2 is the amount of turning in a line from the position PA to the position PB. If the line PA were drawn on a piece of card pivoted to a table by a pin through P, and if it were made to turn completely round, as shown by the arrow, until it came back to its original position, the end A

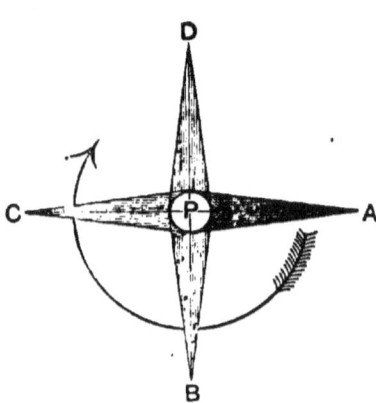

FIG. 2.—Four right angles.

would have pointed in turn all round the room; or if the table were in the open air, all round the horizon. The space of a whole turn is divided into 4 equal quadrants or quarters, each of which is called a right angle, and the amount of turn in a right angle is divided into 90 equal steps called degrees (°), each degree being the 360th part of a whole turn. Every degree is subdivided into 60 equal parts called minutes ('), and each minute into 60 parts called seconds ("). An angle of 1" is thus simply a short name for "the 1,296,000th part of a whole turn," and small though this is, $\frac{1}{10}$ of a second or less can be measured in fine instruments. The amount of turning from the horizon or skyline of a plain to the zenith or point directly overhead is one-quarter of a complete turn or a right angle, *i.e.* 90°. Degrees, minutes, and seconds are thus simply fractions of the unit which is a turn; and a turn is the same whether the turning line sweeps round the horizon, the Earth's equator, or a watch dial.

32. **Position by Angles.**—By fixing points from which to begin the reckoning it is evident that two sets of angles will enable one to define the position of any object on a sphere, such as the sky. In the case of a star, for example, by taking the north point of the horizon as a zero, one first measures the number of degrees, minutes, and seconds of turn until directly under the star, noting in which direction (toward east or west) the turn is taken. Then from the horizon at that point one measures the number of degrees, minutes, and seconds of turn toward the zenith to the star. Angular distance round the horizon is called the azimuth of a point; angular distance toward the zenith from the horizon is called its altitude. The instrument which is most convenient and most generally used for measuring angles of all kinds is the Sextant.

33. **Measurement of Distances by Angles.**—Every one must have noticed in passing a church clock that if, when standing directly opposite it, he sees the long hand pointing exactly to XII, yet from a little distance on one side he sees it to be a few minutes before, and from the same distance on the other side a few minutes past the

hour. This is because we look at the hand from different points of view and at different angles with the direction in which we are going. By measuring the angles, and the distance between the two points of observation, it is possible to calculate the distance of the object by trigonometry—the part of mathematics dealing with triangles. Suppose that in Fig. 3 the distance from A to B, which is called the base-line, is 100 yards, and that by means of a sextant the angles PAB and PBA are measured, then since all the angles of a triangle are equal to two right angles, the angle at P can be got by a simple subtraction, and an easy calculation would give us the distance of P from the eye. The more nearly

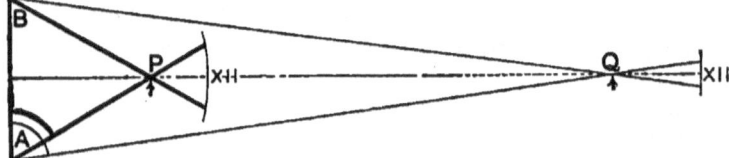

FIG. 3.—Angular measurement of distance. AB, base-line, P, Q, vertical angles.

equal the three angles of APB, the more accurately can this distance be found. For example, from the same base-line the angles to the hand of a much more distant clock would scarcely differ from right angles; the angle at Q would be so minute that the least mistake in measuring the two large angles would put the calculation all wrong. The more nearly the base-line is equal to the other sides of the triangle, the more exact is the trigonometrical measurement of distance. In this illustration the angle at P might be measured without a sextant by noting the amount of displacement of the long hand of the clock on the dial. In the distant clock the displacement would be too slight for the eye to detect.

34. **Exclusiveness** is a term descriptive of the way in which matter occupies space. It means that when one portion of matter is in a certain space no other portion of matter can be in the same space. The fact that a quantity of water can be absorbed by a sponge without much increasing the volume is no argument against this statement,

for the water occupies only the cavities between the sponge fibres. The particles of many kinds of matter are packed loosely together so that vacant spaces or pores occur. Porous bodies, like unglazed earthenware, sandstone, and charcoal, apparently allow air or water to pass through them; really, however, the fluid passes through the otherwise empty pores. The exclusiveness of the space-occupation thus holds good for the smallest particles of matter only (§ 48). The term *impenetrability* is often used for this property.

35. **Stresses and Strains.**—When the form or volume of a body is altered the body is said to be *strained*, and the set of forces which produce a strain is called a *stress*. Stresses act always in two opposite directions, either as a push or a pull. *Rigidity* is the resistance that a solid body offers to shearing stress. Extremely rigid substances, such as steel, require the action of powerful stresses in order to change their form; while less rigid substances may be readily deformed or strained, as a rod of lead is bent or a piece of sandstone pounded into dust. When uniform pressure is applied all solid substances, and still more all liquids and gases, are reduced in volume, the matter in them being compressed into smaller space and the density being of course increased. The amount of compression which the same pressure effects is called *Compressibility* and it differs in various kinds of matter, being greatest of all in gases (§§ 72, 148). The tendency of a body to recover from strain and return to its previous form and volume when the stress ceases to act is termed *Elasticity*. A steel watch-spring is said to be elastic, because after being coiled up tight it returns to its former size and shape. Air is said to be elastic, because when it has been compressed and the pressure is removed it returns at once to its previous volume.

36. **Gravitation.**—*Every portion of Matter attracts or tends to approach every other portion of Matter in the Universe with a force proportional to the masses and inversely as the square of the distance.* This is Newton's Law of Universal Gravitation, and is established beyond

doubt (§§ 19, 20), yet no one understands what gravitation is nor how it produces its remarkable effects. The greater the mass of two bodies, the more strongly do they attract; if the total mass is doubled the attraction is doubled. The nearer they are the more strongly do they attract in the proportion that halving the distance increases the attraction fourfold, reducing the distance to one-third increases the attraction ninefold. Fig. 4 illustrates the law of inverse squares as applied to central forces.

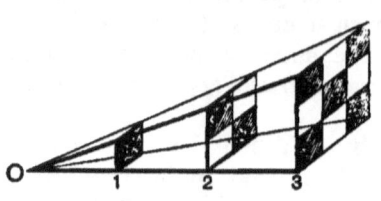

FIG. 4.—Inverse Squares. The gravitational force of O acting on the square at 1, is spread over four times the area at 2, and nine times the area at 3, so that the force acting on a unit square at 3 is ¼ and at 2 is ¼ of that at 1.

37. **The meaning of "Down."**—If two distant bodies equal in mass could be left free to follow the attraction of gravitation, they would approach each other and meet midway. But if one of the distant bodies had a much larger mass than the other it would move a shorter distance, because the result of attraction is to give the same amount of motion or momentum to each (§ 50). If one body is very large and the other very small, the small body seems to fall to the larger, while the latter does not apparently leave its place. This is the case of a stone outside the Earth's surface. It falls directly toward the centre, and the word "down" is used to designate this direction. The movement of the Earth to meet the stone is so slight that it cannot be detected, nor very easily expressed by figures. Still the attraction of gravitation is equal and opposite, the stone attracting the Earth as much as the Earth attracts the stone.

38. **Weight.**—The attraction of the Earth would draw an external body down to the centre, but the rigidity of the Earth's crust resists distortion. Those parts of the surface which possess no rigidity (the oceans) allow any body denser than water to pass through, or sink in obedience to the pull of gravity until it reaches the solid crust below. The pull of gravity which is counteracted by the push of

rigidity is of course greater for greater masses, and the amount of the pull in any case may be measured by pulling against it. Weight is the name given to the pull of the Earth upon some other body. At any definite distance from the Earth's centre the weight of a body is proportional to its mass, and hence it is that when we want one pound mass of tea we ask for one pound weight. If any mass is removed to a greater distance from the Earth's centre the pull upon it is diminished, or, in other words, its weight is less; if it is brought nearer the centre (without passing inside the Earth) the pull upon it is increased, or the weight is greater. Weight, or "Earth-pull," is measured by means of the spring-balance or by the pendulum. On account of the uniform pull of the Earth's gravitation, liquids, which have no rigidity, assume a level surface, or rather a surface parallel to that of the Earth. One of the necessary conditions for equilibrium in a liquid is that all points in the same plane are subject to the same pressure, hence the level of water in a series of connected vessels is always the same. Hence also if the height or the density of a column of liquid is altered equilibrium is destroyed, and the liquid moves under the influence of gravity until it again becomes homogeneous and of level surface (see § 238). Gravitation is a property which affects every kind of matter alike, and it binds together the great masses of the Universe into a firm and flexible whole.

39. **Cohesion.**—When the distance between particles of matter is very minute—too small to be measured—the force of attraction is very great, and binds the particles together very firmly. In this case it is called cohesion. It is by the powerful attraction of particles of matter at very minute distances that a stone is wetted or covered with a thin liquid film when dipped in water. These forces are also shown at work when a liquid rises in a narrow tube, or in a porous body like a sponge, a lump of sugar, or a piece of sandstone. This raising of liquids is called *capillarity* because it is best seen in tubes whose bore will just admit a hair, but it is quite visible on the sides of a tumbler. Another manifestation of the same force is seen in *surface*

tension, or the tendency all liquid surfaces have to become as small as possible. A small portion of a liquid when thrown off as a drop shrinks into a little sphere, because a sphere has the smallest surface possible containing a given volume. A soap-bubble blown on the wide end of a glass funnel contracts and creeps up to the narrowest part of the tube when left to itself. Surface tension accounts for such phenomena as the rapid spreading of a film of oil over a wide surface of water, and the extraordinary gyrations of a piece of camphor floating on clean water.

40. **Analysis and Synthesis.**—If we wish to find out for ourselves of what parts a piece of mechanism, such as a watch, is composed, we must begin by unloosening the parts from one another and taking the watch to pieces. So when we wish to find of what parts a piece of matter, such as a rock, is made up, we must unloosen its parts and take it to pieces. This process is called by the Greek name of *analysis*. There is another process sometimes employed: we might imagine a watch so strongly made that it could not be taken to pieces, but if we had seen the parts put together to make it, we would know of what it was composed. This putting together is called *synthesis*, and the process is sometimes used for investigating kinds of matter.

41. **Mixtures.**—We may take a piece of *granite* as typical of a pure kind of matter which is easily recognised by its characteristic appearance. On examining it with the eye we see that it is made up of three different substances. One of these is clear and glassy, breaking with a sharp edge, and hard enough to scratch glass. It is called *quartz*. Another is milky and opaque, whitish or pinkish in colour, too soft to scratch glass, and when it is broken it splits into regular smooth-sided blocks of similar shape. It is called *felspar*. The third ingredient is silvery or black in appearance; it forms flakes which are soft enough to be scratched by the nail, and flexible, splitting up into thin transparent scales. It is called *mica*. Granite, then, is a mixture of quartz, felspar, and mica,

and the proportion of each ingredient varies in different specimens. In a mixture each ingredient retains all its own properties, and so can readily be recognised and separated. A mixture of sand, salt, and sawdust, for example, could be separated by throwing it into water, in which the sawdust would float, the sand sink, and the salt dissolve.

42. **Compounds.**—Quartz, felspar, and mica may be examined as closely as the most powerful microscope allows, but no sign of any of them being a mixture will appear. Every one part of quartz is exactly like every other. Quartz, which is also called silica, can be separated into two substances by means of certain processes explained by the science of chemistry. One of these substances is a brown opaque solid called *silicon*, the other an invisible odourless gas named *oxygen*. Silica is not called a mixture but a compound, the distinction of which is that the components lose all their characteristics and unite to form a homogeneous substance, different in its properties from any of the components. For example, the metal magnesium is a tough lustrous solid; oxygen is an invisible gas present in the air; the compound resulting from their union is a soft snow-white powder. The composition of compounds is always exactly the same, the same proportion of each component being always present. Silica is invariably composed of 14 parts by mass of silicon and 16 of oxygen; magnesia always contains 24 parts of magnesium and 16 of oxygen.

43. **Analysis of Granite.**—Felspar may be analysed into silica, alumina, lime, and potash, each one of which is in itself a compound; and Mica can be analysed into silica, alumina, magnesia, potash, water, and iron oxide, all of which are compounds. The ultimate components are termed *elements*, of which some, such as oxygen and silicon, are classed as non-metals, the others as metals. Thus :—

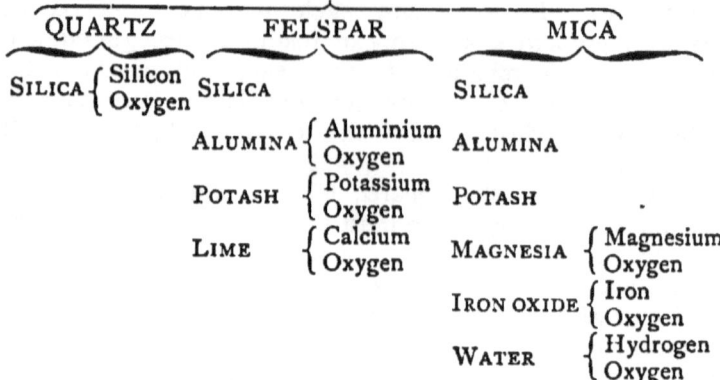

44. Acids and Bases.—Two classes of compounds require to be specially mentioned. The non-metal oxygen when it unites with a metal produces a compound called a *basic oxide*, and this is the case whether we consider the gaseous metal hydrogen, the liquid metal mercury, or any of the solid metals such as magnesium, calcium, or potassium. When oxygen unites with another non-metal, such as carbon, silicon, or sulphur, it produces an *acid oxide*. The main characteristic of basic oxides and acid oxides is that when brought together they unite to form more complicated compounds called *salts*. A certain amount of each acid oxide unites with a certain amount of each basic oxide to form a compound showing neither acid nor basic properties, but in many cases an additional definite amount of acid or of basic oxide takes part in the compound which then shows a more or less distinct acid or basic nature. Other non-metals, such as sulphur and chlorine, unite with metals to form compounds or salts termed sulphides and chlorides. Energy in the form of light or heat is given out when elements combine, and a precisely equal amount of energy must be used up on the resulting compound in order to decompose it. When much energy is involved in the transaction the compound is said to be a firm one.

45. Elements.—The process of analysis ceases when we come to oxygen, silicon, aluminium, etc., for no method yet attempted has been successful in breaking up any of these substances into other kinds of matter, hence they are called the simple substances or elements. There are about seventy elements known to chemists, but those which have been enumerated, together with carbon, appear to make up by far the greater part of the mass of the Earth. Professor Prestwich gives the accompanying estimate of the proportion in which each of the common elements occur in the Earth's crust.

ELEMENTS OF THE EARTH'S CRUST.

Oxygen	50·0
Silicon	25·0
Aluminium	10·0
Calcium	4·5
Magnesium	3·5
Sodium and Potassium	3·6
Carbon, Iron, Sulphur, and Chlorine	2·4
All others	1·0
Total	100·0

46. **Transmutation of Elements.**—For centuries the alchemists firmly believed that one element could be turned into another, and hundreds of men spent their fortunes and their lives in seeking the "Philosopher's Stone" which would bring about the magic change of lead to gold. In more recent times, as the knowledge of the properties of matter has increased, the possibility of such a change has been generally conceded; but although several modern chemists have believed that they got evidence of transmutation, the fact has never been proved. The re-arrangement of the particles with regard to each other in one kind of matter produces great changes in the outward properties. Charcoal and diamond are simply forms of pure carbon, and each has been changed into the other by the action of energy in certain ways. Hence it appears possible that the separate elements may themselves be simply different groupings of the one real thing we call matter, associated with different amounts of the other real thing we call energy.

47. **The Periodic Law.**—Elements are roughly classed into metals and non-metals, but there are intermediate ones which it is not easy to assign to either division. A more natural grouping was discovered by Mr. Newlands in England,

and Professor Mendelejeff in Russia, and is known as the Periodic Law. This states that if the elements are arranged in the order of the mass of their smallest particles, *i.e.* their atomic weight, they will fall into eight groups of about twelve elements each, and the first, second, third, etc., element of each group bears a strong family resemblance to the first, second, third, etc., of each of the other groups. Some of the groups have many gaps, only seventy elements being as yet known; but the atomic mass, the density, the melting temperature, the colour and the nature of the compounds it would form with known elements can be calculated and predicted for each of the elements which are absent. Names have even been given to these hypothetical elements, and in at least two cases the elements were subsequently discovered by chemists and found to correspond very closely to the prophetic description. This fact was the strongest confirmation of the truth of the Periodic Law. If the figures known to chemists as "atomic weights" really correspond to the mass of the atoms of each element, as there is reason to believe that they do, the chief difference between the elements may consist in the fact that their smallest particles contain different amounts of matter; the extreme cases are uranium and hydrogen, the mass of the atom of the former being 240 times that of the latter. We could imagine a great rock to be quarried into blocks of ninety-six definite sizes, the smallest being only $\frac{1}{240}$ of the largest, and ship-loads of these cut and squared stones might be sent to a nation where tools were unknown. These people might use the stones in building houses, but would be unable to change any one size into another until they invented the proper tools. They might be supplied only with sixty or seventy of the sizes, but by studying the weights of these and seeing the order in which they ran they might predict the existence of intermediate sizes. As they could not in the absence of tools change the form or size of the blocks, though recognising their unity of composition, they would look on them as unalterable elements in their building. Similarly modern chemistry has enabled us to understand how it is possible that the elements are merely separate parcels of matter which may be broken up and rearranged when the proper tools are found.

48. Structure of Matter.—Any element or compound appears perfectly homogeneous under the most powerful microscope, but the investigations of scientific men prove that there is a limit to homogeneity. The smallest particles of which matter consists are far too minute ever to become visible—the smallest visible speck is calculated to contain more than 50,000,000 of them. By careful experiments and ingenious reasoning Sir William Thomson has shown that matter is made up of particles so small that if a little cube 1 inch in the side were magnified until it was 8000 miles in the side, neighbouring particles would be 1 inch apart; in other words, there are about 500,000,000 particles in the length of an inch. The study of chemistry has shown that each particle must, in almost every case, consist of at least two, but probably many, parts called atoms which cannot exist separately but always form groups. The atoms of every element are different from those of every other element; but each atom of any element is exactly like all the other atoms of that element. Sir John Herschel compared the immense numbers of exactly similar atoms of hydrogen or of iron or of oxygen that are found on the Earth, in the Sun, and in the remotest regions of space, to manufactured articles all turned out by the same process and all trimmed to exactly the same size and pattern. Sir William Thomson has shown that it is possible to explain the structure of matter as made up of myriads of minute vortex rings or whirlpools set up in a perfect fluid which fills all space.

BOOKS OF REFERENCE

P. G. Tait, *Properties of Matter.* A. and C. Black.
H. E. Roscoe, *Lessons in Elementary Chemistry.* Macmillan and Co.

CHAPTER III

ENERGY, THE POWER OF NATURE

49. **Energy** is the power of doing work. Work, in the scientific sense, is any change brought about in the position of portions of matter against resistance. Change of position implies motion, and thus work may be spoken of as the moving of matter. Lifting water from a well by means of a bucket and rope is work against the resistance of gravity; tearing a piece of paper is work against the resistance of cohesion; pulling a piece of iron from a magnet is work against the resistance of magnetic attraction, and so on. Work is measured by the resistance overcome, and the distance through which it is overcome; the resistance usually chosen for this purpose is weight or the pull of the Earth on matter in consequence of gravitation (§ 38). In English-speaking countries the unit of work usually adopted is the foot-pound, the amount of work necessary to raise 1 lb. weight to the height of 1 ft. The work of raising 10 lbs. 1 ft. is 10 foot-pounds, and the work of raising 1 lb. 10 ft. is 10 foot-pounds also. The work a man of 150 lbs. weight does in climbing to the top of a mountain 10,000 ft. high is 1,500,000 foot-pounds, as much work as lifting 170 tons of coal from the ground up to carts 4 ft. high.

50. **Newton's first Law of Motion** expresses the property of Matter called *Inertia*, thus: *All bodies remain in a state of rest or of uniform motion in a straight line except when compelled by some external power to change that state.* On the Earth friction is always at work retarding motion. A

Sermons

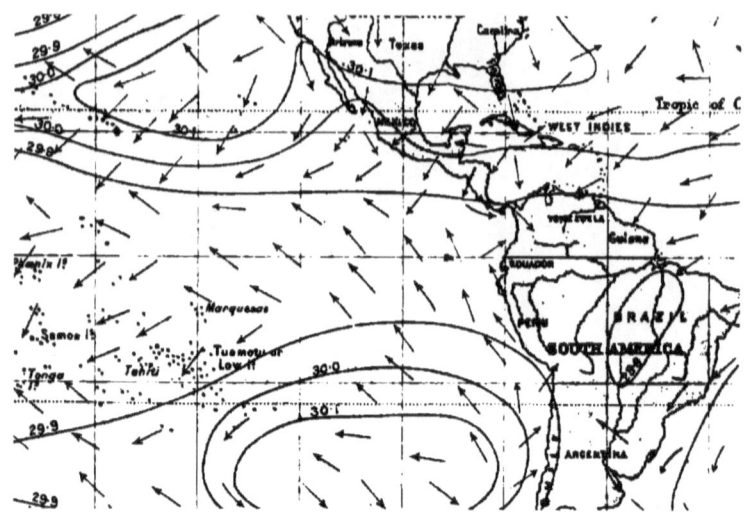

RY.

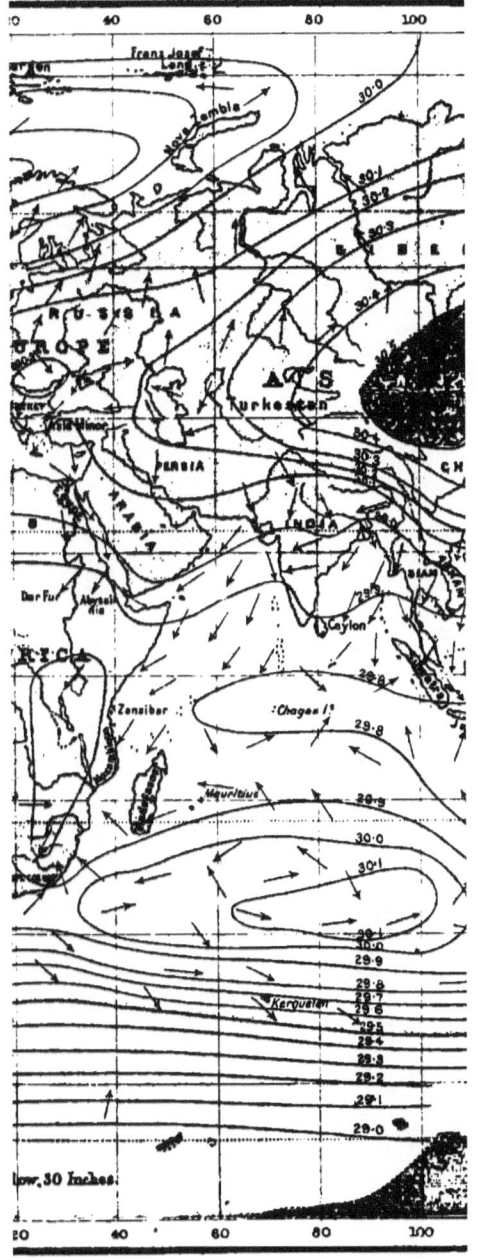

train moving at 60 miles an hour on a smooth level railway only requires the engine to give out enough energy to overcome the resistance of the air and the rails; when that is done the train, however great its mass, continues to move with undiminished speed. When it has to be stopped quickly, shutting off steam from the engine is not enough; great resistance has to be introduced by means of brakes which convert the energy of motion rapidly into heat. The energy expended in setting a mass in motion is preserved in the moving mass when there is no external resistance, and returned unaltered in quantity when the motion is stopped. The amount of motion in a moving body is called its *momentum*, and is measured by the mass and the velocity together. A mass of 1 lb. moving with a velocity of 1000 ft. per second has the same momentum as a mass of 1000 lbs. moving at 1 ft. per second.

51. **The Gyroscope** illustrates the first law of motion. It consists of a heavy leaden wheel turning on an axle in a brass ring. The inertia of the fly-wheel requires to be overcome by imparting a considerable amount of energy to it by means of a cord and a strong pull of the arm; once set in motion it would never stop but for the friction of its axle and of the air. A gyroscope in rotation behaves differently from one at rest. When the experimenter takes it by the stand and attempts to change the direction of its axis of rotation it seems to have a will of its own; it strongly resists any change of position, although when the fly-wheel is at rest its axis may be easily turned in any direction. In the fly-wheel itself there is a struggle going on; the particles tend to move in straight lines, and it is only the attraction of cohesion that compels them to move in a circle. In factories grindstones are sometimes made to rotate so fast that they burst; the tendency of the parts to move in straight lines is too great for the cohesion of the stone to counterbalance. The tendency for bodies to move in a straight line, unless compelled by some power to follow a curve, is often called *centrifugal force*.

52. **Work against Gravity.**—In employing energy to overcome weight there seems at first sight to be a real loss

unlike the case of inertia (§ 50). An exhausted mountaineer, on reaching the summit referred to in § 49, might ask, " Where are my million and a half foot-pounds of energy ?—are they not lost for ever ? " If the mountain were precipitous on one side the climber could answer his question by an experiment, not on his own person, but on a block of stone of equal weight (150 lbs.) Such a block in virtue of its elevated position has acquired the power of doing work. The attraction of the Earth draws the stone downward, and once allowed to fall it moves faster and faster until it strikes the ground with enough energy of motion to do 1,500,000 foot-pounds of work. This energy in a real case would be expended partly in heating the air during descent, and partly in shattering the stone and heating the fragments and the ground. The amount of energy expended and the ultimate form assumed are the same if the stone rolls down a slope as if it falls vertically.

53. **Energy of Motion.**—The faster a body is moving the more work it can do, *i.e.* the more energy it contains. A leaden bullet thrown against a man by the hand might inflict a painful blow, projected from a sling at the same distance it would produce a serious bruise, but fired out of a gun it would pass right through the victim. The greater the velocity of the bullet the greater is its power of doing work. But a small bullet striking a steel target is stopped, while a cannon ball, though moving at the same speed, breaks its way through; hence the greater the mass in motion the greater is its energy. When the mass of a moving body is doubled its energy is doubled, but when the velocity of a moving body is doubled the energy is increased fourfold. For example, a small river flowing at 6 miles an hour could do as much work in turning mills as a river four times the volume flowing at the rate of 3 miles an hour. This is expressed in the form of a Law—*Energy of motion is proportional to the moving mass and to the square of the velocity.*

54. **Potential and Kinetic Energy.**—Energy of position may be termed an expectant, energy of motion an active power of doing work ; or, to use the usual terms, the former

is potential, the latter kinetic. The raised weight or coiled spring of a clock contains potential energy, which is gradually converted into the kinetic energy of moving wheels and hands. The simple *Pendulum* consists of a heavy ball hung by a thin cord. Its practical value depends on the fact that if the length of the cord does not change, the ball swings from one side to the other in exactly the same time through any small arc. If the ball is pulled to one side to A (Fig. 5), since the cord does not stretch A is more distant from the Earth's centre than is B, and when let go its weight makes it swing back toward B. At A the pendulum has a certain amount of potential energy on account of its raised position, and as it falls it loses that potential energy, gaining instead kinetic energy, so that it passes the point B in the full swing of its active movement. The power immediately begins to do work against gravity in raising the ball to C, and the ball rises more and more slowly as its kinetic energy is being used up until at C it comes to rest. Here it possesses as much potential energy as it did at A, and so swings back again. The swings are shorter and shorter and finally it comes to rest only because the friction of the air and of the cord on its point of attachment gradually change all the energy into heat.

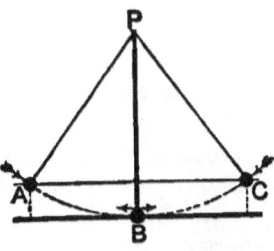

FIG. 5.—Swing of a pendulum. A, C, highest point of swing; B, lowest point.

55. **Conservation of Energy** is the term employed to denote the fact that the total amount of energy in Nature, as in the case of a frictionless pendulum in a perfect vacuum, never varies; that energy like matter can neither be created nor destroyed. Many clever mechanicians have endeavoured to find the *Perpetual Motion*, by which a machine when once wound up and set agoing would not only go on for ever, but would do work as well. In January 1890 an advertisement in the *Times* stated that the discovery had been made, and the inventor wanted pecuniary help to complete it. Knowledge of the laws of energy would have

saved the advertiser much lost time and useless trouble. We know that if a machine could run without resistance it would go on for ever at the same rate in virtue of inertia if energy is once imparted to it. But if a machine could not only keep going but set looms in motion as well, energy must be created at every turn, and experiment proves that this has never taken place. If energy be a real thing the Perpetual Motion is impossible. Energy is always undergoing transformation, visible motion, magnetism, electricity, heat, and light being a few of the many forms which it assumes. But Nature says sternly and unmistakably, " Nothing for nothing." No form of energy can be obtained without paying an exact equivalent in some other form.

56. **Invisible Energy.**—Work can be done and potential energy stored in separating atoms (§ 44) as well as in climbing mountains ; and the union of the separated atoms reconverts potential to kinetic energy as truly as the downward rush of an avalanche. When a stone strikes the ground its energy of motion as a whole is changed into energy of motion of its parts, which we recognise as heat. Three kinds of motion occur both on the great scale, perceptible to the eye, and on the small scale, discoverable by observation and reason. These are simple *translation*, like the movement of falling stones or of the darting particles of gases ; *wave motion*, like the undulations of the sea or the vibrations producing light ; and *vortex motion*, like whirlpools in tidal streams or the disturbances we recognise as magnetism.

57. **Wave-motion.**—Every elastic substance (§ 35) can propagate wave-motion. This motion consists in one particle moving through a comparatively short path and returning to its previous position, after passing on its energy of motion to another particle which also moves a short distance and returns. Waves of to-and-fro or up-and-down motion occur in solids and liquids ; and waves of alternate compression and expansion occur in gases. Waves are measured by the distance between similar parts of successive waves. The distance between crest and crest (CC in Fig. 6) or between trough and trough (TT) of

waves in water, or between succeeding maxima of compression or of rarefaction in waves of air, is spoken of as the wave-length. The amplitude of a wave is the height from crest to trough (C'T), or the difference in degree of compression and dilatation.

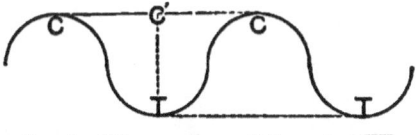

FIG. 6.—Wave-motion. CC, crests; TT, troughs.

58. Sound.—When a wave of alternate compression and rarefaction of air strikes the ear, it produces the sensation of sound; the more rapid the vibration and shorter the wave-length the shriller is the sound, but neither very short rapidly vibrating waves of air nor very long slowly vibrating ones affect the ear at all. The greater the amplitude of an air-wave, the louder is the sound. Waves of compression and rarefaction pass through the air at the rate of about 1100 feet per second when the temperature is 32° F., and travel 2 feet per second faster for every degree that the air is warmer. Sound-waves pass through water with four times the velocity, and through solids with many times the velocity of their passage through air. Air is set into wave-motion by any substance that is vibrating as a whole, such as a tuning-fork, a stretched string, or a column of air in a pipe. A tuning-fork when made to vibrate sets up air-waves that produce the sensation of a particular musical note in the ear; if that tuning-fork is at rest, and air-waves of the same kind as those it can set up strike it, they transfer their energy to the fork and start its vibrations. All other air-waves, longer and shorter alike, pass by with but slight and transitory effects, and, stated generally, the law holds that *Bodies absorb vibrations of the same period as those which they give out.* When certain notes are sung, or struck on a piano, the gas globes in a room absorb the particular waves which they would set up if struck, and ring in response to them.

59. Molecular Vibrations are the minute movements of the smallest particles of bodies, either as a quivering of the particle itself or as quick oscillations to and fro.

As long as there is any kinetic energy associated with a portion of matter the particles will be in motion. The amplitude of the oscillations in solids is very slight, not sufficient to overcome the resistance of cohesion (§ 39). However large a body may be, its particles will in time come to oscillate at the same rate throughout if not interfered with, any more quickly-moving particles passing on some of their energy to their more slowly-moving neighbours. The process of passing on and equalising the rate of molecular vibration is called *conduction*, and takes place, although more slowly, in liquids and gases as well as in solids.

60. **Radiant Energy.**—As the vibrations of bodies, as a whole, set up waves of various length in air which may travel to a distance, and some of which are capable of impressing the ear, so the invisible vibration of the particles of bodies sets up waves of radiant energy which travel to a distance, and some of which impress the senses. The quiverings of particles are very complex, and the particles of each kind of matter seem to quiver and oscillate in a way of their own, setting up waves which, although excessively minute, are far more complex than those of sound. There is much difficulty in understanding how the waves of radiant energy travel, and it is assumed that a very remarkable kind of matter called the *Ether* fills all space, and penetrates freely between the particles of ordinary matter. It is so fine that it offers no perceptible resistance to the movement of the planets through it, or to the movements of the particles of matter; but it is so elastic that it passes on the smallest and swiftest undulations. The undulations travel in straight lines through the ether at the rate of nearly 186,000 miles per second, and all amplitudes of these undulations travel at the same rate, about a million times as fast as the waves of sound in air.

61. **Reflection and Refraction.**—When the waves of radiant energy reach a surface through which they cannot pass, they are turned into a new path, either directly backward or at a definite angle to their former direction. Sound-waves meeting an obstacle are *reflected* in the same

way, giving rise to echoes, and so are the little ripples of a water surface on meeting a straight line of cliffs. When the ripples of the sea pass among a number of half-covered stones their onward path is changed in direction, each little undulation being bent from its course by the obstacle it meets. Similarly, when a ray of radiant energy passes from one medium into a denser, from the ether into air, or from air to glass, for example, the undulations are diverted by the particles of matter, and the path of the ray is bent or *refracted*. Radiant energy is made up of many different vibrations; some are comparatively long and are slow in their vibration, others are very short and much more rapid. The short quickly-vibrating waves are most bent from their straight path by passing into a different medium, and are therefore said to be most refrangible. It is evident that if a beam of radiant energy, in which each ray corresponds to a definite wave-length, travelling straight on, enters a denser medium, the separate rays will be spread out like the ribs of a fan, those of the shortest waves being most turned from the straight line, those of the longest waves least.

62. **The Spectrum.**—When the undulations which come from an intensely vibrating solid enter a triangular glass prism (Fig. 7, P) through a narrow slit, they are spread out by refraction and arranged side by side in perfect order from those of shortest wave-length, v, to those of longest, r forming a *spectrum*. The waves shorter than $\frac{1}{67000}$ of an inch have a peculiar power of affecting certain substances and producing chemical changes, but they have no effect on the senses. The waves between $\frac{1}{67000}$ and $\frac{1}{28000}$ of an inch in length (vr) affect the sense of sight through the eye, producing the sensations of light and colour, hence they are termed light-waves. Waves of longer wave-length

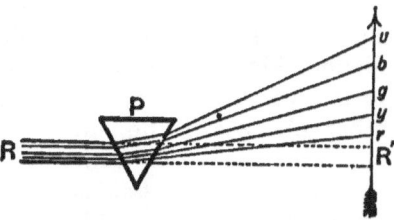

FIG. 7.—Prismatic refraction. RR′, straight path of light ray; Rvr, refracted path.

set the particles of bodies in vibration when they fall on them; they are invisible to the eye and are known as heat-waves. The shortest of the light-waves (v) produce the effect of violet light, longer ones (b) blue, still longer (g) green, longer yet (y) yellow, and the longest that produce any effect on the eye (r) red. Thus when one looks at a glowing solid body through a spectroscope, an instrument containing one or more prisms, the colours red, yellow, green, blue, violet are seen ranged in a row as in the rainbow (Fig. 8, which gives a detailed view of the range vr of Fig. 7), but the eye sees nothing of the short wave-length rays beyond the violet, nor of the relatively long wave-length rays beyond the red. Still longer waves can be detected by their electro-magnetic action. In fact, all radiation is essentially electro-magnetic.

63. Radiation and Absorption.—The different wave-lengths of sound in air correspond to different musical notes, the different wave-lengths of light in the ether to different colours. The molecules of each of the elements vibrate in a way of their own when set in motion, and produce waves in the ether of one or more definite lengths only. Sodium vapour, for example, when intensely heated sets up only rays the wave-length of which is $\frac{1}{43000}$ of an inch, and these produce the sensation of yellow light in the eye. A spectroscope sorting out the light from glowing sodium shows only a strong double yellow line (D in Fig. 8). The molecules of calcium vapour produce several distinct kinds of quivering, originating rays corresponding to definite colours of light. The same is true of all the other elements; the spectra of the radiant energy sent out from them are distinctive in every case. But, as in the case of sound, bodies absorb the same kind of radiations as they emit. If a beam of white light, which includes rays of all wave-lengths, is passed through sodium vapour, the particles of sodium are set vibrating by the waves $\frac{1}{43000}$ of an inch in length, and the energy of these waves is absorbed, so that when the beam is examined by the spectroscope, and the rays are spread out side by side, the peculiar double yellow ray is missing and in its place there is a blank or black line. The same is true with the vapours of all the other elements, the particular waves absorbed differing in each case. *Spectrum Analysis* is a term used to describe the discovery of the elements whose vibrations give out a certain kind of light. It is not only analysis or unloosening; it is also a method of seeing

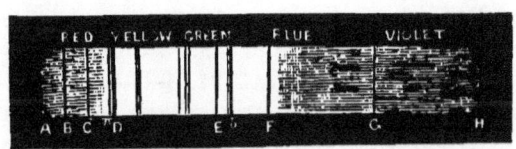

FIG. 8.—Diagram of the solar spectrum, showing the order of colour and the position of the principal absorption lines.

through a compound when taken apart by the action of heat. However distant a body may be, if it gives out light, the light tells its own tale as to the matter whose quiverings sent waves through the ether, and as to any other kinds of matter which may have exercised absorption on it in intermediate space.

64. **Light and Colour.**—White light is produced when waves of radiant energy corresponding to all or nearly all the wave-lengths that affect vision strike the eye together. When waves of light fall upon any object, some of them are absorbed and the others are reflected; the report these reflected rays convey through the optic nerve to the brain names the colour of the object. Thus when sunlight falls on grass the rays whose vibrations produce the effect of red, yellow, blue, and violet are almost all absorbed, their energy being set to do work in the plant (§ 399), and only those which produce the sensation of green are sent back to the eye. Similarly when light falls on a sliced beetroot the yellow, green, blue, and violet-producing vibrations are absorbed and only the red-producing rays sent back. When light falls on a piece of charcoal it is all absorbed, and as none is reflected the body appears devoid of light, or black. A sheet of paper, on the other hand, absorbs very little of the light and reflects white light as white. The fact that

colour comes from the light, not from the object, may be illustrated by sprinkling salt on the wick of a burning spirit-lamp. The sodium of the salt gives out light of one wavelength only, producing the sensation of yellow. Objects which reflect all kinds of light and those that reflect yellow appear yellow, but such things as beetroots and grass absorb all the yellow light and appear black, like charcoal, which absorbs all light whatever, and the most brilliant painting appears in tones of black and yellow only.

65. **Heat and Temperature.**—The action on matter of radiant energy, particularly of the comparatively long and slowly vibrating waves known as heat, is to make the particles oscillate more rapidly. When the particles of matter vibrate rapidly they send out waves of radiant energy, and thus a heated body radiates heat. Two bodies are said to be at the same temperature when each communicates the same amount of heat to the other as it receives from it. If one body by conduction (§ 59) or radiation (§ 63) gives to another body more heat than it receives from it, the former is said to be at a higher temperature. The hand plunged into water (§ 7) lets us know whether the water is at a higher or lower temperature than the hand. If the water is at a higher temperature, heat passes into the hand which feels warmth, if the water is at a lower temperature heat passes out of the hand which feels cold. The amount of heat which gives a small body a great rise of temperature imparts to a large body a much smaller rise of temperature. Heat is the total amount of molecular motion in the mass, while temperature depends on the rate of that motion. The unit of heat used in this volume is the amount required to raise the temperature of 1 lb. of water 1° F. Temperature is measured by the thermometer (§ 440).

66. **Capacity for Heat.**—Heat bears to temperature exactly the same relation as volume of a liquid does to level. When a large quantity of liquid must be poured into a vessel to raise the level one inch, we say that the vessel has great capacity; while if only a few drops are required to raise the level one inch, the vessel is said to

have small capacity. It is level alone that decides the direction in which the liquid will flow when two vessels are connected by a pipe. Similarly there are some kinds of matter one pound of which requires a great deal of heat to raise its temperature by one degree, while an equal mass of others is raised in temperature to the same amount by very little heat. The former class of substances are said to have a great *capacity for heat*, or, as it is sometimes called, a high *specific heat*. Thirty times as much heat is required to raise the temperature of 1 lb. of water 1° as to raise the temperature of the same mass of mercury by the same amount. Water, indeed, has the greatest capacity for heat of any substance known. On the same fire, if other conditions are the same, mercury becomes as hot in a minute as an equal mass of water does in half an hour; but then as a necessary consequence heated mercury cools as much in a minute as an equal and equally heated mass of water does in half an hour.

67. **Expansion by Heat.**—When the temperature of matter is raised the oscillations of the particles are not only more rapid but of greater amplitude. Each particle occupies a greater space in its longer swing, and consequently the volume occupied by the matter is increased and the density diminished. Expansion of volume by heat takes place in solids, liquids, and gases alike, though its amount is different in each kind of matter and is always greater for gases and liquids than for solids. The lengthening of a bar of iron when heated or its contraction when cooled takes place with nearly irresistible force. The rails of the railway 400 miles long between London and Edinburgh are nearly 1000 feet longer on a summer afternoon than on a winter night. The expansion of a metal rod is often used as a measure of temperature; but thermometers (see § 440) are usually constructed by taking advantage of the greater expansion of liquids or gases. If heat is applied to the lower part of a vessel containing liquid the layer next the source of heat is raised in temperature, expands, and becoming less dense rises to the surface, allowing the denser liquid above to subside to the bottom and get heated in its

turn, thus setting up complete circulation throughout the mass. This transmission of heat by the translation of heated portions is called *convection*, and in consequence of it the temperature of a liquid heated from beneath becomes much more rapidly uniform than that of a solid. The conduction (§ 59) of heat in liquids is very slow, and when the upper layer is heated the vibrations of its particles are passed on by conduction to the mass below very slowly indeed (§ 229), as the expanded upper layer tends to remain in its position.

68. **States of Matter.**—If the particles of any kind of matter were absolutely at rest, that is to say if they possessed no kinetic energy, it is usually assumed that the body would be absolutely cold, or at the absolute zero of temperature. This total absence of heat has never been actually observed. The difference between the same substance in the solid, liquid, and gaseous states is due to the rate of motion of the particles alone, and the work of moving the particles may be readily expressed in terms of heat. Thus in solids which contain relatively little heat the particles move so slowly that cohesion confines them to excessively minute paths, and the substance possesses rigidity (§ 35). In liquids there is much more internal movement or heat, and the particles having a longer path and greater rapidity of motion partly overcome cohesion and show the property of fluidity. Gases contain so much heat that their particles are in very rapid motion through comparatively long paths and the power of cohesion is quite overcome. When the pressure remains the same, every additional degree of temperature makes the particles of a gas move more quickly through a longer path, and the volume occupied by the gas is increased by $\frac{1}{490}$ ($\frac{1}{273}$ for each centigrade degree). A fall of 1° reduces the volume by $\frac{1}{490}$. Hence a fall of 490° of temperature in a gas at 0° should reduce its volume to nothing, which is impossible; hence it is believed that no gas or liquid can exist at $-490°$ F. In other words the particles of solid matter would be motionless, that is, absolutely without heat or at the *Absolute Zero* of temperature.

69. **Action of Heat on Ice.**—We may follow the action

of heat on matter by supposing radiant heat to be supplied to a mass of 1 lb. of ice at 0° F. Each unit of heat raises the temperature of the mass by 2° (hence the capacity for heat of ice is only half that of water), and by the time 16 units of heat have been absorbed, the mass of ice has expanded considerably, and its particles are vibrating with increased energy so that the temperature is 32°. The next 144 units of heat which enter the mass produce no effect on the temperature, which remains at 32°. But the energy is doing other work, for when the 144 units have been absorbed we are dealing with water, not ice. Those 144 units have been expended in work against cohesion and are stored up as potential energy. The heat employed in doing this work of separating particles is sometimes said to become latent, and the *latent heat* of water, *i.e.* the amount of heat necessary to change 1 lb. of the solid into 1 lb. of the liquid substance, is 144 F. heat-units. This is higher than the latent heat of any other substance known.

70. **Action of Heat on Water.**—The volume of 1 lb. of water at 32° is 8 per cent less than the volume of 1 lb. of ice. This is a very significant fact, for almost all other substances occupy a greater volume in the liquid than in the solid state. When 7 heat-units are absorbed by 1 lb. of water at 32° the temperature rises to 39°, but the volume continues to diminish, a state of things which appears to show that in water, unlike almost all other liquids, the faster moving particles fit in a smaller space. But after 39° is past each fresh unit of heat raises the temperature by about 1°, and the volume of the liquid increases faster and faster. From 32° the addition of 180 heat-units raises the temperature to 212° at ordinary atmospheric pressure; but here another change takes place, and the water is said to boil. No less than 967 units of heat must be supplied before the temperature of 1 lb. of water rises above 212°, and at the end of that operation there is not water but 1 lb. of steam or water-vapour at 212°. A real experiment would not proceed so regularly, because at all temperatures water, and even ice, are partly converted into vapour, to produce which a certain amount of heat is used up.

71. **Action of Heat on Water-vapour.**—The work done by 967 heat-units on 1 lb. of water at 212° was done once more against cohesion. The vibrating particles have been enabled to increase the amplitude of their oscillations to a great extent, the volume of the gaseous steam being 1700 times as great as that of the water from which it was derived, and every particle of the water-vapour is darting with the speed of nearly 1 mile per second. When heat is supplied to steam every unit raises the temperature by 2° (its specific heat being only half that of water); the rise of temperature means increase in the velocity of the darting particles and brings about an increase of volume by $\frac{1}{490}$ part for each degree if the pressure upon the vapour remains the same, or a corresponding increase of pressure on the sides of the containing vessel if expansion is prevented. When water-vapour is raised to a very high temperature the heat begins to do the work of breaking up the molecules of water into its components oxygen and hydrogen, thus doing work against chemical attraction and storing up potential energy in the separated gases.

72. **Pressure and Change of State.**—Under pressure ice melts at a lower temperature than 32°, and the few other bodies which contract when they liquefy also have their melting-points lowered by pressure. Bodies which expand when they liquefy—like mercury, rocks, and most other substances—have their melting temperatures raised by pressure so that more heat is required to liquefy them. The effect of pressure on the temperature at which the change from liquid to gas takes place is much more marked. In every case an increase of pressure delays complete vaporisation or boiling until a higher temperature is reached. Water, for example, cannot be heated in the liquid state to a greater temperature than 68° if the atmospheric pressure is one-fortieth of its average amount, but to 176° at half the usual pressure, and to 250° if the usual pressure is doubled. The boiling-point of a liquid may thus be used to measure atmospheric pressure.

73. **Heat-energy.**—The changes which take place when heat is withdrawn from matter are the exact opposite of

those accompanying the application of heat. When oxygen and hydrogen unite, the potential energy of separation is changed into kinetic heat-energy, as already explained (§ 44). When 1 lb. of hot water-vapour radiates out its heat-energy its temperature falls gradually to 212° at ordinary pressure; but then, in assuming the liquid state, 967 heat-units are given out as the particles rush together under the influence of cohesion. One pound of steam at 212° if passed into 4 lbs. of water at 32° gives out heat enough in liquefying to warm up the whole 5 lbs. of water to 212°; hence the great value of condensing steam as a heating agent. One pound of water cooling from 212° to 32° gives out 180 heat-units, and as the particles come fully under the influence of cohesion and group themselves into solid crystals of ice, the energy that held them apart is changed into 144 units of heat.

74. **Mechanical Equivalent of Heat.**—The great task of measuring the quantity of heat-energy which is equal to a certain amount of work (§§ 25, 49), and so of comparing the invisible motion of molecules with the visible motions of masses, was attempted and triumphantly accomplished by Joule in 1843, when the modern theory of energy was founded. He showed that 1 heat-unit was equal to 772 foot-pounds. In other words, if a mass of 1 lb. were to be pulled down by gravity through 772 feet, and the whole of its kinetic energy changed into heat in 1 lb. of water at 32°, the temperature of the water would be thereby raised to 33°. Thus we can measure the work done by heat in melting 1 lb. of ice at 32° (§ 69) and find it to be equal to 111,000 foot-pounds, while that done in evaporating 1 lb. of water at 212° (§ 70) is 747,000 foot-pounds. It appears that the heating of $1\frac{1}{2}$ lbs. of ice at 32° until it becomes steam at 212° requires as much energy as the feat of mountain-climbing described in §§ 49, 52.

75. **Degradation of Energy.**—It is always possible and easy to change work or electricity or light into heat, and 772 foot-pounds of work will always yield the full heat-unit. The inverse operation is different, and from 1 unit of heat the best machine it is possible to imagine could only

obtain a small fraction of its equivalent of work. As water tends to flow to the lowest level, so in Nature energy of every kind tends to assume the least available form, which is that of heat. This process is called the degradation of energy, and in course of time, if it continues to act, all the energy of the Universe will be reduced to the form of heat-vibrations in one uniform mass of matter at one uniform temperature, and although present in full amount quite unavailable for doing work. Viewing the past of the Universe in the light of the degradation of energy, Sir William Thomson has shown that there was a time when the distribution of heat was such as could not have been derived from any conceivable previous distribution ; in other words, that there was a beginning or a creation and that ever since the Universe has been like a machine running down.

76. **Electrical Energy** is not yet sufficiently understood to admit of its nature being simply explained. It seems to be the energy of any form of stress or motion of the ether. Electricity is often spoken of as a fluid, but this is simply the survival of a more dense ignorance of its nature. Electrical energy appears to take part in nearly every change of matter as to composition or state. It has the power of decomposing many chemical compounds which resist the action of every other form of energy, and it can also make some elements combine together which do not unite by any other means. As heat is transmitted from matter at a high temperature to matter at a low temperature, so electricity passes from matter at a high electrical potential to matter at a lower potential. This passage of electricity is called an electric current.

77. **Conductors and non-Conductors.**—Electricity passes readily through some substances, such as copper, silver, metals of every kind, sea-water, damp earth, etc., and these are called conductors. Other substances, such as dry air, glass, sealing-wax, allow it to pass with such difficulty that they are called non-conductors. There is no perfect conductor, nor any absolute non-conductor. Even copper and silver offer a certain resistance to the passage of electricity, and if the difference of potential is sufficiently

great, electricity will overcome the greatest resistance of glass or air. The energy expended by electricity in overcoming resistance is changed directly into heat or light vibrations, as in the case of an electric glow-lamp.

78. **Disruptive Discharge.**—When the amount of electricity on the surface of a small body increases, the potential rapidly rises, and a transference of electricity takes place along the path that offers least resistance. With high potential, electricity can force its way across an interval of air, and as the resistance of air is very great much of the electrical energy is transformed into heat in the process, and the particles of air are set in such violent vibration that they become luminous. Such a transference is called a disruptive discharge, or when it occurs in Nature a flash of lightning.

79. **Magnetism.**—An oxide of iron which exists naturally in considerable quantities has the power of attracting to itself pieces of iron, this attractive force being much more powerful than gravitation. When a bar of this mineral is cut, and so uniformly shaped that no difference in appearance can be found between its two ends, the ends still differ, much as the right hand differs from the left. If the bar be balanced on a pivot it will turn and come to rest with one end pointing toward the north. On this account the mineral is called the *lodestone*. If two similar bars are balanced in this way the north-seeking end of each can be found and marked. The effect of one such lodestone on another emphasises the difference between the two ends. If the north-seeking end of a lodestone is brought near the south-seeking end of another which is balanced the latter is strongly attracted, but if brought near the north-seeking end of the balanced lodestone there is as strong repulsion. The property of two-endedness in bodies outwardly similar is called polarity, and the ends are termed poles. The rule of magnetic attraction and repulsion is very simple—*Unlike poles attract, like poles repel.* The lodestone imparts all its properties to steel when rubbed upon a bar of that metal, and such steel bars are then termed magnets.

80. Electro-magnetism.—The properties of magnets would be inexplicable had not an accidental discovery shown the close relation of magnetism and electricity. It was found that when electric energy is passing through a wire placed above a balanced magnetic needle, the needle swings round and sets itself at right angles to the wire. It was found later that when a coil of copper wire traversed by electricity surrounds a bar of iron, the iron becomes a powerful magnet and retains its properties of polarity and attraction as long as the electricity passes, losing them the instant the current ceases. A coil of copper wire without any iron in the centre was subsequently found to possess polarity, and to exert attraction and repulsion as long as an electric current flowed through it. Hence magnetism can be produced by electricity, and the reverse also holds good. A magnet placed inside a coil of common wire generates a momentary current of electricity. By merely making a coil of wire move in the field of a powerful magnet electricity can be produced in the wire, and thus work can be changed directly into electric currents.

In Nature nothing is so simple as has been represented in this and the last chapter. We do not know how particles vibrate and oscillate, and only guess at the real nature of the forms of matter and energy. Authorities differ in their interpretation of many of the facts, and we have only presented a few of the simpler conclusions in order to assist the student who does not know much of physics and chemistry to follow the chapters which come after.

BOOKS OF REFERENCE

Balfour Stewart, *Elementary Physics.* Macmillan and Co.
P. G. Tait, *Recent Advances in Physical Science.* Macmillan and Co.

CHAPTER IV

THE EARTH A SPINNING BALL

81. The Earth a Sphere.—The field of view at sea or on a level plain is always bounded by an unbroken circle called the horizon; and in all parts of the Earth when one watches a receding object at sea or on a level plain the horizon appears slowly to swallow it up, and it disappears like a traveller over a hill. In all parts of the Earth if the eye is placed 5 feet above sea-level the lower 5 feet of any object are concealed when 4 miles away. Across a lake 4 miles wide, two men of ordinary height standing erect and looking at each other with telescopes can see only the head and hands of the other apparently floating on the water, their bodies being entirely concealed from view (Fig. 9). So from the sea-shore the hull of a ship 10 feet above the water vanishes at 5 miles' distance, and its masthead

FIG. 9.—Curvature of the Earth, exaggerated 400 times.

100 feet high sinks out of sight at 12 miles. Since the same length of an object is concealed by the horizon at the same distance from the observer in all parts of the Earth, it is evident that the dip of the horizon, as it is termed, is practically the same everywhere, and that the surface of the Earth is uniformly curved in a convex form. The only figure which has uniform convex curvature is a sphere, and

the Earth is hence generally spoken of as being a sphere or globe. From 5 feet above sea-level the horizon is only 3 miles distant; from a height of 4000 feet it is 80 miles, so that an observer can see to a distance of 80 miles all round; while from 24,000 feet it is more than 200 miles distant, and in each case a perfect circle.

82. **The Earth an Ellipsoid.**—If the Earth were a perfect sphere its size could be measured by measuring the length, in miles or yards, of the arc of a great circle (*i.e.* a circle the centre of which is at the centre of the Earth) subtending one degree, and multiplying by 360 to give the circumference, for each degree subtends an equal arc on a sphere. It is easy by observations of the stars (§ 92) to tell exactly how many degrees one has advanced along a great circle; and parts of great circles (arcs of the meridian) have been measured in many parts of the Earth with much exactness. In Great Britain 1° was found to be almost exactly 365,000 feet long; but in Peru 1° was found not quite 363,000 feet in length, and in the north of Sweden 1° was found to measure about 366,000 feet. These measurements are undoubtedly correct to within a few feet; and the only conclusion that can be drawn from them is that the Earth is not a sphere, but a figure the curvature of which is less than that of a sphere in some parts and greater in other parts. It resembles a sphere slightly compressed along one diameter, and correspondingly bulged out in the direction at right angles. The length of the shortest diameter has been calculated as 7899·6 miles (about 500,000,000 inches), and the diameter at right angles as 7926·6 miles. The circumference is about 24,000 miles. The form is very nearly that known as an ellipsoid, or oblate spheroid of revolution —a figure that could be made in a turning-lathe, with the axis of rotation in the lathe as the shortest diameter.

83. **The Earth a Ball.**—The most exact measurements which have been made, show that the figure of the Earth is not a true ellipsoid. It appears to be compressed to a slight extent at right angles to the shortest diameter, so that the equatorial diameters vary in length by one or two miles. The exact form of the Earth is being gradually discovered

by very careful measurements of the force of gravity (§§ 38, 252) by means of a pendulum or fine spring-balance. The weight of a given mass on the Earth's surface depends only on its distance from the centre, and thus as the strength of gravity at different places is found, the figure of the Earth is gradually felt out. The form of the Earth is termed by mathematicians a *geoid* or earth-like figure; and it is more accurate to speak of it as a ball than as an ellipsoid or sphere. Yet the difference in shape is so slight that if a geoid or ball, exactly like the Earth, an ellipsoid and a sphere were made each a foot in diameter, it would be quite impossible to tell which was which by the eye or touch.

84. **Structure of the Earth.**—The Earth is a structure composed of three divisions—(1) a vast stony ball termed the *lithosphere* with an irregular surface, part of which forms the dry land; (2) a liquid layer resting in the hollows of the lithosphere, a great part of which it covers; this is termed the *hydrosphere* or water-shell; and (3) a complete envelope of gas surrounding the whole to a considerable height and known as the *atmosphere* or air.

85. **Mass and Density of the Earth.**—To weigh the Earth, all that is necessary is to measure the attraction of gravity between a large block of metal and a small block set at a measured distance. Then (making allowance for the distance of the small block from the Earth's centre) the attraction of the large block on the small one bears to the weight of the small one, *i.e.* the attraction of the Earth on it, the same proportion as the mass of the large block bears to the mass of the Earth. Cavendish, who first carried out this experiment a hundred years ago, employed a cumbrous apparatus in which the large attracting mass took the shape of two leaden balls a foot in diameter. The small block consisted of two small leaden balls fixed to the ends of a light rigid rod, which was hung by a fine silver wire. This arrangement is termed a torsion balance, because when the small spheres were attracted by the large ones and moved slightly toward them the wire was slightly twisted, and the force required to twist the wire to that

extent having been found by experiment, was a measure of the attraction between the small and large spheres. The weight of the small balls is the measure of the attraction of the Earth upon them, and as the distance of the small balls from the centre of the Earth is known, the mass of the Earth can be calculated from the known mass of the large leaden spheres. Mr. Vernon Boys has recently succeeded in making a very fine elastic thread of quartz which acts as an extremely sensitive spring, and can be used to measure the force of attraction between bodies as small as ordinary bullets.[1] As the result of several independent methods, the mass of the Earth has been found to be the same as if it were a globe of homogeneous substance $5\frac{1}{2}$ times as dense as water; the mean density of the Earth is thus said to be 5·5.

86. **The Earth in Motion.**—On a clear morning the bright disc of the Sun appears somewhere on the eastern horizon, rises slowly and wheels round the sky, then, as slowly sinking, it disappears somewhere on the western horizon. When the Sun is visible its light fills the whole sky, which appears as a bright blue dome unless clouds interrupt our view of it. Sometimes a glimpse may be had of the Moon, as a ghostly white broken disc like a little fleecy cloud; very rarely, indeed, the bright light of a planet is visible, or the weird form of a comet. At night the curtain of the Sun's excessive light is dropped, and we see that the whole sky is really gemmed over with bright points or stars, as if a dome or hollow sphere of black paper pricked with innumerable holes had been wheeled between us and the Sun. This star-dome appears to revolve round the Earth, the various marks on it preserving an unaltered arrangement. The stars have been grouped into fanciful constellations, which are easily recognised and serve as a rough-and-ready way of naming any definite part of the sky. By a curious mixture of guessing and of reasoning on the observations which they made, Copernicus and Galileo and their followers came to the conclusion that the regular changes in the appearance of the sky from hour to hour and month to month could only be accounted for by the Earth having at

least two different kinds of motion. The first convincing proof of the Earth's motion was the discovery that a weight dropped from the top of a high tower did not reach the Earth's surface perpendicularly under the point from which it was let go, but always a little to the east (§ 93).

87. Rotation of the Earth.—The old difficulty that the Earth could not be moving because we do not feel it, and that the star-dome could not be fixed because we see it move, no longer troubles people who are familiar with the imperceptible motion of a well-started train, and the apparent gliding away of the platform in the opposite direction. The Earth spins uniformly and regularly from west to east, as may be inferred from the uniform and regular apparent rotation of the starry sky at night. The first Law of Motion (§ 50) enables us to understand how the rotation of the Earth has been actually proved, and what the immediate consequences of rotation are. The French physicist Foucault showed how a large pendulum once set swinging changed the plane of its swing slowly and regularly. If started, for instance, swinging above a table from north to south, at the end of twelve hours it would be found swinging from east to west, and in twenty-four hours it would have changed its plane still farther and be swinging from south to north again. Since the only force which could act on a moving pendulum hung from the solid roof of a building is the rotation of the Earth, this change in the direction of the pendulum proves it. The pendulum does not really change its direction of swinging in space; it remains in a state of uniform motion, and the apparent twisting is produced by the house and the whole Earth turning while the pendulum marks out its invariable line.

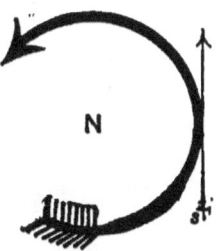

FIG. 10.—Direction of rotation round North Pole, and direction of deviation of moving bodies in southern hemisphere.

88. Polarity.—A ball at rest has no ends or natural points from which to reckon position, but as soon as the ball is made to spin two opposite points on its surface become different from all others, although there may be no visible mark or sign of the fact. These points, which are called ends or poles, are relatively at rest like the centre of a wheel, and the rate at which a point on the surface of a spinning ball moves is greater in proportion to its distance from them. A body spinning uniformly turns round the axis (NS in Fig. 12) or line joining its poles as a wheel spins round an axle. The two poles of a spinning body are distinguished from each other by the apparent direction of rotation about them. Looking down on the Earth from above one pole, an observer would see the surface rotating in a direction opposite to that of the hands of a watch, as shown by the thick arrow (Fig. 10), while if he were to look down similarly on the other pole the surface would appear to rotate in the same direction as the hands of a watch do (thick arrow, Fig. 11). The end first mentioned is called the North Pole, and the opposite is named the South Pole. The Earth always rotates in one direction, from west to east (arrows in Fig. 12); the apparent difference at the poles is due to our looking from opposite sides. The arrow of Fig. 10 appears turning to the left in its flight, that of Fig. 11 appears turning to the right, but on holding the page up to the light they are seen to be one and the same. The student should, if possible, make himself familiar with these facts by actual observations on a terrestrial globe.

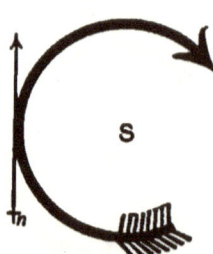

FIG. 11.—Direction of rotation round South Pole, and direction of deviation of moving bodies in northern hemisphere.

89. **Ferrel's Law.**—On a steamer at rest or moving steadily straight forward a passenger has no difficulty in walking in a straight line parallel to the planks of the deck, or in any other direction. But if the steamer is turning rapidly to the right, the promenader, trying to keep in a straight line, has the greatest difficulty in preventing himself from deviating to the left and running against the bulwarks; or if the steamer is turning to the left he can hardly help deviating to the right with reference to the planking. The passenger tends to continue walking in a straight line

with regard to objects outside the ship all the while, and the real motion of the deck toward the right gives an apparent motion of the passenger toward the left. The same thing is true of everything moving rapidly on the surface of the rotating Earth, whether the moving body be a shot from a cannon, a railway train, a river, or simply wind. This fact is thus stated by the American meteorologist, Professor Ferrel: *If a body moves in any direction on the Earth's surface, there is a deflecting force arising from the Earth's rotation, which deflects it to the right in the northern hemisphere, but to the left in the southern hemisphere.* The moving body has a tendency to keep on in a straight line; it is the Earth that changes its direction, as in Foucault's pendulum experiment. Fig. 10 represents the apparent deviation of a body moving in the southern hemisphere, Fig. 11 that in the northern—the thin arrow showing the original direction, the thick arrow the deviation.

90. **Position of the Axis.**—The axis of the Earth about which it rotates is the shortest diameter (§ 82). If the Earth was once much hotter than now and in a semi-fluid condition (as we shall subsequently see reasons to believe), the mere fact of its rotation would make it bulge out along the line farthest from the poles, and that to the precise degree which is found to be the case. As in the case of the rapidly spinning gyroscope (§ 51), and for the same reason, the axis of the Earth preserves its direction practically unchanged in space; and consequently the ends of the axis always point to opposite parts of the starry sky. As the Earth rotates, these points—the poles of the heavens—appear to be at rest, while the sky with its constellations appears to revolve round them from east to west. The north pole of the Earth points very nearly to a bright star which has received the name of the Pole Star or *Polaris*, and is of the greatest importance as a guide to direction and position on the Earth in the northern hemisphere.

91. **Direction on the Earth.**—On account of the Earth's rotation it is possible to fix direction and position on its surface. The line which we may imagine to be traced round the Earth equally distant from both poles is termed

the *Equator*, and it is the only great circle the plane of which cuts the axis at right angles. The half of the globe in which the north pole is situated is termed the northern hemisphere; the half whose centre is the south pole is the southern hemisphere. Great circles running through the poles, and therefore having a north and south direction, are called *meridians*. The direction toward which the Earth turns is called the east, that from which it turns the west. East and west thus indicate merely a direction of turning, and do not refer to fixed points. Small circles traced round the Earth, their planes cutting the axis at right angles, have thus an east and west direction and are called *parallels*. They are, of course, smaller and smaller as the poles are approached. The equator, meridians, and parallels are well shown on the map of the world in hemispheres (Plate XIV).

92. **Latitude** is the name given to the angular distance at the centre of any point on the Earth's surface from the equator measured toward the poles. The equator is chosen as 0° of latitude, and as the distance of the poles is a quarter turn or right angle (§ 31) the north pole has latitude 90° N., the south pole latitude 90° S. The latitude of any place, except the poles, merely refers to the distance from the equator of a small circle, or parallel of latitude, passing through the place in question. Latitude is always measured astronomically by observing the altitude of the pole of the heavens, directly or indirectly. The altitude of the pole, or its angular distance above the horizon of an observer, is equal to the angular distance of the observer from the Earth's equator. Standing on the equator an observer (if the effects of refraction are not considered) would see the north pole of the heavens close to the pole star on the northern horizon, and the south pole of the heavens on the southern horizon, while all the stars would appear to rise in the eastern half of the sky, to describe vertical semicircles, and sink on the western side. If the observer were to journey farther north he would lose sight of the south pole of the heavens, while the north pole would rise higher and higher above the horizon. By the

time he had got half-way from the equator to the pole (45°
N., at O Fig. 12) the pole star would appear to have
risen half-way from the northern horizon toward the zenith,
an elevation of 45°. All the stars within 45° of the pole
would remain in sight all night, never rising or setting, but
circling round the pole; a star exactly 45° from the pole
would describe a circle, passing through the zenith at its
highest point, and touching the northern horizon at the
lowest. Stars beyond that limit would rise in the eastern
part of the sky, describe oblique arcs, and set in the
western; while stars more than 135° from the north pole
of the heavens would never become visible. Finally, if it
were possible to reach the north pole of the Earth, the pole
of the heavens would appear in the zenith (altitude of
90°). All the stars within 90° of the pole would be visible,
but no others. They would never rise nor set, but always
wheel round in horizontal circles, once in twenty-four hours.
Measuring with a sextant the altitude of the pole of the
heavens above the horizon thus gives the latitude of the
observer. In practice the altitude of some bright star or of
the Sun when at the highest point of its daily apparent
path is observed, and the relative position of the Sun
or star being given with proper corrections in the *Nautical
Almanac*, it is easy to calculate the latitude. Thus the
position of an observer on the Earth with respect to the
poles can be found by observations of the stars without
any measuring of distances on the surface, and the position
of a degree of the meridian can be fixed. A degree of the
meridian varies a little in length (§ 82) but averages 69.09
miles; the sixtieth part of this, or one minute of latitude,
measures nearly 6000 feet, and is called a sea-mile, or
nautical mile; the second of latitude measures about 100
feet.

93. Angular and Tangential Velocity of Rotation.
—The Earth turns on its axis uniformly, and the rate of
turning or angular velocity is the same at all parts. A
line drawn perpendicularly from the equator to the Earth's
axis at C describes a whole turn in the same time as a
line drawn perpendicular to the Earth's axis at A from a

point B in 60° latitude. But the line CE is nearly 4000 miles long, while the line AB is not 2000 miles; therefore during the time of one rotation the point E is carried through more than 24,000 miles, while the point B is carried through little over 12,000 miles, and the points N and S are at rest.

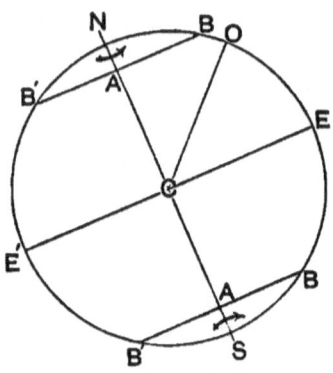

FIG. 12.—Diagrammatic Section of the Earth. C, centre; CE, CO, CN, radii of the Earth; NS, axis; AB, perpendicular to axis; O, a point in 45° N. lat.; B, B, points in 60° lat.

The rate of movement of the Earth's surface by rotation is called its tangential velocity, and diminishes from over 1000 miles an hour at the equator to 500 miles an hour at 60°, and 0 at the poles. A body resting on the Earth's surface has a tendency to fly away at a tangent, like a stone in a sling, and the force of gravity is partly employed in preventing this. The centrifugal force (§ 51) makes bodies weigh less at the equator than at the poles, reinforcing the change due to the fact that the equator is more distant than the poles from the Earth's centre (§ 38). If the Earth rotated seventeen times more rapidly than it does the centrifugal force would be equal to gravity, and if it rotated in the least faster the equatorial part of the Earth would split off like the edge of a burst grindstone. The increase of tangential velocity with length of radius enabled the fact of the Earth's rotation to be proved in the seventeenth century by dropping a weight from the Leaning Tower of Pisa, and observing the distance of its fall to the east of the perpendicular line. The weight was moving eastward on the top of the tower more rapidly than the base of the tower, and retained its original motion in consequence of inertia.

94. **Measurement of Rotation.**—The period which elapses between the Sun crossing the meridian or north and south line of a place on two successive occasions is called a day, and is divided into 24 equal parts or hours;

this is the apparent time occupied by the Earth in making one rotation. It is in many ways more convenient, and also more exact (§ 111), to determine the period of rotation of the Earth by observing the successive transits of conspicuous stars. By this means the exact period of the Earth's rotation has been fixed as 23 hours, 56 minutes, 4 seconds. The name Sidereal Day is given to the rotation period of the Earth as measured by the stars, and astronomers divide it into 24 hours, subdivided into minutes and seconds of sidereal time.

95. **Time.**—The uniform rotation of the Earth is the only standard of time which is practically employed, and for common purposes the solar day of 24 hours is everywhere used as the unit. The Sun crosses the meridian of any place midway between its hour of rising and of setting, and the name meridian (mid-day) was given to the north and south line on this account. Mid-day or noon can be determined exactly by measuring the altitude of the Sun by a sextant or transit circle, or roughly by watching the shadow cast by a stick or a pillar. As the Sun is rising the shadow gradually becomes shorter, and at noon the Sun being at its highest the shadow is at its shortest, and marks out on the ground the north and south line or meridian of the place. The movement of mechanism actuated by a falling weight or an uncoiling spring, and regulated by a pendulum or a balance-wheel, is universally employed for time-measuring; but all clocks, watches, and chronometers must be adjusted according to astronomical determinations of the rotation period of the Earth.

96. **Local Time.**—As the Earth turns, the Sun appears successively on every meridian. It is always noon somewhere, but it can never be noon on two meridians at the same moment. The rate of angular rotation is 360° in 24 hours, or 15° in 1 hour, or 1° in 4 minutes. Thus when the Sun is on the meridian of Greenwich it is 12 hours since it shone on the meridian of the Fiji Islands (180°), where it is consequently midnight. Two towns 15° apart differ 1 hour in their local noon, so that it is necessary in describing

the time of any occurrence to specify by what meridian the time is regulated. The local time in different parts of the world at Greenwich noon is shown on Plate XIX. Greenwich time is used throughout all Great Britain, although at Greenwich noon it is 12.7 local time in the east of Norfolk and 11.37 in the west of Cornwall. In Ireland, Dublin time is employed, the clocks there showing 11.35 at Greenwich noon. Throughout the United States and Canada the time is changed by 1 hour at every 15° of longitude; so that in each belt of that width the same time is shown on all the clocks, and between the Atlantic and Pacific there are five changes of this kind. Travelling eastward or toward the sunrising has the effect of making the Sun rise earlier each day and set earlier each night; passengers on an eastward-bound steamer in the North Atlantic have their meals 20 minutes or half an hour earlier each day according to the speed of the vessel, and the clock appears to go slow. Going right round the world in an easterly direction the few minutes cut off each day by meeting the Sun before the complete rotation of the Earth amount to one whole day extra, so that, for example, in 100 Earth rotations the traveller has seen 101 noons, and recorded the doings of 101 days (each 1 per cent shorter than a day at home) in his diary. Similarly going in a westerly direction the rising and setting of the Sun are delayed by an equal interval of time, and on going round the world westerly in 100 Earth rotations there have been only 99 noons and the doings of only 99 days recorded, each "day" of course being 1 per cent longer than a day at home. In order to keep the dates right a day is dropped out of the reckoning of all vessels sailing eastward when they cross the meridian of 180° from Greenwich, and a day is added on to the reckoning when they cross the same meridian bound westward.

97. **Longitude.**—The longitude of a place is the angular distance of its meridian from some prime meridian, that of Greenwich being usually adopted. In order to find the longitude of a place from the meridian of Greenwich it is only necessary to know the local time and Greenwich time at the same moment. Local noon is easily ascertained by direct

observation of the Sun, or by observing when the Sun attains equal altitudes, before or after crossing the meridian, and halving the interval of time. To get Greenwich time in remote places is more difficult. Accurate chronometers, very carefully regulated and rated, are usually relied on, the average time shown by two or three instruments being taken as correct. If at noon local time, when the Sun is on the meridian, the chronometer shows that it is 11 A.M. Greenwich time, it is evident that an hour must elapse before the Earth has turned sufficiently far toward the east to bring the meridian of Greenwich under the Sun. The interval between the local meridian and that of Greenwich is therefore 1 hour's turning or 15°; and since the Earth is turning toward the east the local meridian must lie 15° E. of that of Greenwich. If at local noon in another place the chronometer showed 2 P.M. Greenwich time, it is evident that the Earth has been turning for 2 hours toward the east since Greenwich was under the meridional sun, and the place of observation lies 2 hours of turning or 30° W. The apparent position of the Moon on the star-dome at successive intervals of Greenwich time is given in the *Nautical Almanac*, the Moon thus serving as a clock-hand pointing to the hour. But seen from different parts of the surface of the Earth the Moon is displaced to one side or another, and it is necessary to calculate the angular distance of the Moon from certain stars as it would appear if measured from the centre of the Earth, just as correct time is only shown by a clock when the observer stands in front of it (§ 33). When this *correction for parallax*, as it is termed, is made, the lunar distances give the Greenwich time by a simple calculation and the longitude can be found at once. Since the great circle of the equator, the circle of only half the size of the parallel of 60°, and the minute circle immediately surrounding the pole are all divided into 360° of longitude, it is evident that while the arc subtending 1° on the equator is equal to that of a degree of latitude, a little over 69 miles, the arc subtending 1° of longitude at the parallel of 60° is only $34\frac{1}{2}$ miles, and that close to the pole only a few feet or inches. The parallels of latitude are equidistant from each other, but the meridians of longitude converge and all meet at the poles.

98. **Terrestrial Magnetism.**—The rotation of the Earth is probably the cause which confers on the globe as a whole the properties of a great magnet (§ 79). The poles of the Earth-magnet are near the poles of rotation, but do not coincide with them; the north magnetic pole lies in 70° 51' N. 96° 46' W. and the south about 73° S. 146° E. (see map Plate I.) When a small straight magnet is hung by a fine thread so that it can move freely in all directions, it takes up a position which in most parts of the world is nearly north and south, hence its use in the mariner's compass (§ 438) as a ready means of finding directions. A suspended magnet when free from any disturbing attraction points due north and south in all places, marked in the map by the curves of 0° or agonic lines. The angle between the meridian and the direction of a suspended magnetic needle is called the *declination*, or by sailors the *variation* of the needle. Between the agonic lines over almost all Europe, Africa, the Atlantic and Indian Oceans, the needle points west of north, the lines in the magnetic chart showing the number of degrees in different places. In the north-west of Greenland the declination is 90°, or the needle points due west; while northward of the magnetic pole it is 180°, or the north-seeking pole turns due south. Over most of Asia, America, the Pacific and Indian Oceans, the declination is to the east of north. After a freely suspended steel needle, balanced so as to rest horizontally upon its pivot, is magnetised one end is found to be drawn downward by the magnetic attraction of the Earth. This phenomenon is called the *Dip* of the needle. Along a certain line on the Earth's surface there is no dip; this line is termed the magnetic equator and is shown in the map. North of it the north-seeking pole dips more and more until at the north magnetic pole it points vertically downward. South of the magnetic equator the south-seeking end of a suspended magnetic needle dips downward. The intensity of magnetic force varies from place to place, being nearly proportional to the dip. In certain regions the rocks beneath the surface

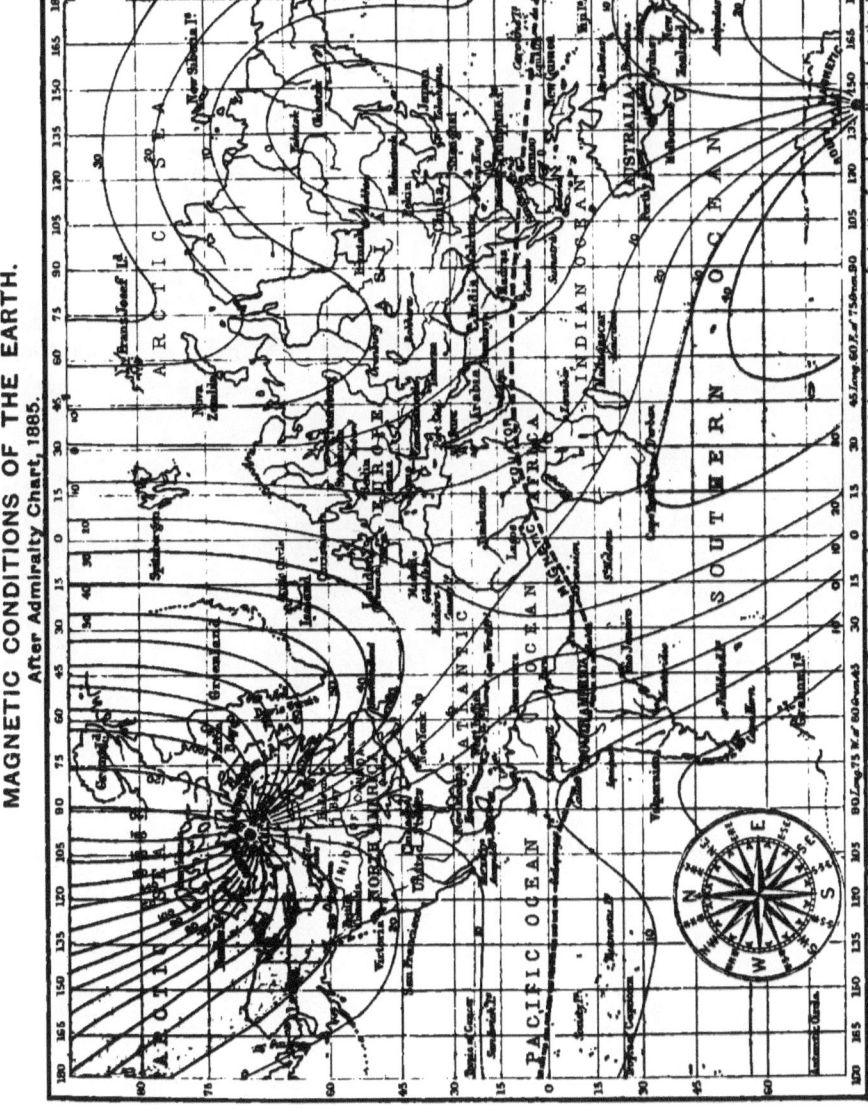

MAGNETIC CONDITIONS OF THE EARTH.
After Admiralty Chart, 1885.

of the Earth exercise a powerful attraction on a suspended magnet (§ 348).

99. Periodical Magnetic Changes.—In 1576, when the declination of the magnetic needle was first measured in London, the north-seeking pole pointed 11° east of north, but the easterly declination gradually diminished until in 1652 the needle pointed due north, and, the change still continuing, in 1815 it pointed $24\frac{1}{2}$° west of north. Since then the declination has gradually diminished, being only 17° W. at London in 1891, and decreasing about 9' per annum. The dip is subject to a similar slow change. These changes were formerly accounted for by supposing that the magnetic poles changed their position on the Earth's surface. Recent observations indicate that this is not the case; they rather suggest that the alteration of declination and dip may be produced by geological changes taking place in the Earth's crust. Commander Creak, as the result of the "Challenger" observations, states that the change is most rapid at several points in a line drawn from the North Cape along the Atlantic to Cape Horn, and that the British Islands are situated in the region where the rate of change is greatest of all.[2] Regular changes of shorter period also occur, the needle daily swinging perhaps 5' or 6' to E. and W. of its average position and back again; and there is a yearly periodicity as well. Irregular variations of much greater extent, sometimes amounting to one or two degrees, are called magnetic storms, and are closely connected with the appearance of the aurora (§ 174). Auroras and magnetic storms are most frequent at intervals of about 11 years, corresponding to the periods of greatest frequency of sun-spots. It is remarkable that whenever a great uprush of heated gas takes place in the Sun, producing solar prominences (§ 116), there is a simultaneous disturbance of all the delicately-hung magnetic needles on the Earth. Thus it appears that while the Earth's magnetism resides in the massive rocks of its crust, and is probably produced and maintained by the Earth's rotation, the Sun's energy exercises a regulating or disturbing influence upon it.

REFERENCES

[1] See *Nature*, xl. p. 65 (1889).
[2] Summary of Creak's Report on "Challenger" Magnetic Observations, *Nature*, xli. p. 105 (1889).

BOOKS OF REFERENCE

See end of Chapter V.

CHAPTER V

THE EARTH A PLANET

100. **The Moon.**—So far we have looked on the heavenly bodies as convenient marks blazoned on the hollow dome of space around the spinning Earth. In § 97 it was implied, however, that the Moon at least was free to change its position on the star-dome. The Moon appears to move amongst the stars, from west to east, so fast that if we observe it rising due east at the same moment as a star, it will be seven times its own diameter behind the star on crossing the meridian, and the star will have set about half an hour before the Moon reaches the western horizon. The Moon often passes between us and a star, and occasionally it passes in front of the Sun, causing an eclipse. These facts prove that the Moon revolves round the Earth from west to east, and that it is the nearest of all the heavenly bodies. The diameter of the Earth affords a sufficiently long base-line (§ 33) to measure the distance of the Moon accurately, the vertical angle at the Moon of the triangle of which the radius (or semi-diameter) of the Earth is the base being 57'. This angle is called the horizontal parallax of the Moon, and shows that the diameter of the Earth as seen from the Moon would be 1° 54'. The parallax varies somewhat during a month, showing that the distance of the Moon is not always the same; but from its average value the average distance of the Moon is found to be 238,793 miles, or in round numbers 240,000. The apparent, or angular, diameter of the Full Moon as seen from the Earth is about

F

30′; that is to say, 180 full moons, one above another, would extend from the horizon to the zenith. The diameter of a body subtending this angle at a distance of 240,000 miles must be about 2000 miles, or, to be exact, 2153 miles. The mass of the Moon has been estimated to be $\frac{1}{80}$ of that of the Earth; its mean density is about 3 times that of water.

101. **The Moon's Surface.**—The Full Moon appears to be diversified with patches of unequal brightness, but observations with powerful telescopes prove that it is simply a lithosphere surrounded by neither water nor air. Ring-shaped mountains closely resembling volcanic craters may be easily seen by using an ordinary field-glass, especially when the Moon is so placed that sunlight illuminates only part of the surface. The Moon shines by reflecting sunlight, and even when most brilliant its light is so feeble that if the whole visible sky (a surface equal to 105,000 moons) were to shine as brightly the effect on the Earth would only be equal to one-fifth that of the Sun. As the Moon revolves round the Earth we see the side turned toward us wholly lit by the Sun once a month and call it Full Moon; a fortnight later the Sun is shining only on the side turned from us and we see the Moon dark, calling it New Moon. Between these periods the illuminated area wanes or dwindles down to a crescent, and again waxes or grows into the full round.

102. **Period of the Moon.**—The Moon revolves round the Earth in 27 days, 7 hours, 43 minutes; but the interval of time between successive new moons or full moons (the lunar month) is rather more than two days longer. The Moon always presents the same aspect to the Earth—only one half, and always the same half, is to be seen, although now and again slight irregularities in its motion reveal a narrow additional strip at one edge or another. The fact that no one has seen the other half proves that the Moon rotates on its axis in exactly the same time as it revolves round the Earth. If it had no rotation we should see all round it. To prove this, pass a loop of thread over a drawing-pin fixed in a horizontal board or table and the other end of the loop round a pencil. Keep the cord stretched, and, holding the pencil between the finger and thumb facing in the direction

of the arrows (Fig. 13), trace a circle without allowing the hand to rotate. The diagram shows that the drawing pin, A, if endowed with vision, would see all sides of the pencil (represented by the arrows) in succession. Next trace a similar circle, holding the pencil firmly but keeping one side of it, say that covered by the thumb, toward the centre, so that the

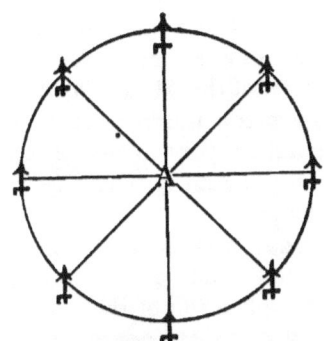

FIG. 13.—Revolution of a non-rotating body; presenting all sides consecutively to the centre.

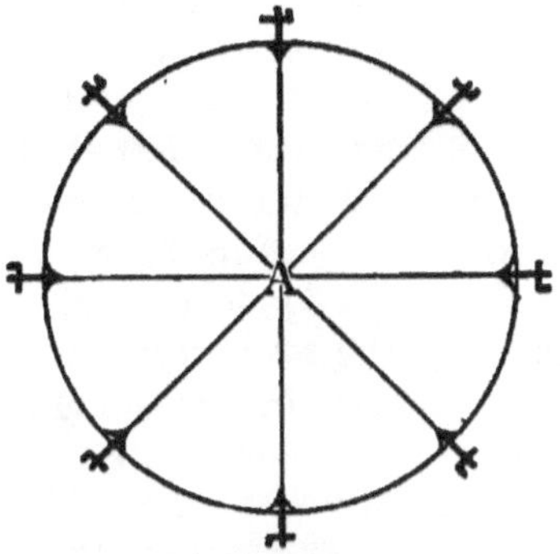

FIG. 14.—Revolution of a body rotating once in the same time as it revolves; presenting always the same side to the centre.

drawing-pin can only see the thumb-nail (arrow-head in Fig. 14). When the circle is complete the cramped position of the hand will prove that there has been rotation at the wrist. The fact of rotation is shown in the diagram by the arrow pointing successively in every direction.

103. **Differential Attraction and Tides.**—Since attraction varies inversely as the square of the distance between the attracting bodies (§ 36), it follows that the Moon must exert a greater attractive power on the side of the Earth which is nearest to it than on that which is 8000 miles farther away. In consequence of this, the Earth is subjected to a stress tending to lengthen it out toward the Moon. The rigid lithosphere is not perceptibly strained; the gaseous atmosphere is so readily disturbed by other causes acting irregularly that only the slightest effect from this cause can be detected in it; but the liquid hydrosphere responds readily and swells into a long low wave, the crests

of which are on opposite sides of the Earth, and equal troughs between them. As the Earth rotates, high water and low water succeed each other regularly, from east to west, as the crest and trough of the wave pass at intervals of about $6\frac{1}{2}$ hours. Without mathematical reasoning it is impossible to explain how the tidal wave, pulsating round the world, is related to the actual position of the Moon in its orbit and in the sky (§ 218). On account of the formation of tidal currents, the hydrosphere is very gently pressed like a brake on the lithosphere by the differential attraction of the Moon; and as the energy of the currents comes from the Earth's rotation, the rate of rotation at the end of each century is slower by the fraction of a second, and the time of rotation, or day, is longer in the same minute proportion.

104. **The Tidal Romance of the Moon.**—Millions of years ago the Earth must have rotated much more rapidly than now, when it suffers from long application of the brake. At that remote epoch the Moon was much nearer than now, for it is a property of revolving bodies, which cannot be explained here, that any reduction in the rate of the Earth's rotation is necessarily accompanied by an increase in the Moon's distance. The nearer Moon must have raised far greater tides than those we now know, in the more extensive and denser hydrosphere of those ancient days. In the remotest past on which this argument casts light the Moon must have been close to the Earth, whirling round its little orbit in the same time as the Earth spun round on its axis, which was then only a few hours. The Moon, indeed, seems to have been originally part of the semi-fluid Earth whirled off by the furious rotation (§ 93) of the earliest times. As the Moon receded from the Earth in its slowly widening spiral path it also had a hydrosphere in which the Earth's differential attraction raised tides, the friction of which gradually brought the rapid rotation of our satellite to correspond with its period of revolution round the Earth.

105. **The Sun** even more conspicuously than the Moon, separates itself from the other heavenly bodies, which are dim by contrast with its brilliance, and when the Sun

rises vanish from sight like tapers when an electric arc is turned on. The altitude of the Sun at noon, observed at any place, varies throughout the year, increasing day by day until a certain maximum is reached, and then decreasing gradually to a minimum. The period from highest Sun to highest Sun, as observed in regions outside the tropics, is about 365 days. The angular diameter of the Sun when measured daily is found to gradually increase from a minimum of 31' 32" to a maximum of 32' 36", and then to diminish again to its former value, and this change also takes place in about 365 days. Unless with the aid of a very powerful telescope we cannot see the constellations in daylight so as to be able to tell amongst what group of stars the Sun appears at noon; but we know that these stars are just opposite those which cross the meridian at midnight. In the course of 365 days all the constellations of the star-dome successively cross the meridian at midnight, and from this fact we know that the Sun, like the Moon, moves amongst the stars from west to east, although in a year instead of a month.

106. **Problem of the Earth and Sun.**—The most natural explanation of the Sun's annual path amongst the stars is that the Sun, like the Moon, revolves round the Earth, but in a year instead of in a month. Another hypothesis, that the Earth revolves round the Sun, would also explain the facts. In Fig. 15 both hypotheses are illustrated. S represents the sun, E the earth, the arrow ESN shows where the Sun appears amongst the stars at noon, and the arrow EM shows what stars cross the meridian at midnight. The dark circle is the hypothetical orbit of the Earth round the Sun, the lighter circle the hypothetical orbit of the Sun round the Earth. The arena is so vast that the gyrating pair of globes are practically at the same distance from the amphitheatre of stars. Whether we assume that the arrow ESN, passing through the Sun, turns round the centre E, or that the arrow ESN, passing through the Earth, turns round the centre S, the arrow would point successively to the same parts of the star-dome, and observation of the stars would not decide which is the

correct hypothesis. The law of gravitation explains that two revolving bodies circle round the centre of gravity of the pair. In the case of the Earth and Moon the centre of gravity of the system lies within the Earth, hence the Moon

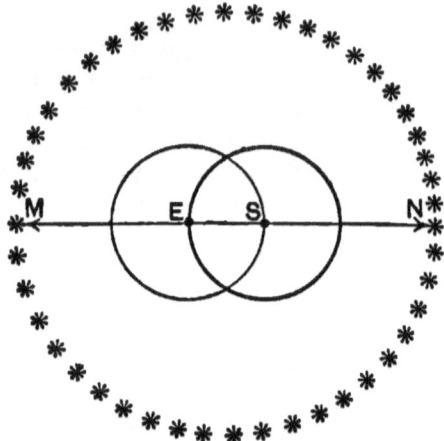

FIG. 15.—Problem of the Earth and Sun. Showing how observation of the Sun's place amongst the stars cannot tell whether the Earth (E) goes round the Sun (S), or the Sun goes round the Earth.

appears to revolve round it. It remains to inquire where the centre of gravity of the Earth and Sun lies; in other words, whether, and by how much, the Earth or the Sun is the greater body.

107. The Sun's Distance and Mass.—The horizontal parallax (§ 100) of the Sun is not quite $9''$; and being so minute it is not easily measured accurately. Since the Sun's parallax is about $\frac{1}{380}$ of the Moon's (§ 100), it follows that the Sun must be about 380 times more distant from the Earth than is the Moon. Accurate determinations give the average distance as 92,700,000 miles. Since the Sun subtends as large an angle to our eye (about 30′) as the Moon does, it follows that the Sun, being 380 times as distant, must have a diameter 380 times as great as that of the Moon, that is to say, about 800,000 miles. The Sun's volume is thus more than 1,200,000 times that of the Earth. By the attractive force it exerts the Sun's mass is proved to be more than 300,000 times that of the Earth. The centre of gravity of the Earth-Sun System must, indeed, lie within the Sun, and it is therefore as certain that the Earth goes round the Sun as that a weight of 50 lbs. will cause 1 grain to fly up if the two are placed in the opposite scales of a balance.

108. **Proof of Revolution.**—If a man, sitting in a dog-cart on a dead-calm day while a steady downpour of rain is falling, finds the raindrops driving against his face instead of falling straight upon his hat, he concludes correctly that this *aberration* or wandering of the raindrops from their normal path is due to the fact that the dogcart is not at rest but in rapid motion. By estimating the angle at which the rain strikes he may even calculate the rate at which he is being carried along. The astronomer, sitting in his observatory, detects a similar aberration in the light-rays from each of the stars. He finds the light reach him at a different angle at various times of the year, so that each star traces out a minute annual curve on the sky the greatest radius of which is about 20″. No other cause can account for this aberration of the starlight except the fact that the observatory and the Earth itself are rushing with tremendous velocity through space in a closed curve which takes one year to complete. The rate of motion can be calculated from the angle of aberration, when the velocity of light is known.

109. **The Earth's Orbit.**—The regular change in the angular diameter of the Sun seen from the Earth (§ 105) proves that the annual orbit is not a circle, as the two bodies are sometimes nearer and at other times farther apart. The form is an ellipse (Fig. 16), of which the Sun occupies one focus (S); but the ellipse is very like a circle, the ratio of the longest to the shortest diameters being as 100,000 to 100,014. Indeed, if a circle 3 inches diameter were drawn with a very sharp pencil making a line $\frac{1}{5000}$ of an inch thick, it would represent the orbit correctly, the difference between the ellipse and the circle being concealed by the thickness of the line. The place of the Sun would, however, require to be represented $\frac{1}{40}$ of an inch from the centre of the circle. Certain slow changes take

place in the form of the orbit on account of the perturbation of the Earth by other planets. The eccentricity, or distance of the Sun from the centre, increases to a very marked degree, diminishes until the orbit becomes almost a circle, and then begins to increase again. The time elapsing between successive maxima of eccentricity is about half a million years. The Earth moves round this orbit with varying speed, moving fastest when nearest the Sun (or in perihelion, p), and slowest when most remote (or in aphelion, a); the average velocity is about $18\frac{1}{2}$ miles

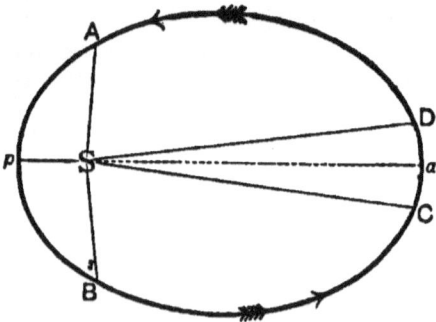

Fig. 16.—Ellipse, representing the Earth's orbit enormously exaggerated in ellipticity and eccentricity. S, the sun; a, aphelion; p, perihelion.

per second or 66,000 miles an hour. Before Newton proved that the power of gravity would produce precisely this effect, Kepler had discovered the nature of the motion, and expressed it in his "Second Law" thus: *The radius vector, or line joining the centres of the Earth and Sun, sweeps through equal areas in equal times.* In Fig. 16 the figure SAB is equal in area to the triangle SCD; S being the sun, SA, SB, SC, SD, being successive positions of the Earth's radius vector. Hence, since the radius vector sweeps through the angle SAB in the same time as it takes to sweep through SCD, the Earth traverses the long part of its orbit from A to B through perihelion in the same time as it traverses the much shorter distance from C to D through aphelion.

110. **The Year.**—The period in which the Earth

accomplishes one revolution round the Sun is called a year, and is the unit for long intervals of time. The unit for shorter intervals of time is the solar day or apparent period of the Earth's rotation. Unfortunately these two natural units are incommensurable; the revolution period of the Earth with regard to the stars is not made up of an even number of rotation periods or of solar days, but consists of 365 days, 6 hours, 9 minutes, $9\frac{1}{2}$ seconds. The *tropical year* or time of apparent revolution is 365 days, 5 hours, 48 minutes, 46 seconds; and it is in order to fit in the extra 5 hours and odd minutes that the plan of having an extra day every fourth year (leap year), and omitting it once a century, is adopted. If this were not done, the same period of the year would not occur in the same part of the Earth's orbit at each successive revolution.

111. **Solar and Sidereal Time.**—The revolution of the Earth round the Sun once in a year accounts for the interval between two successive transits of the Sun across the meridian, the solar day being nearly 4 minutes greater than the Earth's rotation period or sidereal day. While the Earth is turning once round on its axis it advances so far upon its orbit that nearly 4 minutes of turning more than a complete rotation are necessary to bring the Sun once more on the meridian. Since the Earth moves with unequal velocity in different parts of its course, and its axis is not perpendicular to the plane of its orbit, the day, as measured from noon to noon, varies slightly in its length throughout the year. The average solar day is taken in order to calculate the solar mean time which is always used in ordinary affairs.

112. **The Ecliptic.**—The Earth's orbit lies always nearly in the same plane, because there is no force competent to change its direction. That is to say, the Earth goes round the Sun in limitless space as a boat sails round a ship on the surface of a calm sea. We may imagine the plane to extend beyond the Earth's orbit through all space so that it intersects the dome of stars. The line of intersection is the apparent yearly path of the Sun amongst the stars, and is called the *ecliptic*; the constellations it traverses are the

well-known twelve "signs of the zodiac." The plane of the ecliptic in space serves as a standard level, to which other directions may be referred for comparison. It seems most natural that the Earth's axis should be perpendicular to the plane of the ecliptic, but, as has been said, this is not the case. The axis is inclined about $23\frac{1}{2}°$ from the perpendicular. We have thus to picture the Earth sailing round the Sun, not "on even keel" but with a list or inclination of $23\frac{1}{2}°$, and with the north end of the axis always pointing toward the same bright star on the celestial dome. This inclination is not absolutely constant, but like the eccentricity of the orbit is subject to slight increase and diminution in long periods.

113. **Eclipses.**—Instead of saying that the Earth revolves round the Sun we should, in order to be accurate, say that "the Earth-Moon System" does so; for the Moon shares the annual revolution of the Earth as a point on the tire of a wheel shares the onward movement of the centre. If the Moon's orbit lay in the plane of the ecliptic, the Moon would pass between the Earth and Sun once every month, and a fortnight later the Earth would cut off the sunlight from the Moon. In other words, at every New Moon there would be an eclipse of the Sun, at every Full Moon there would be an eclipse of the Moon. But the Moon's orbit is inclined at an angle of about 5° to the ecliptic, and it is only when the Moon happens to be at one of the nodes, or points on the orbit where its plane intersects the ecliptic, that an eclipse can take place. From this fact the ecliptic gained its name. Eclipses of the Moon are common occurrences, for they happen several times in a year and are visible from a large area of the Earth's surface, as the Earth's shadow is wide compared with the angular diameter of the Moon. Eclipses of the Sun are more frequent, but are more seldom seen at a given place, being visible only for a comparatively short time and over a limited tract of the Earth's surface, since the Moon's shadow thrown by the Sun is a comparatively narrow cone. When the Moon is at its nearest point to the Earth, in the course of its elliptical orbit, its angular diameter is great enough to entirely conceal the Sun, and the eclipse is said to be total. But when the Sun

is at its nearest, its disc appears larger than that of the Moon at its farthest; and if an eclipse occurs in such conditions it is said to be annular, the black disc of the Moon being surrounded by a narrow bright ring of the Sun, like a penny lying on a half-crown.

114. **Solar Tides.**—The differential attraction of the Sun on the opposite sides of the Earth has a tide-raising power like that of the Moon (§ 103). But the Sun is so distant that in spite of its vast mass the difference in its attracting power on opposite sides of the Earth, due to the distance of 8000 miles, is only two-fifths as great as the difference in the attracting power of the nearer Moon. At New Moon and at Full Moon the tide-raising power of Sun and Moon is exerted in the same direction, and produces Spring-tides in the ocean; the tidal wave rises highest and sinks lowest or has its greatest amplitude. At the quarters, on the other hand, the Sun is raising high water where the Moon is producing low water, and consequently the amplitude is much less, the tidal wave not rising to the average height nor sinking to the average depth. These are called neap-tides, and represent the difference, as spring-tides represent the sum, of the tide-raising power of Sun and Moon, the relative amplitudes being as 3 to 7.

115. **Precession of the Equinoxes.**—The tropical year or apparent time of the Sun's circuit of the heavens is 20 minutes shorter than the Earth's revolution period (§ 110); in other words, if the Sun starts from that point of the ecliptic known as the vernal equinox it will reach it again 20 minutes before completing the annual circuit of the heavens. Thus the equinox seems to be moving slowly along the ecliptic to meet the Sun, and so every year it precedes or comes before its former position, the phenomenon being known as the precession of the equinoxes. The star-dome, not sharing the movement, appears to rotate about an axis at right angles to the plane of the ecliptic, but so slowly that 25,000 years are required for a single turn. Consequently the constellations on the zodiac have ceased to correspond with the "signs" of 30° each which formerly included them. This apparent movement of the heavens must be produced by a

real rotation of the Earth in 25,000 years round an axis perpendicular to its orbit. The axis of diurnal rotation thus describes a slow conical motion like the mast of a boat which is pitching and rolling equally, and the north pole, instead of pointing steadily to the pole star, traces out a circle on the star-dome about 47° in diameter in the course of 25,000 years. The horizontal axis of a gyroscope at rest is at once drawn into a perpendicular position by attaching a light weight to one end. But if the fly-wheel is in rapid rotation, the angle which the axis makes with the perpendicular remains constant, and the weight attached merely sets up a slow rotation of the gyroscope about the perpendicular, the axis of spinning tracing out a circular cone (§ 51). The differential attraction of the Sun and Moon on the protuberant region about the Earth's equator (§ 82) exerts a force tending to pull the equator into the plane of the ecliptic and make the axis of diurnal rotation perpendicular. Rotation sets up resistance as in the gyroscope, and the attempt to make the Earth sit upright results in the very slow rotation about the perpendicular, to which the axis of diurnal rotation preserves the nearly constant angle of $23\frac{1}{2}°$.

116. **The Sun's Surface.**—The bright disc of the Sun which we see is termed the *Photosphere*, and although it appears uniform in texture to the eye, the telescope shows that it is finely mottled with brilliant granules separated by a less luminous network. The Sun rotates in about 25 days, but not like a solid globe, and the fact that marks on different parts of the surface move at different rates proves that the photosphere is the surface of a dense and intensely heated atmosphere in which the bright granules are vast luminous clouds. During a total solar eclipse red flames of fantastic form are usually seen projecting beyond the black disc of the Moon, and these *Prominences* may also be observed without an eclipse by an ingenious arrangement of the spectroscope. They consist of great outbursts of intensely heated gas, mainly hydrogen. Prominences have been seen rising to the height of 400,000 miles above the Sun's surface in a few hours, against gravity 27 times as powerful as that of the Earth. This gives us some idea of the terrific violence

of the manifestations of solar energy. Down-rushes of comparatively cool gases from the upper regions of the Sun's atmosphere are believed to be the cause of black marks which are often seen on the photosphere and termed *sun-spots*, although sometimes many thousand miles in diameter. Though apparently black, compared with the intense glow of the rest of the surface, sun-spots really shine with a light brighter than that of the electric arc lamp. Photographs of the Sun's disc are taken daily in some observatories in order to preserve a record of the number and movements of sun-spots, and in this way much information has been obtained on the subject. It has been observed that spots usually originate at some distance on either side of the Sun's equator, and for a time they increase in size; then beginning to diminish they travel toward the equator and gradually vanish, being succeeded by others, which are smaller and fewer. Finally, after about twelve years or so, the whole set fades away, and a new series of larger size appear and go through the same changes. Periods when sun-spots are at a maximum succeed each other at intervals of about eleven years, and relations have been traced between them and the influence of the Sun's radiant energy on the Earth. During total eclipses a halo of silvery light, sometimes circular, sometimes spreading out like great wings, surrounds the Sun. It is called the *corona*, and is probably composed of fine particles of dust either thrown off by the Sun or being attracted toward it and shining, in part at least, by reflected light.

117. **The Spectrum of Sunlight** is a continuous band of colour crossed by an immense number of black lines (the more conspicuous of which are named in Fig. 8, § 63), showing that the light from some glowing solid or liquid has reached us after traversing an expanse of cooler vapour. Every year more of the lines in this spectrum are identified, and those which are produced in the Earth's atmosphere are being distinguished from those due to the Sun's. The lines produced by absorption of light in the Earth's atmosphere are best recognised by comparing the spectrum of the Sun low in the sky, when they are strongest, with that

at noon, when they are faint. When a body giving out light is in rapid motion toward the observer, the wave-length of the light is apparently shortened and the lines of its spectrum are shifted toward the violet end. In the light of a rapidly receding body the lines are similarly shifted toward the red end. At its equator the Sun's surface is moving 70 miles a minute, toward an observer on one side —from him on the other. By causing a small image of the solar disc to flit across the slit of the spectroscope several times in a second, an observer analyses in quick succession the light from the approaching and receding edges. Consequently the most distinct solar absorption lines are seen to oscillate slightly from side to side, being displaced alternately toward the red and toward the violet, while the lines produced in the Earth's atmosphere remain motionless and can be readily distinguished. The elements which have been detected in the Sun are identical with those found in the Earth, but the spectrum shows that they are at an enormously high temperature, so much so that some of the solar lines not yet identified may be due to matter of a simpler form than any elements known on the Earth (§ 47).

118. **The Heat of the Sun.**—The temperature of the Sun is higher than any that has been produced on Earth, and it does not perceptibly differ from year to year. If the Sun were a heated solid or liquid globe it would be falling in temperature as it radiated heat, unless the supply were kept up in some way. There is no external source of heat that is sufficient to account for the vast solar expenditure. The collision of meteorites and many other theories have been suggested, tested, and rejected, and we must look to the Sun itself for an explanation. Sir William Thomson[1] and Professor von Helmholz have shown that as the solar atmosphere loses its heat the power of gravity draws its particles closer together, and this shrinking transforms the potential energy of separation (§§ 54, 56) into heat, which is sufficient to maintain the diminished volume at the same or even a higher temperature. The process will go on, loss of heat being compensated, or more than compensated,

by shrinkage, as long as the Sun remains mainly gaseous. If this theory is correct, Sir William Thomson estimates that twenty million years ago the substance of the Sun was so diffused and cool that it had not begun to give out light such as we now enjoy, and that five or six million years hence the sphere will have grown solid, cold, and dark.

119. **The Earth's Share of Sun-heat.**—Since the Sun's parallax is less than 9" it follows that, viewed from the Sun, the Earth only occupies $\frac{1}{2000000000}$ of the sky, or a disc 18" in diameter. The Earth consequently receives less than $\frac{1}{2000000000}$ of the radiant energy sent out by the Sun. If the Sun were expending, instead of energy, money at the rate of £18,000,000,000 a year, the Earth's annuity would be only £9. This endowment, however, is payable continuously, and at the same rate throughout the year, in the proportion of 6d. every day or ¼d. every hour. Minute as the energy which reaches the Earth appears in view of what streams away into space, it is stupendous when compared with the power of the greatest steam-engine ever constructed, and is, indeed, the source of all the work and all the wealth of the world actual and prospective.

120. **Effects of Inclined Axis.**—If the Earth's axis of rotation were perpendicular to the plane of the ecliptic the Sun's radiant energy would be dispensed for an equal time each day over the whole surface—every place would always have 12 hours of daylight and 12 hours of darkness. The Sun would always be in the zenith at noon on the equator, but never elsewhere; at the poles the Sun would always be half above the horizon, and at every intermediate point the meridian altitude would always be (as in fact it is at the equinoxes) the complement of the latitude, *i.e.* 90° minus the latitude. In consequence of the inclination of the axis the distribution of radiant energy on the Earth is unequal and varies at different times of the year, giving rise to the difference of the *seasons*.

121. **Vernal Equinox.**—The position on 21st March (Fig. 17) is such that the equator lies in the plane of the Earth's orbit as viewed from the Sun, and the Sun appears in the zenith at noon viewed from the equator. Sunlight

reaches both poles simultaneously, and as the Earth rotates, every place on the surface is lighted up for twelve hours and plunged in darkness for the other twelve, day and night being equal everywhere. This period is therefore called the vernal or spring *equinox*, and happens at that point in

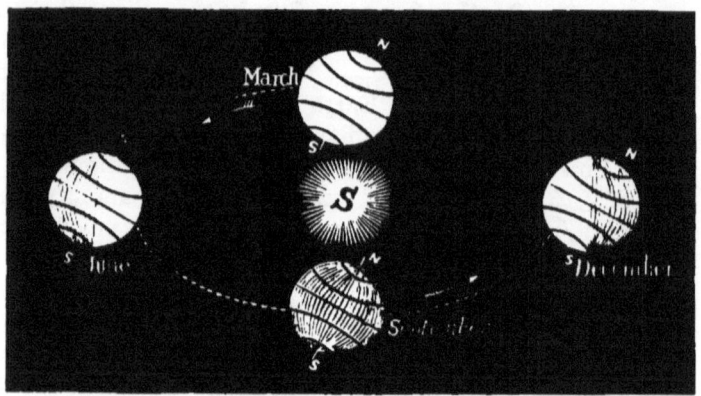

Fig. 17.—Diagram illustrating the cause of the seasons.

the Earth's orbit from which the Sun appears projected on the star-dome in the sign of Aries. This season is spring in the northern and autumn in the southern hemisphere.

122. **Summer Solstice.**—In three months, the Earth having advanced along one quarter of its path, the equator dips $23\frac{1}{2}°$ S. of the plane of the ecliptic when viewed from the Sun, hence the Sun viewed from the Earth appears at noon in the zenith on the parallel of $23\frac{1}{2}°$ N.; and as at this time the Sun is projected on the star-dome in the sign of Cancer, this parallel is called the Tropic of Cancer. This is the highest northern latitude for a vertical Sun, and is called a *tropic* because the Sun appears to *turn* southward after reaching it. Sunlight reaches $23\frac{1}{2}°$ beyond the north pole, and falls short of the south pole by $23\frac{1}{2}°$. As the Earth rotates the whole region for $23\frac{1}{2}°$ round the north pole keeps in sight of the Sun, the whole region round the south pole rests in darkness, and the period of daylight diminishes while that of darkness increases over the world

from north to south, being 12 hours each at the equator. The Sun being vertical at noon, $23\frac{1}{2}°$ north of the equator, its meridian altitude from the south point of the horizon in the northern hemisphere is equal to the complement of the latitude plus $23\frac{1}{2}°$. In the southern hemisphere the Sun's greatest altitude is equal to the complement of the latitude minus $23\frac{1}{2}°$. This period is termed the summer *solstice*, as the Sun *stops* in its northern path. It is the middle of the northern summer and of the southern winter. The parallels of $66\frac{1}{2}°$ ($23\frac{1}{2}°$ from the poles) are termed the *Arctic* and *Antarctic Circles*, and these are the lowest latitudes in which sunlight or darkness can last for 24 hours at a time.

123. **Autumnal Equinox and Winter Solstice.**—In three months more it is the autumnal equinox; the equator comes again into the plane of the Earth's orbit, day and night are equal from pole to pole, and the Sun's meridian altitude is again equal to the complement of the latitude. The Sun is projected on the star-dome in the sign of Libra, and it is the autumn of the northern hemisphere and the spring of the southern. Another period of three months brings the Earth into such a position that the equator is $23\frac{1}{2}°$ N. of the Sun's place in the ecliptic, and consequently the Sun is seen vertically overhead at noon from the parallel of $23\frac{1}{2}°$ S., which is termed the Tropic of Capricorn after the sign in which the Sun is projected on the star-dome. This is the highest south latitude for a vertical Sun. The Sun is visible everywhere within the antarctic circle, but all within the arctic circle is in daylong darkness. In all parts of the southern hemisphere the Sun's meridian altitude above the north point of the horizon is $23\frac{1}{2}°$ greater than the complement of the latitude; in the northern hemisphere it is $23\frac{1}{2}°$ less, and the days grow shorter and the nights longer from south to north, day and night being equal on the equator. This is the winter solstice, midwinter in the northern hemisphere and midsummer in the southern.

124. **Altitude of the Sun.**—The altitude of the Sun and duration of daylight are described above for a globe without an atmosphere. On account of refraction (§ 150)

G

the Sun always appears higher in the sky than its true position; the period of daylight is thus increased and the period of darkness diminished, the effect being greatest in high latitudes.

LENGTH OF THE LONGEST DAY.

Latitude	0°	10°	20°	30°	40°	50°	60°	70°	80°	90°
Hours	12	12h. 35	13h. 12	13h. 56	14h. 51	16h. 19	18h. 30	65 days	161d.	186d.

(Refraction slight and not allowed for)   (Refraction allowed for)

Between the tropics the Sun is vertical in every latitude twice in the year; outside the tropics never. Even in summer the altitude of the Sun is low in high latitudes; it can never be more than $23\frac{1}{2}°$ at the poles, nor more than $53\frac{1}{2}°$ in 60° latitude. The amount of radiant energy falling on the surface varies with the altitude of the Sun. Fig. 18 shows that the same beam of light which, falling vertically, covers 1 sq. ft. of surface, will, when falling at an angle of 30° cover 2 sq. ft., and so produce on each square foot only one half of the effect of vertical light; at a lower angle the heating effect of sunlight is very slight. Oblique rays of light also pass through a thicker layer of the Earth's atmosphere, and so are more absorbed than vertical rays.

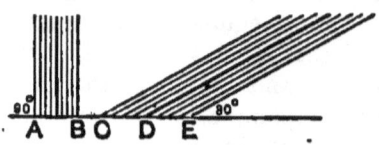

FIG. 18.—Angle of Light rays. The breadth of the beam CE is the same as that of AB, but striking at an angle of 30° the length CE is twice the length AB where the rays fall perpendicularly.

125. Zones of Climate.—It follows that the region between the tropics receives most of the solar energy, higher latitudes sharing it in smaller and smaller proportions. The Earth has consequently been divided into zones of climate—a word originally meaning *inclination* of the Sun's rays. The areas within the polar circles, poorest in radiant energy, are termed the *Frigid Zones*, those between the polar circles and the tropics, where there is a tolerable abundance of radiation, the *Temperate Zones*, and the wide belt between the tropics which is overflowing with solar wealth the *Torrid Zone* (Fig. 19). If the Earth were a smooth lithosphere, either free from water or surrounded by a continuous hydrosphere and atmosphere, this unequal distribution of solar energy would give rise to a regular system of redistribution by currents streaming from the equator to the poles in the upper regions of the atmosphere, and from the poles to the equator in the lower, their paths curved in consequence of the rotation of the Earth; and in this way the tropical warmth would be distributed with some approach to uniformity over the whole surface. The actual redistribution is much more complicated (§ 178 and following).

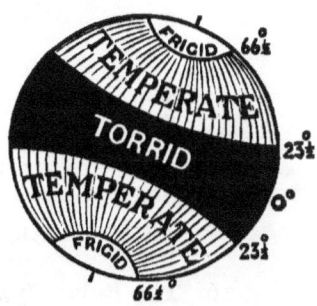

FIG. 19.—Zones of Climate on the Earth.

Reference

[1] Sir Wm. Thomson on "The Sun's Heat," *Nature*, vol. xxxv. p. 297 (1887).

Books of Reference

J. F. W. Herschel, *Astronomy*, Cabinet Cyclopædia. (The most perfect description of simple mathematical astronomy.)

R. S. Ball, *Time and Tide: A Romance of the Moon*. S.P.C.K.

James Nasmyth, *The Moon considered as a Planet, a World, and a Satellite*. John Murray. (Unique illustrations of the surface of the Moon.)

See also list at end of Chapter V.

CHAPTER VI

THE SOLAR SYSTEM AND UNIVERSE

126. The Solar System.—The Sun and Moon are not the only celestial bodies which pass between our eyes and the dome of stars. Several bright objects, which, unlike the stars, shine without twinkling by light reflected from the Sun and show a distinct disc in the telescope, were long ago called *planets*, or wanderers, for they pursue a devious track among the constellations, changing in position on the star-dome from night to night. All the planets are related to each other, as their wanderings are all confined to the belt of sky termed the zodiac, extending only a few degrees on each side of the ecliptic. The distances of these bodies from the Earth have been measured, and it has been proved that like the Earth they all rotate and revolve round the Sun in elliptical orbits, the planes of which are, as a rule, only slightly inclined to the plane of the ecliptic. Some of the statistics of the members of the solar system are given in the following table.

127. Inner Planets.—The four planets next the Sun are often called the inner planets. Mercury and Venus are never seen very far from the Sun, and Mercury is rarely visible to the naked eye. **Venus**, visible sometimes as the evening star shortly after sunset, and at other times as the morning star shortly before sunrise, is a magnificent object, its light being often strong enough to throw a distinct shadow. These two planets exhibit phases like the Moon, those of Venus being clearly visible by the aid of an opera-glass. Signor Schiaparelli has recently proved that the period of rotation of Mercury is equal to its period of revolution round the Sun; and this is probably true of Venus also. Solar tidal friction has evidently acted on these planets as the tidal friction of the Earth has acted

THE PLANETS.

Name.	Symbol.	Mean Distance from Sun. Million Miles.	Periodic Time. Solar Days.	Diameter of Planet. Miles.	Rotation Period.	Satellites.
Mercury	☿	35.9	88	2,992	Days. 88	...
Venus	♀	67.0	224.7	7,660	224.7	...
Earth	⊕	92.7	365.3	7,918	Hrs. Min. 23 56	1
Mars	♂	141	687	4,200	24 37	2
ASTEROIDS	...	...	...	...	...	...
Jupiter	♃	482	4,332	85,000	9 55	4
Saturn	♄	884	10,759	71,000	10 14	8
Uranus	♅	1780	30,687	31,700	...	4
Neptune	♆	2780	60,127	34,500	...	1

on the Moon; and it is interesting that the two planets nearest to the Sun, and receiving enormously more heat and light than the Earth, have perpetual day in one hemisphere, and perpetual night with a cold approaching the absolute zero in the other. Mercury and Venus occasionally pass between us and the Sun, the planet appearing to pass across the solar disc like a small black spot. A transit of Venus affords the best opportunity of measuring the solar parallax, and hence the Sun's distance, by noticing how far the path of the planet across the disc is altered when viewed from distant parts of the Earth.

128. **Mars**, the first planet beyond the Earth, most resembles it. The rotation period is nearly the same, and the surface is diversified by marks which evidently indicate continents and seas, while at each pole a gleaming white patch increases and decreases as the planet wheels round

the Sun, suggesting the forming and melting of great areas of snow. Until 1877 Mars was supposed to have no satellites, but in that year Professor Hall of Washington discovered two. One is very small, very near the planet, and races round it, from west to east, in little more than 7 hours, making three complete revolutions whilst the planet rotates once; the other, farther away, revolves in 30 hours.

129. **Asteroids.**—It had been observed even before Kepler's time that there is a certain symmetry in the placing of the planets. This relation was subsequently formulated by the German astronomer Bode in the end of the eighteenth century, and has since been termed *Bode's Law*. It is as follows: If 4 be added to each member of the numerical series—

0	3	6	12	24	48	96

we get—

4	7	10	16	28	52	100
Mercury.	Venus.	Earth.	Mars.	—	Jupiter.	Saturn.

These figures represent very nearly the relative distance of the planets from the Sun, *e.g.* Saturn is 10 times farther than the Earth. There is a gap between Mars and Jupiter, and although no physical reason was, or is, known for this arrangement, the whole system seemed so orderly that Kepler supposed this gap to represent the place of a missing planet. Bode and several other astronomers were so impressed by the gap in this law that they agreed to examine the sky very minutely for the missing planet. While their search was in progress the Italian Piazzi (who was not one of the number) discovered on the first night of the nineteenth century a small planet occupying exactly the position prescribed by this law, and gave it the name of Ceres. Next year another little planet was discovered, and when half the century had elapsed no less than fifteen had been found. A more systematic search was then commenced by many astronomers, and the small stars made visible only by powerful telescopes were followed individually night after night, with the result that a great many were found to have no fixed place on the star-dome, and to show the movements of planets. They are so like stars that the name Asteroid (star-like) is usually given them. No. 311 was discovered on 11th June 1891. These minor planets are all very small, the largest being probably only 300 miles in diameter; the orbits of some are very long ellipses, and lie far out of the plane of the ecliptic (see § 132).

130. **Outer Planets.**—Beyond the asteroid ring the giants of the solar system, each attended by a train of satellites, rotate with amazing speed, and are surrounded by thick atmospheres loaded with heavy clouds. **Jupiter**, the largest of all, with four satellites, has a temperature so high that dense layers of cloud, arranged in belts parallel to the equator by its rapid rotation, completely obscure the body of the planet. The spectrum of its light shows some dark bands which are not due to reflected sunlight, and it is generally assumed that Jupiter is only now cooling down from being a self-luminous body. **Saturn**, although somewhat smaller, is unique in being accompanied by a series of rings or thin flat discs surrounding its globe parallel to the equator, and reflecting sunlight like the planet itself. These rings can only be accounted for on the assumption that they are composed of orderly crowds of innumerable minute satellites. Outside the rings there are eight separate satellites of various sizes, one being larger than the Moon.

131. **Uranus and Neptune.**—Uranus has been known as a planet since 1781, when it was discovered by Herschel. One astronomer had observed it previously twelve times, and only the careless way in which he kept his notes prevented him from recognising it as a new member of the solar system. This remote body is remarkable for its four satellites revolving in apparently circular orbits in a plane at right angles to that of the planet's orbit, and from east to west, whereas the satellites of all planets nearer the Sun revolve, like the Moon, from west to east. The movements of planets in their orbits under solar attraction is calculated from Kepler's Laws (§ 109), but allowance has always to be made for the perturbations or deviations produced by the attraction of other planets. After all possible allowances

were made, and the path of Uranus along the star-dome calculated, it was found that the planet did not keep to its time-table. The English astronomer Adams and the French Leverrier made calculations on the assumption that this irregularity was produced by an unknown planet beyond Uranus. In 1846 their work was finished almost simultaneously, and each predicted the position of the hypothetical planet in the sky. The very day that the information from Leverrier reached the observatory of Berlin, the German astronomer Galle turned his telescope to the part of the sky indicated, and there discovered the new planet which was named Neptune. Like Uranus it had previously been recorded as a star, and it was only by mistrusting his observations that an earlier astronomer failed to detect its true nature. One satellite has been observed which revolves, like those of Uranus, from east to west.

132. **Comets.**—Occasionally a luminous body appears in the sky, brighter in some cases than the planets, and usually enswathed in a long flowing tail of gauzy texture, from which peculiarity it is called a comet. Many comets have been found to travel in elliptical orbits, much more elongated than those of the planets, but like them with the Sun in one focus. As a comet pursues its path, it approaches the Sun with increasing velocity, sweeps round and sometimes almost touches the solar surface, and then flies on with ever diminishing speed to its aphelion. Halley's comet was the first the regular return of which was noticed; its period is 76 years, and it should next return to perihelion in 1910. It will then pass within the Earth's orbit, but its aphelion lies outside the orbit of Neptune. Several comets have their farthest points from the Sun near the orbit of Neptune; others show a similar relation to Uranus and to Saturn, while quite a number of comets of short period are associated with the orbit of Jupiter. Many of the grandest comets that have been seen pursued a path shaped like a parabola or hyperbola, and after passing the Sun swept out of the solar system for ever. It is supposed that the orbits of comets are naturally parabolas, but when the comet happens to pass near enough to a planet the

path is changed by attraction either into a closed curve—an ellipse—or into a hyperbola. Comets are thus viewed as the carriers of new stores of matter and energy into the solar system from remoter realms of space. Halley's comet is believed to have been captured by the attraction of Neptune when it was sweeping through the solar system, and the other periodic comets are similarly the slaves of the great planets. The planes of the orbits of comets show no relation to that of the ecliptic, sometimes indeed being perpendicular to it. To revert to a former simile (§ 112), if the Sun be compared to a large ship, and the ecliptic to the surface of the ocean, steam-launches manœuvring round the ship represent the planets, all nearly in the same plane, though the swell of the ocean causes them to be above the mean level at one part of their evolutions and beneath it at another. A comet would be represented by a diving bird going round the ship by diving under the keel and flying above the deck.

133. **Nature of Comets.**—The tail of a comet, sometimes several million miles long, is greatest when near the Sun, away from which it points whether the comet is approaching or receding. Comets shine, according to the spectroscope, partly with reflected sunlight and partly with the light of glowing vapour. The density of their substance is very slight, and they were long supposed to consist of masses of glowing gas. Recent observations, however, make it almost certain that they are swarms of very small solid bodies far enough apart to let starlight pass between them, and these when heated by approach to the Sun give off vapour at first composed of a compound of carbon and hydrogen, latterly, as the temperature is higher, of metals such as sodium and iron. The particles which make up comets may be only a few inches, or possibly only the fraction of an inch in diameter, and they are known as meteorites.

134. **Meteors.**—Attentive observers may see a few meteors or "falling stars" on any clear night. A star apparently detaches itself from its neighbours on the star-dome and silently glides downward, sometimes leaving an

evanescent track of light. At certain times, particularly about 10th August and 13th November, this phenomenon is so common that showers of shooting-stars are seen. At those dates the Earth crosses the orbits of two comets. The November shower is sometimes marvellously magnificent, and the grandest displays recur at intervals of about 33 years. The last is still remembered in 1866, and a similarly fine spectacle may be looked forward to in 1899. Meteors are not falling stars, for the stars are as numerous after a meteor shower as before. They are produced by small solid bodies, on the average perhaps as large as a pea, which enter the Earth's atmosphere with enormous velocity. The energy of motion is converted into heat by the friction of the air, and the solid is immediately driven into vapour and vanishes, being condensed into fine invisible dust (§§ 161, 277). Meteors usually begin to glow at the height of about 80 miles above the Earth's surface, and die out at the height of at least 50 miles.

135. **Meteorites.**—It has occasionally happened that meteoric masses of considerable size, weighing several pounds or even hundredweights, have fallen on the Earth, and in about a dozen cases this has happened in the sight of intelligent witnesses. Meteorites, as such masses are termed, are of at least two classes, either metallic composed mainly of iron and nickel, or stones resembling volcanic rock, although frequently associated with minerals not known in terrestrial rocks. They often contain carbon, and almost always considerable quantities of various gases absorbed in their pores. When a powdered meteorite is heated in a tube from which the air has been exhausted, and through which an electric current is passed, it glows with a faint light, the spectrum of which is very like that of comets, strongly confirming the meteoric theory of those bodies (§ 133). The close relation of meteors and comets was proved very forcibly in 1861 when the Earth dashed through the tail of a comet; again in 1872, and in 1885 when Biela's comet was calculated to cross the Earth's orbit close to the Earth's place at the time. The only sign of the

collision on these occasions was a fine shower of shooting-stars, through which the Earth sailed as safely as a locomotive passes through a cloud of dust. Meteorites of all sizes, from an invisible granule to masses of several tons and moving in various directions, seem to be scattered in infinite numbers through all space, and occasional denser swarms moving together form comets.

136. **The Stars.**—The Sun, surrounded by its orderly family of planets and an irregular host of attendant comets and meteorites, is practically alone in the centre of the star-sphere, forming one system isolated by inconceivable expanses of space from the fixed stars. But the Sun and its train are sweeping with tremendous velocity in the direction of the constellation Hercules. The number of stars or fixed points of light on the star-dome which are visible at any one time to the unaided eye of an observer on the Earth is about 3000. More people in fact assemble to hear a popular concert than there are stars in the heavens, so far as our vision can tell. By the aid of an opera-glass more than 120,000 stars, too feeble in their light to be seen by the unaided eye, spring into sight. A million may be seen through a small telescope; in a large telescope the number is enormously increased, and with every instrumental improvement smaller specks of light crowd in myriads on the view. Some stars, invisible in the most powerful telescopes to the eye, have been discovered by their effect on a sensitive photographic plate. Altogether the existence of something like 100,000,000 stars has been ascertained. The telescope, no matter how powerful, fails to make even the brightest star appear as a disc; but it often shows that what we see as a single star is actually double, triple, quadruple, or multiple. In some cases this is an accidental result of stars, perhaps very distant from one another, lying nearly in the same line as seen from the Earth; but there are many "physical doubles" the associated stars of which are seen to revolve round one another. This discovery proves that these stars are subject to gravitation. Several stars vary in their brightness at definite intervals, at one time blazing out with extraordinary brilliance and then

fading down to invisibility. This happens so regularly in some as to suggest that a dark body revolving round the star comes between it and us. In other stars the increase in brightness is accompanied, according to the spectroscope, by a change in chemical constitution and a great increase of temperature, as if perhaps swarms of meteorites flying in opposite directions had come into collision.

137. **Distance of the Stars.**—The stars are so remote that when corrected for aberration (§ 108) there is, as a rule, no apparent parallax. This means that the displacement of our eye by 186,000,000 miles from one side of the Earth's orbit to the opposite does not alter their apparent position on the star-dome. In several cases a minute parallax has been measured. The largest, barely 1", was found in the case of a Centauri, one of the brightest stars visible in the southern hemisphere. The parallax of Sirius, the brightest star in the sky, is $\frac{1}{7}$ of a second, that of the Pole Star only $\frac{1}{15}$ of a second. Light which travels at 186,000 miles per second requires 8 minutes to flash from the Sun to the Earth, and would require 9 hours to traverse the diameter of Neptune's orbit. Yet the light from a Centauri, the nearest star, has been more than 3 years on its way to us. We see Sirius by the rays sent out more than 17 years ago, and for nearly half a century the light-waves which are now arriving from the Pole Star have been shooting with lightning speed across the awful voids of space. Other stars are perhaps a hundred or a thousand times more remote than these. Although the star-dome may be spoken of as a vastly remote whole with reference to the solar system, it is really made up of remotely isolated objects placed at different distances and seen by us at different dates. For all our sight can tell us to the contrary, every star that shines placidly in the sky may have grown cold years or centuries ago, and snapped the thread of light the end of which may now be fast approaching our Earth.

138. **Stars as Suns.**—For classifying the stars the spectroscope has entirely superseded the telescope. By its means great differences have been detected in the chemical composition and physical states of various stars, and the

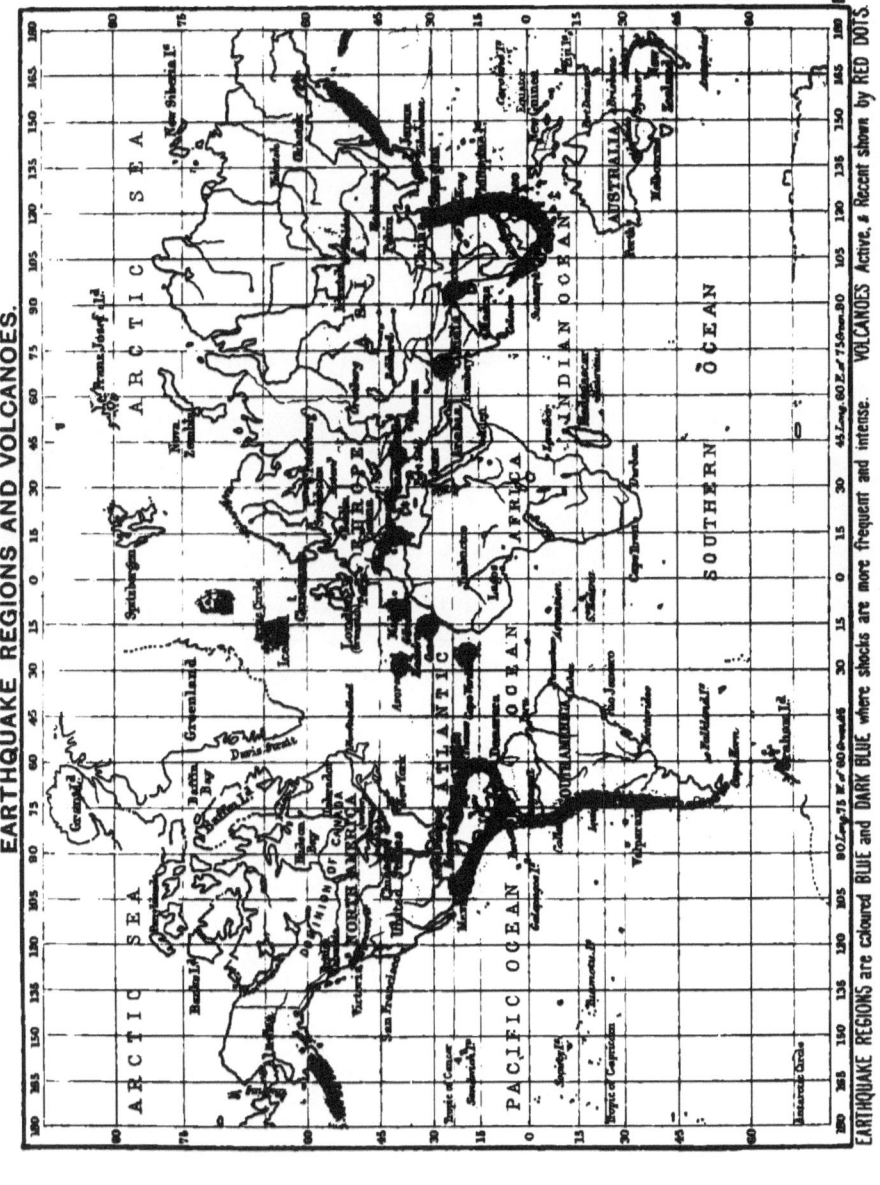

classification now viewed with most favour is of a biographical character, referring the star to its position in the long evolution or series of changes through which our Sun is passing (§ 118). In arranging the stars in the order of their evolution their state at the period their light left them is of course referred to. Stars of youth, or the earlier stages, are comparatively cool and diffused agglomerations of matter gradually condensing and rising in temperature. Stars of middle life, or the central stages, are intensely hot, invested with a glowing atmosphere of gas which gives bright lines in the spectra of their light. Stars of old age, or the later stages of evolution, have survived the period of maximum temperature and are steadily consolidating and cooling down. There is reason to believe that many stars are invisible to us because they have ceased to glow. We may infer, from their general similarity to our Sun, that stars of the central and later stages at least are accompanied by systems of planets. Some double stars present much the same appearance as the Sun would have done at a similar distance when Jupiter was still brilliantly incandescent. Many of the stars have a rapid motion through space as shown by the displacement of their spectral lines. This is termed their proper motion, to distinguish it from the various apparent movements, but though it is inconceivably swift it has produced very little change in the appearance of the constellations in 2000 years.

139. **Charting the Heavens.**—Although the constellations remain of the same form as when first described by astronomers, some change must be taking place. Common star-maps fail to let the changes appear, but a series of large photographic charts of the sky would probably show a definite alteration of position amongst the stars on account of their proper motion in a few years. An International Astronomical Congress held in 1891 decided that in several observatories such photographs should be taken with the ultimate object of completing a photographic survey of the entire star-dome. In order to prevent confusion from chance specks and to detect asteroids, a device has been suggested by which the photographic plate is exposed in the telescope

in three long stages with a slight shift of position in each. Each star thus prints itself as a little triangle of three points, while in consequence of its relative motion an asteroid presents its record in one little blurred streak and can thus be readily detected.

140. **Form of the Universe.**—On a clear moonless night a luminous gauzy band called the Milky Way may be seen spanning the sky like a wide but ragged and colourless rainbow. As this is visible from all parts of the Earth it evidently forms a complete girdle round the star-dome. A telescope of moderate power shows that the Milky Way is really a dense pavement of stars. There is no reason to believe that any two of these stars are nearer each other than the Sun and α Centauri, and the apparent crowding is simply an optical effect due to their great number. If we were led blindfolded into a regular pine plantation, and on looking round found that to east and west the tree trunks stood out sharply against the sky, affording a glimpse of diversified country beyond them, while to north and south the trunks were crowded so closely that they formed merely a reddish mist under the dark green leafage, we would naturally conclude that the wood was planted in a long narrow belt running north and south. So from our station in the Universe the Milky Way appears as the direction in which the extent of star-sown space is greatest; the widely strewn stars indicate the sides on which it is least. The form of the Universe, if this mode of reasoning be correct, is that of a vast disc, the edge of which, as shown by a division in the Milky Way, is partially split and bent back. Within this expanse the great family of 100,000,000 or more stars is supposed to be clustered together, and separated by incalculable distances of vacancy from other universes which may exist.

141. **Star-clusters.**—As one might catch glimpses of other forests through the tree trunks on either side of the long plantation, so we catch glimpses of remote universes through the thinly star-sown regions remote from the Milky Way. These are faint patches of light, which were long called *Nebulæ* from their cloudy appearance. Genera-

tions of astronomers have laboured to discover the nature of these cloudy tracts, and in many cases they have succeeded in showing them to be clusters of immeasurably remote stars. The forms of these star-clusters or remote universes are in many cases wonderfully beautiful—ring-shaped, oval, rod-like, or resembling dumb-bells or spirals of much complexity.

142. **Nebulæ.**—The old observers were accustomed to find that many nebulæ which their telescopes only showed as a gauzy cloud were resolved into star-clusters when a more powerful instrument was brought to bear on them. Consequently it was long believed that all unresolved nebulæ were simply star-clusters that larger telescopes could make plain. When Mr. Huggins first succeeded in observing the spectra of the unresolved nebulæ in 1864 he detected bright lines unlike those of stars, and doubtless coming from intensely heated gases. The nebulæ were therefore supposed to be distant masses of glowing gas. Professor Norman Lockyer has recently suggested a somewhat different explanation of the spectrum. He points out that the spectra of nebulæ and of comets' tails and of meteorites in a vacuum tube (§§ 133, 135) are all so much alike that they are probably produced by the same materials. Following an earlier suggestion of Professor P. G. Tait, he views a nebula as a vast swarm of meteorites moving in different directions, and by dashing against each other producing heat enough to drive a part of their substance into luminous vapour.

143. **The Nebular Hypothesis.**—The Prussian philosopher Kant and subsequently the French astronomer Laplace accounted for the origin of the solar system by supposing that at one time in the remote past it consisted merely of a vast nebula or cloud of intensely hot gas extending far beyond the orbit of the outermost planet. As this cloud cooled and contracted it acquired a whirling motion from west to east, and formed a rotating gaseous disc which gradually condensed at the centre to form the embryo Sun. The edge of the whirling disc was thrown off as a ring by centrifugal force, and the ring ultimately condensed into the planet Neptune. The gaseous disc

continuing to contract and to spin more rapidly threw off another ring which gave rise to Uranus, and so on with the other planets, which themselves by a similar process threw off rings to persist like those of Saturn or to condense into satellites. The ring thrown off after the formation of Jupiter, instead of condensing into one planet, consolidated, perhaps on account of perturbation by its great neighbour, into separate bodies—the asteroids. The residue of the original nebula remained as the great globe of the Sun.

144. **Meteoritic Hypothesis.**—Recently Norman Lockyer has pieced together the facts discovered by modern astronomers, and he believes them to countenance the theory that originally all space was filled with matter in its simplest or primary form, that this matter commenced to aggregate under the influence of gravity and chemical affinity, producing a fine moving dust of the elements and latterly of their compounds. This dust further condensed and gave rise to meteorites in great moving swarms separated by tracts of empty space. As the meteoritic swarms shrank by gravity, collisions between the individual meteorites became more frequent, and some of their energy of motion was changed to heat which partly vaporised them, giving rise to the bodies we recognise as nebulæ or as variable stars. These swarms of moving meteorites present many of the properties of a gas on a very large scale, and the motion and equilibrium of a meteoritic nebula would be very similar to those of a gaseous one. Meteoritic material is supposed to pass from the nebular state into that of separate and much denser suns surrounded by families of planets somewhat in the manner Laplace suggested. Lockyer differs from Laplace in making gravitation and molecular attraction the primary cause rather than heat, and so including in the theory the heating up as well as the cooling down of the Universe.

BOOKS OF REFERENCE

J. Stuart, *A Chapter of Science*, S.P.C.K. (A thoroughly scientific and attractively simple explanation of the movements of the solar system.)

S. P. Langley, *The New Astronomy*. Boston : Ticknor and Co. (Remarkable for its fine illustrations.)

R. S. Ball, *Star-Land*. Cassell and Co. (Simple and racy sketch of elementary astronomy.) *The Story of the Heavens*. Cassell and Co.

J. N. Lockyer, *The Meteoritic Hypothesis*. Macmillan and Co. See also list at end of Chapter V.

CHAPTER VII

THE ATMOSPHERE

145. The Ocean of Air.—We live and move at the bottom of a shoreless ocean of invisible fluid to the surface of which we are powerless to rise. The existence of this ocean is revealed to us by its power of exercising pressure, but the substance composing it was long supposed to have no weight, and the phrases "light as air," "an airy nothing" are remnants of that idea. The simple experiment of inverting a tumbler over a cork floating in a basin of water shows that air can exert pressure and that it occupies space. By means of the air-pump a glass vessel can be nearly emptied of air, and on weighing it before and after emptying, it is ascertained that a pint of air has the mass of about 10 grains, or a cubic foot that of $1\frac{1}{5}$ ounce.

146. The Barometer.—Torricelli, an Italian mathematician of the seventeenth century, when investigating the action of the common sucking-pump, made a discovery which laid the foundations of scientific knowledge of the atmosphere. He took a tube closed at one end and about 33 inches long, filled it with mercury, and placing his thumb on the open end inverted it (Fig. 20) in a basin of mercury. The column of mercury in the tube sank gradually and stood just 30 inches above the level of the mercury in the basin. Mercury placed in a tube open above and below and set in the same manner would immediately run out by its own weight. Torricelli argued that the only difference in the mercury in the closed tube was that the weight of the atmosphere could not press upon it. He knew that in a liquid at rest every point in the same horizontal plane must be at the same pressure, so he argued that every point in the line *a b* (Fig. 20) must be at the same pressure. The points between *c* and *d* were pressed upon by the weight of 30 inches of mercury, but were free from the weight of the air, while the points from *a* to *c* and *d* to *b* were free from the weight of mercury, but subject to the pressure of the weight of the air. Thus the pressure of the atmosphere on a given area is equal to the weight of 30 inches of mercury, or $14\tfrac{3}{4}$ pounds on a square inch. This reasoning proved that the atmosphere presses as heavily on the Earth's surface as if it were an ocean of mercury 30 inches deep, or, since mercury is about $13\tfrac{1}{2}$ times denser than water, an ocean of water 34 feet deep. Exact observation shows that the column of mercury balanced by the atmosphere at sea-level over the whole Earth averages 29.9 inches, and it is calculated from this that the whole mass of the atmosphere is 5500 million millions of tons. Since the mercury tube enables one to measure the weight of the atmosphere it has been called the Barometer (see also § 439).

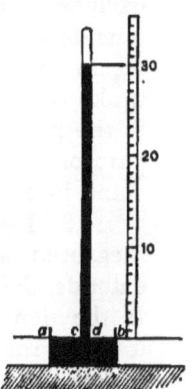

FIG. 20.—Mercurial Barometer and yard measure.

147. Pressure of the Atmosphere.—Torricelli's experiment made it clear that the piston of a common suction-pump lifts the atmosphere from above the piece of water in which the pipe dips, and that the pressure of the atmosphere on the rest of the surface forces up the water over that space until the weight of the column is equal to the pressure on an equal area of the free surface: this height never exceeds about 34 feet, which is the limit of lifting power in a pump. Air, and fluids generally, exert pressure equally in all directions; and on account of this uniform pressure of the air all round us and through the tissues of our bodies, we do not feel the pressure to which we are always subjected of $14\tfrac{3}{4}$ pounds on every square inch, or 14 tons for the whole body of a man of ordinary size. A common limpet weighing perhaps half an ounce sticks to a smooth level rock as if its weight were from 10 to 15 pounds, because the soft tough foot is planted so closely on the stone as to exclude all air from below and the pressure comes from the outside only. The limpet sticks as firmly to a vertical or an inverted surface as to a level one. The vacuum brake is a powerful illustration of the pressure of air, for by it the pressure of the atmosphere applied to a very small part of the surface of a rapidly moving train brings it to a stand in a very few minutes.

148. **Density of Air.**—The mass of the air has been measured with great accuracy, but the height to which it extends, the depth of our aerial ocean, is difficult to estimate. If the density of the air ocean were uniformly the same as it is at the Earth's surface (about $\frac{1}{800}$ of the density of water), its height would be five miles. That this is not the case was proved by Mr. Glaisher, who once ascended more than seven miles in a balloon and still found air around him, though of much less density than at the Earth's surface. But the fact was known by theory two centuries earlier. Boyle, in 1662, announced the discovery of the law known by his name :—

The density of any gas is proportional to the pressure it supports.

The pressure of the atmosphere produced by its own weight is greatest on the Earth's surface or in a mine, where the density is accordingly greatest also. As one ascends in the atmosphere the pressure falls steadily, because less air remains above, and the density of the remaining air is consequently less. Thus the barometer can be used to measure heights: near sea-level a fall of one inch in the barometer corresponds to a rise of 1000 feet. One half of the atmosphere lies beneath the height of $3\frac{1}{2}$ miles, or 18,500 feet, from the Earth's surface, and the half which is above this height can exert a pressure only equal to about 15 inches of mercury at that level. Another rise of $3\frac{1}{2}$ miles (to 7 miles) leaves half of the half atmosphere below, and only one quarter above, the pressure being equal to $7\frac{1}{2}$ inches. At

$10\frac{1}{2}$ miles above the Earth's surface $\frac{1}{8}$, at 14 miles $\frac{1}{16}$, at $17\frac{1}{2}$ miles $\frac{1}{32}$, and at 21 miles only $\frac{1}{64}$ of the atmosphere lies at a higher level: at 21 miles the barometer would stand at half an inch. Thus, if Boyle's law holds good the atmosphere has no definite limit, but extends with diminishing density throughout infinite space. It has however been proved that this law does not hold for gases of very small density, which behave like very light liquids and have a definite surface, so that the atmosphere has an upper limit, beyond which the particles of gas do not stray.

149. **Height of the Atmosphere.**—Observations of twilight (§ 162) show that the atmosphere is not less than 45 miles high. The aurora, which is produced in the upper atmosphere (§ 174) has been measured at more than 100 miles above the Earth, and meteors (§ 134) sometimes become visible at 200 miles. Hence it is probable that the atmosphere extends at least 200 miles beyond the Earth's surface; but in consequence of its compressibility nearly three quarters of the air lies between sea-level and the summit of the loftiest mountain.

150. **Atmospheric refraction.**—When light from any of the heavenly bodies enters the atmosphere, it traverses denser and denser layers, and is consequently bent downward from a straight line as it approaches the surface (§ 61). The amount of this bending or *refraction* is proportional to the obliqueness of the rays of light—thus when the light falls perpendicularly from the zenith there is none, but when it comes parallel to the horizon the refraction is great. A person

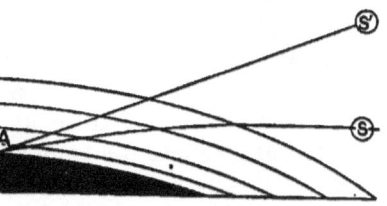

FIG. 21.—Atmospheric Refraction. A, observer; S, true position; S', apparent position of Sun. The density of the atmosphere is indicated by the closeness of the lines.

always refers an object to the direction from which the light enters the eye. When the Sun is near the horizon its light is bent into the curve SA (Fig. 21) and as the light reaches the eye of an observer at A from the direction S'A, he sees the Sun's image at S', considerably higher in the sky than it really is. In astronomical observations it is necessary to correct this error, and tables of refraction at every altitude of a star and for different temperatures of the air have been compiled. The atmosphere, by raising the apparent position of the Sun, thus serves to lengthen the period of sunlight by about four minutes on the equator, and by several hours and even days in high latitudes (§ 124). For the same reason the midnight sun is visible in places where it would not appear above the horizon if there were no atmosphere. Thus at Archangel in lat. 64° 32', nearly 2° south of the Arctic Circle, there is perpetual sunlight for several days at midsummer. When from unequal heating or other causes the distribution of density in the atmosphere becomes irregular, light is reflected and refracted by the layers of air in such a way as to make objects at a great distance visible as if near at hand. This effect, which is most marked in deserts and at sea, is called *mirage*. All our knowledge of the outer regions is obtained by looking through the window pane of air which encloses the world, and allowance must always be made for its imperfections.

151. **Composition of Air.**—The experiments of Priestley, Black, and Rutherford at the close of the eighteenth century proved that common air is a mixture of several different airs or gases, and at that date it ceased to be considered an element. Innumerable analyses of air have since been made which show that in all parts of the world the atmosphere has almost the same composition. Traces of nearly every gas which exists naturally, or is produced artificially in large quantities, have been found in air, but the main constituents are few. A rough analysis of air may be made thus :—(*a*) A large tightly-corked flask of warm air when chilled by being covered with snow or ice is seen to become dewed with liquid drops on the inside. These drops are *water*, and their appearance proves that water-vapour is a constituent of air. When a person wearing spectacles steps from the frosty night into a warm room he is the victim of an irritating variation of

this experiment, for the cold glasses immediately condense a blinding film of dew-drops from the air. (*b*) When a little clear lime-water is shaken in a flask of air the liquid becomes milky from the formation of solid carbonate of lime, a compound of carbonic acid with lime. Hence, *carbonic acid* is one of the constituents of air. (*c*) When a candle, or a piece of charcoal, or of phosphorus is allowed to burn in a limited quantity of air under a tumbler or bell-jar inverted in water, the flame soon goes out, and another bit of burning charcoal, or phosphorus is extinguished the moment it is introduced; moreover, the water rises until it fills about one-fifth of the jar, showing that about one-fifth of the atmosphere is a gas which is consumed by burning substances. This gas is *oxygen*. (*d*) The residue from which burning phosphorus has extracted the oxygen is a gas with no striking properties called *nitrogen*. (*e*) When a sunbeam traverses a darkened room, or when strong sunlight streams through an opening in a thick cloud, immense multitudes of motes may be seen dancing in the light. Thus *dust* is an ingredient of the atmosphere. The amount of water-vapour is variable, and the amount of dust is still more uncertain; but the other constituents occur always very nearly in the proportions :—

	By weight.	By volume.	
Nitrogen . .	76·80	79·00 or	$\frac{4}{5}$
Oxygen . .	23·14	20·96 or	$\frac{1}{5}$
Carbonic acid .	0·06	0·04 or	$\frac{1}{2500}$
Total	100·00	100·00 or	1

152. Nitrogen.—The most abundant gas of the atmosphere has no colour, no taste, no smell, no tendency to combine with other elements, no poisonous effect on living creatures, and no power to keep them alive. From the last circumstance it is sometimes called *Azote*. Its service in the atmosphere is mainly to dilute the other ingredients, and to produce mechanical effects. Most of the pressure of the atmosphere, the strength of wind, the refraction of light, and the buffer-action which breaks the force of meteorites and drives them into dust, are due to nitrogen.

When an electric discharge passes through air, a small quantity of nitrogen is always caused to combine with hydrogen and oxygen to form salts of ammonia.

153. **Oxygen** was originally known as *Vital Air*, for it is the ingredient of the atmosphere which sustains life, and by its ready combination with other elements supports combustion. The oxygen of the atmosphere is a great store of potential energy when taken into account with the uncombined substances in the Earth (§§ 56, 44). Oxygen in the pure state combines very energetically with carbon, hydrogen, and almost all the other elements; but when it is diluted with four times its volume of inert nitrogen, combustion is slower and quieter, although the same amount of energy is ultimately set free as would be the case if no nitrogen were present. Under the influence of electric discharge, and of the growth of some trees, oxygen is partly changed into a condensed form called *ozone*, and partly combined with water to form peroxide of hydrogen. These substances exist in the air in very minute proportions, but when either of them is present it is believed to increase the healthfulness of a neighbourhood. Oxygen in small quantities is a colourless and transparent gas, but in the atmosphere it absorbs a good deal of sunlight, giving broad black bands in the red part of the spectrum. The blue tint of the sky may be due in part to the true colour of oxygen. The proportion of oxygen in the free air of the country is a very little greater than in crowded towns.

154. **Carbonic Acid**, though present in small amount, has an important part to play in the economy of the atmosphere. Green plants in sunlight absorb it, decompose it, retain the carbon to build up in their own substance, and breathe back the oxygen into the air. Animals and also plants (§§ 399, 400) breathe in air, absorb the oxygen, which is ultimately combined with carbon and breathed out as carbonic acid. There is a large proportion of carbon in coal, oil, wood, fat, and almost all combustible substances, which thus produce carbonic acid as the principal result of their union with oxygen. The amount present in the atmosphere varies considerably; 3 parts in 10,000 is the proportion in

ISOTHERMS
After

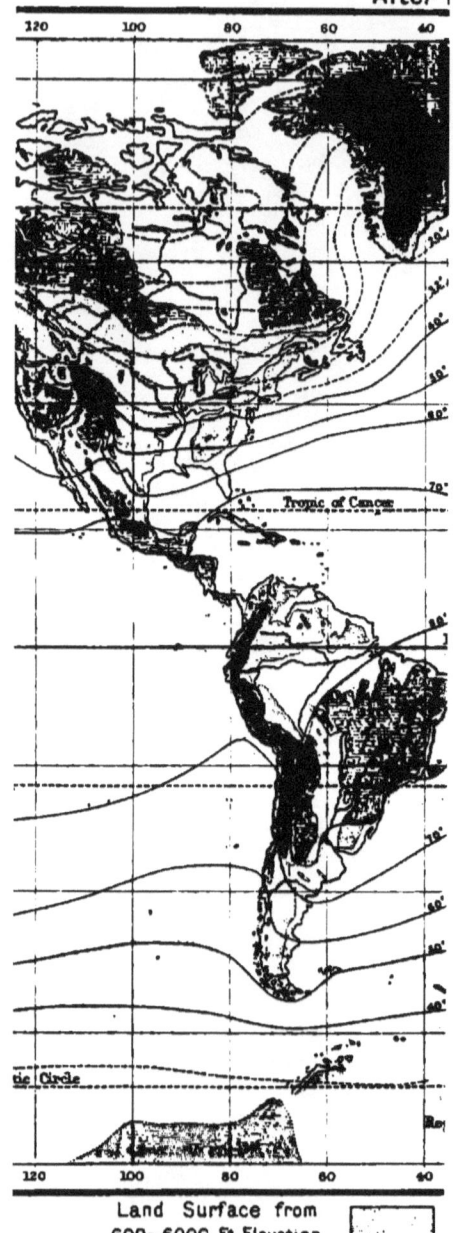

Land Surface from 600–6000 Ft Elevation.

R JANUARY,
ıchan.

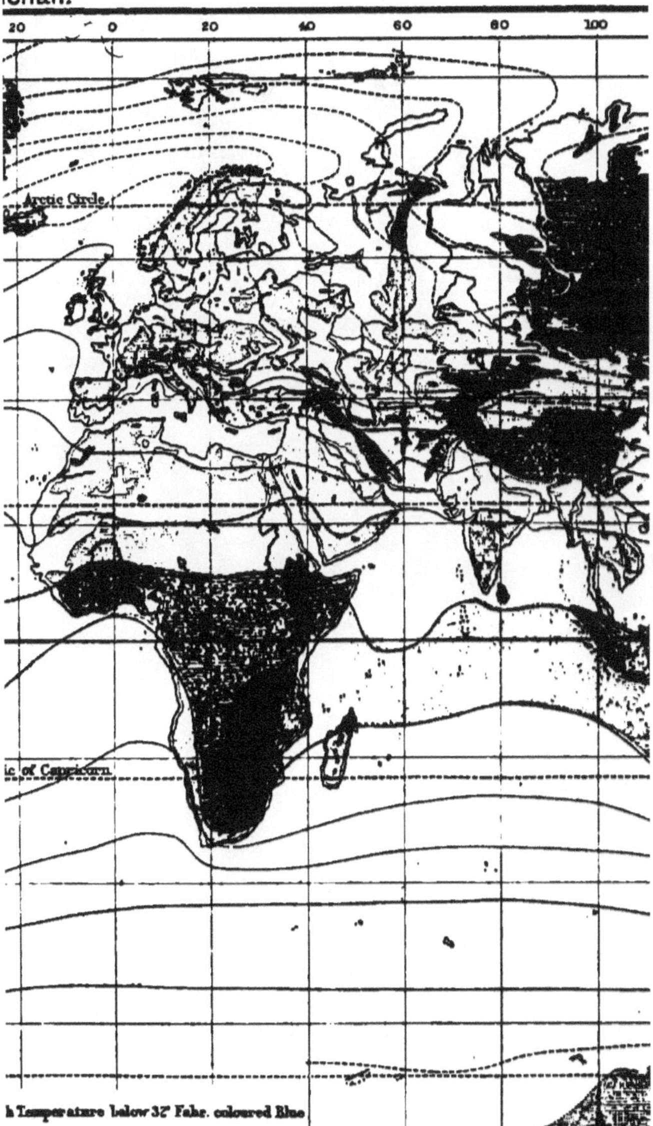

20 0 20 40 60 80 100
Arctic Circle
ic of Capricorn
h Temperature below 32° Fahr. coloured Blue
20 0 20 40 60 80 100

Land Surface Above
6000 Ft Elevation.

the open country, 5 parts is common in towns, and as much as 30 parts of carbonic acid in 10,000 of air may be found in badly-ventilated overcrowded rooms. More than this proportion acts poisonously on animal life. Carbonic acid is the most soluble of the atmospheric gases, water at 60° F. and under ordinary pressure absorbing its own volume.

155. **Mixture of Gases.**—One consequence of the nature of gases is that when two or more different kinds are mixed, each one acts as if it alone were present. This is known as Dalton's Law. Thus there is an atmosphere of nitrogen surrounding the globe, and exerting the pressure of its weight upon the Earth's surface, and an atmosphere of oxygen pressing upon the surface with its weight, which is rather less than one quarter of the pressure exerted by nitrogen, and a very thin atmosphere of carbonic acid exerting a very feeble pressure. There is also an atmosphere of water-vapour pressing with its independent weight on the Earth's surface, and all these partial pressures together make up the pressure exerted by the whole atmosphere. The particles of the different gases pass each other freely, without interfering, like crowds moving in different directions across a market-place. Thus it is that the composition of the atmosphere as a whole remains constant so far as regards the three gases, nitrogen, oxygen, carbonic acid, and the proportion of each of them is the same at all heights.

156. **Water-vapour.**—Next to oxygen, water-vapour is the most important ingredient of the atmosphere. The other gases are a long way above their liquefying point, so that the addition or withdrawal of heat only affects their temperature and their volume. But water-vapour in the atmosphere is near the temperature at which it becomes liquid or solid, and is nearly always in the presence of liquid water, hence changes of temperature greatly affect the amount of vapour present. Let us suppose for a moment that the atmosphere consisted of water-vapour only, and that the hydrosphere covered the Earth uniformly with a liquid layer. The amount of this atmosphere, and consequently its pressure, would depend upon the temperature. Evaporation takes place from cold water, or even ice, but

at every temperature when the vapour exerts a certain definite pressure upon the liquid, evaporation is stopped, and the vapour is said to be saturated at that temperature.

157. **Water-vapour and Temperature.**—At the freezing-point (32°) water-vapour is saturated, *i.e.* presses sufficiently to stop evaporation, when its pressure is equal to that of 0·18 inch of mercury; at 50° it must exert twice this pressure, or 0·36, before evaporation ceases; at 70° it must exert a pressure of 0·73, and at 90° a pressure of 1·45 inches, in order to be saturated. These figures show that at 50° twice as much vapour is required to form a saturated atmosphere as at 32°, and at 70° twice as much as at 50°, and at 90° twice as much as at 70°. If an atmosphere of water-vapour saturated at 50° is warmed up to 70°, evaporation is at once allowed to commence and will continue until the amount of vapour present above the water is doubled. Then the vapour will exert pressure sufficient to stop further change, and will be saturated. Again, if the temperature of the saturated vapour is reduced from 70° to 50°, half the vapour must return to the liquid state or become condensed in order that the pressure may fall to that which is just sufficient to prevent further evaporation. Hence it is plain that every rise of temperature is accompanied necessarily by evaporation, every fall of temperature is accompanied necessarily by condensation, until the vapour exerts the pressure proper to its new temperature. Precisely the same thing happens, as explained by Dalton's law, when there are atmospheres of nitrogen, oxygen, and carbonic acid surrounding the Earth. The pressure of saturated water-vapour at 50° is still equal to 0·36 inches of mercury,—the only difference is that it takes a longer time for the pressure to readjust itself to a change of temperature, as a party of excursionists crossing a broad railway platform reach their carriages, whether the platform is left to themselves or is thronged by crowds moving in different directions, only in the latter case the transference takes a longer time. On account of the low temperature at great elevations, water-vapour, although its density is only half

that of air, is almost entirely confined to the lowest region of the atmosphere.

158. Vapour Pressure and Humidity.—The fraction of atmospheric pressure exerted by the water-vapour it contains is often termed *vapour tension*, but preferably vapour pressure. The amount of water-vapour in the atmosphere at any place as measured by the hygrometer (see § 441), and expressed in the pressure it exerts in inches of mercury or by the number of grains weight in a cubic foot of atmosphere, is called the *absolute humidity*. In the case of saturated vapour this depends only on the temperature. The vapour in the atmosphere has seldom an opportunity to become saturated, for the air is never at rest. Suppose, for example, that air containing water-vapour saturated at 32°, and therefore exerting a vapour pressure of 0.18 inches, is carried inland to a waterless place and heated up to 50°. Or suppose simply that its temperature is raised so rapidly that the somewhat slow process of evaporation has not had time to produce its full effect. The absolute humidity or vapour pressure is consequently only 0.18 inches, but evaporation could continue if time and opportunity were given until the amount of vapour would be doubled. Hence this portion of the atmosphere has only one half, or 50 per cent, of the water-vapour it could contain at its temperature. If the same portion of air were cooled without other change to 32° it would contain all the vapour possible at that temperature, or 100 per cent, and have no tendency to evaporate more. If it were heated to 70° it would contain only one quarter, or 25 per cent, of what might be present at that temperature, and evaporation would go on rapidly from free surfaces of water. The term *relative humidity* is applied to the percentage of the whole possible amount of water-vapour which is present at any particular temperature. When the relative humidity is low the atmosphere is "drying" or has a tendency to raise more vapour from water or damp soil; when on the other hand the relative humidity is high, there is little tendency to evaporation, and a slight fall of temperature leads to saturation and condensation.

159. Thermal Changes in Evaporation and Con-

densation.—The change of a pound of water into a pound of vapour requires the same expenditure of energy (§ 70), whether it takes place in a kettle boiling on a fire, or over the surface of a freezing pond. The work of evaporation uses up heat, and produces a lowering of temperature. On the other hand, when vapour is condensed to the state of water, the potential energy stored up is reconverted into heat; thus condensation produces a rise of temperature (§§ 70-73). When air resting over water is heated by the Sun's rays, evaporation begins actively and diminishes the rate of rise of temperature in the air. On the other hand, when a portion of the atmosphere containing saturated vapour is cooling down by radiation, the vapour begins to condense, giving out heat, and so retarding the rate of fall of temperature. In both cases the tendency is toward moderation and slowness of change. The cooling of air containing unsaturated vapour goes on unchecked until the temperature of saturation is reached.

160. **Absorptive Power of Air.**—The water-vapour of the atmosphere is not transparent to all light; it absorbs certain rays from sunlight, producing black lines or bands in the spectrum, particularly a set in the yellow known as the *rain-band* (π in Fig. 8). The rain-band in the spectrum increases in width and darkness as the amount of vapour in the slice of atmosphere looked through increases, and the probability of rain occurring within a certain time may be judged from the darkness of the band. The heat rays of the Sun pass readily into the atmosphere, but heat does not so readily pass out through the air into space. The atmosphere thus acts toward the Earth as a great blanket, or rather a heat-trap allowing radiant heat to enter freely but greatly retarding its escape. Water-vapour has usually been considered the chief heat-entrapping agent, because the chilling by radiation at night is always greatest when the proportion of water-vapour in the air is least. But there is now reason to believe that condensed water and solid dust-motes are more powerful in producing the effect.

161. **Dust.**—Solid dust is always present in the at-

mosphere throughout its whole depth. Twenty million meteorites are calculated to reach the Earth every day, and most of these are broken up by the friction of the air, furnishing a supply of *Cosmic dust* (§ 134), which being excessively fine, and even invisible, settles down very slowly. Terrestrial dust is carried into the atmosphere by ascending currents of air and is of many kinds, resulting from the wearing down of rocks, from volcanic explosions (§ 297), from flowers in the form of pollen, from minute organisms either plants or animals (§ 401), from burning fuel, from factories, mines, flour-mills, and from the spray of the sea. The number of motes is almost incredible. Every puff of smoke from a cigarette contains about 4000 million separate granules of dust. Dust appears to float in the atmosphere, and the motes of a sunbeam seem to be rising as often as falling. This is, however, a result of currents of air. In still air, dust always falls, but the large motes fall most rapidly under the pull of gravitation, and against the resistance of the friction of the air. When a cube of stone one inch in the side is falling, its mass drags it down, and the friction of the air on its six square inches of surface resists the fall. If the cube were cut into ten slices $\frac{1}{10}$ of an inch thick, each of these into ten bars, and each of these into ten cubes $\frac{1}{10}$ of an inch in the side, there would result 1000 little cubes drawn down by the same force as had acted on the one; but the atmosphere would now have sixty square inches of surface to act on. If each of these little cubes were cut into 1000, the downward attraction of the Earth on the whole million would be the same as for the one-inch cube, but the air-brake would be applied to no less than 600 square inches of surface, so that their fall must be very slow indeed. The average dust-motes of the air are much smaller than these, hence it is not surprising that even the stillest air is never free from dust.

162. **Quantity of Dust in Air.**—Mr. John Aitken, the discoverer of the importance of dust in Nature, invented an ingenious piece of apparatus by which he was able to count the number of invisible dust-motes in any sample of

air.[1] His numerous experiments show that in one cubic centimetre of the air of great cities there are hundreds of thousands of motes; in the air of small villages there are thousands, and there are hundreds even in the open country far from towns or factories. The purest air met with was on one occasion on the summit of Ben Nevis where one cubic centimetre contained only one dust-mote, the mean of ten observations. These minute motes catching and scattering the sunlight are the agents by which the whole atmosphere is so illuminated that not even the brightest of the stars is visible by day. If the air were free from dust we should probably see the Sun shining from a perfectly black star-filled sky, and one side of a house would be dazzlingly illuminated, the other in a shadow of absolute darkness. The blue colour of the clear sky (§ 153) is largely due to the scattering of sunlight by the dust-motes of the higher layers. The red tints produced at sunrise and sunset (§ 297) and the lingering twilight of high latitudes have a similar origin. Twilight is produced when light from the Sun, while still below the horizon, strikes on the upper atmosphere, too obliquely for refraction (§ 150) to bend the rays down to the surface; then the illuminated dust-motes of the upper air light up the sky for hours with a soft shimmer.

REFERENCE

[1] J. Aitken, "On the Number of Dust Particles in the Atmosphere." *Transactions Roy. Soc. Edin.* xxxv. p. 1 (1888).
See also *Nature*, xxxvii. 428 (1888) and xli. 394 (1890).

BOOKS OF REFERENCE

R. Angus Smith.—*Air and Rain.* Longmans.
See also lists at end of Chapters VIII. and IX.

CHAPTER VIII

ATMOSPHERIC PHENOMENA

163. Solar Energy in the Atmosphere.—All the changes in the atmosphere are directly or indirectly due to the radiant energy received from the Sun (§§ 119-125), the whole of which must pass through the air before reaching the Earth's surface. Thermometers placed in specially contrived screens are employed to measure the temperature of air. On lofty mountains, where the atmosphere contains little water-vapour and few dust-motes, the air is heated so slightly by the Sun's rays passing through, that it remains bitterly cold, although the Sun's direct heat blisters the traveller's face and hands. At an elevation of 11,000 feet, water has even been boiled by exposing it in a blackened bottle to the sunshine. On account of the low pressure of the air at great heights, air from sea-level rising as a heated current expands greatly, as explained by Boyle's law (§ 148). But the work of expansion against the attraction of gravity consumes heat, and the temperature of the expanded air, if unsaturated, falls 1° for every 180 feet of ascent. When cold air from a great altitude is carried toward the Earth's surface by a descending current, the pressure upon it is continually increasing, and its volume is being reduced. The work thus done on the air by gravity is changed into heat, and the temperature of the air rises 1° for each 180 feet it descends. The actual rate of change of temperature in the air near the Earth's surface is not so great as this, for the Sun has a certain heating effect. Several years of continuous observations on the summit of Ben Nevis, and at sea-level at Fort William have shown that the actual falling off of temperature with height is 1° for every 270 feet of ascent. Thus, whatever the temperature may be at sea-level, there is a certain height where the air has an average temperature of 32° F., no matter how much sun-heat passes through; and snow which falls above that height does not melt. This limit is termed the *snow-line*. It is sea-level in the extreme arctic regions, about 5000 feet at latitude 62° in Norway, about 9000 feet in latitude 46° in Switzerland, and above 16,000 feet at the equator (see figure 63 and section in Plate VIII.)

164. **Heating and Cooling of Air.**—Near sea-level the dense air is charged with water-vapour and dust which, during the day, absorb solar radiant energy and pass on the heat to the air. The ground also is rapidly heated, as its specific heat is only about one quarter that of water, and its temperature therefore rises four times as much as water does for the same amount of heat. Once heated, the ground is effectual in heating up the air in contact with it. In the case of water, the Sun's rays penetrate to a great depth, the temperature of the surface is very slightly raised, and transfers little heat to the air over it. Hence in sunshine a land surface heats air greatly, a sea surface heats it only slightly. After sunset the hot land radiates its heat through the atmosphere, and falls to a low temperature, thereby chilling the air in contact with it, and were it not for the dust-motes and condensed water catching and retaining most of this heat (§ 160) the radiation of a single clear night would chill down the land far more than the solar energy received during the day could heat it up. The temperature of the dust-motes is also lowered by radiation from the particles at night, and this is not fully compensated by the heat radiated from the earth, so that the air temperature falls greatly. From a water surface heat is radiated slowly at night, and the air over water is not greatly chilled.

165. **Dew and Hoar Frost.**—On a clear night, when

ISOTHERM
After

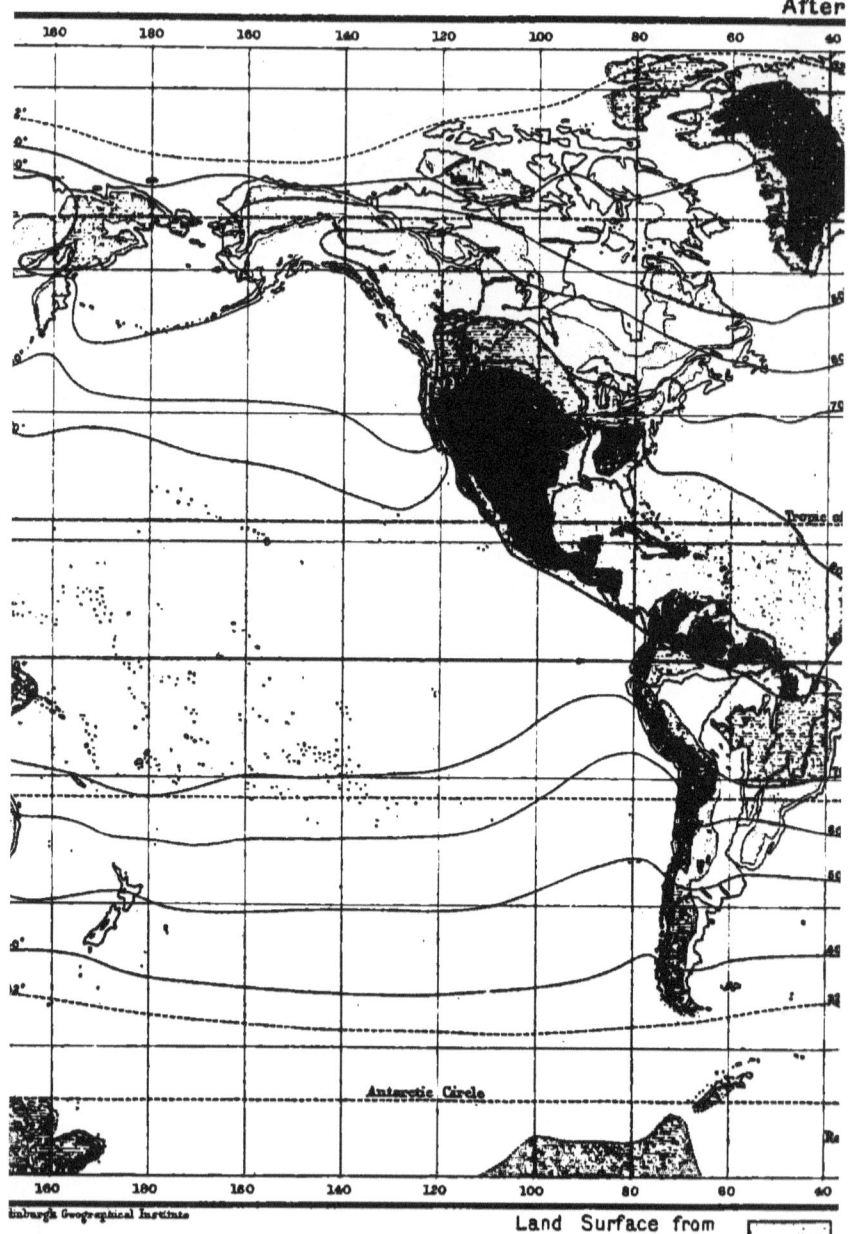

Land Surface from
600-6000 Ft Elevation.

OR JULY.
chan.

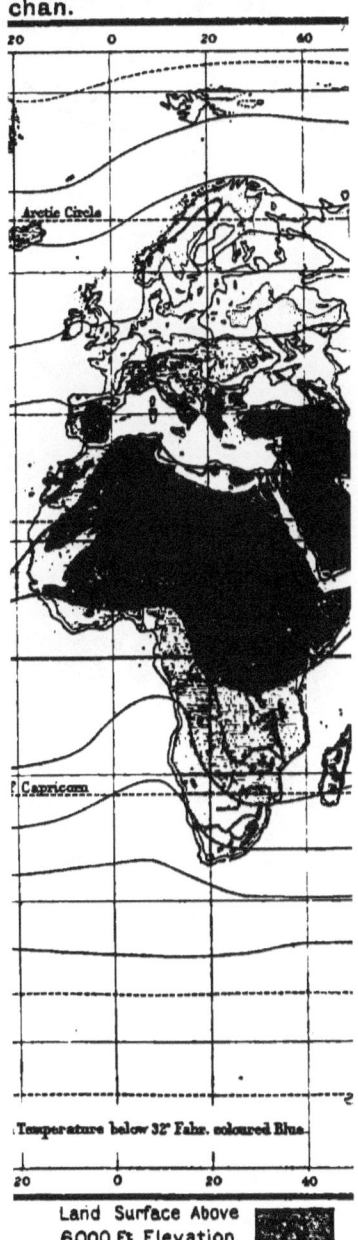

Temperature below 32° Fahr. coloured Blue.

Land Surface Above 6000 Ft Elevation.

the temperature of the land surface falls to the point at which the water-vapour present becomes saturated, moisture is deposited on all exposed objects in the form of drops of dew or as small crystals of ice, called hoar-frost. The temperature of saturation of water-vapour is hence called the *dew-point*. The deposition of dew, or of hoar-frost, liberates heat (§ 159), and so diminishes the subsequent fall of temperature. In last century, Dr. Wells made a number of experiments on the cause of dew. He showed that it was only deposited when the sky was clear, and on objects which had become greatly cooled by radiation, and he proved that these in turn chilled the air below the dew-point, and so condensed the water-vapour on their surfaces. On a cloudy night radiation is checked, the water spherules of the clouds retaining and radiating back the heat lost by the Earth, so that dew is not formed. Mr. John Aitken has recently shown that though the chilling by radiation of exposed objects is certainly the cause of dew, only a small part of the moisture is extracted from the air. Indeed, on a still night when there is no wind the air resting over a cabbage, for example, could never have contained the quantity of water found on the leaves in the morning. This is really condensed from the water-vapour always being breathed out by plants. On a gravelled road also, the under side of the gravel and not the upper, is often wet with dew, the stones chilled by radiation condensing the water-vapour which is always rising from the ground.[1]

166. **Condensation and Dust.**—It is remarkable that water-vapour never condenses except upon a solid substance. In air quite free from dust, water-vapour has been cooled far below the dew-point without condensation; but the instant a puff of common dust-laden air is admitted, each dust-mote becomes a nucleus, and a globule of water is formed upon it. All condensation of water-vapour in the air, whether it appears as rain, mist, fog, cloud, or snow, takes place on a nucleus of dust.

167. **Fog and Mist.**—When dust-motes are very numerous, and the temperature of air falls suddenly below the dew-point, each mote can receive only a small coating

of water. The minute globules formed in this way fall very slowly, and in the absence of wind may remain suspended in the air for a long time. This accounts for the black winter fogs of great cities where the specks of soot are very numerous and are only thinly coated with water. Over the open sea, when a broad stream of warm air carrying saturated water-vapour crosses a cold current of water or meets an iceberg, the sudden cooling of the vapour necessitates an enormous condensation, and the dust which is abundant even far from land, enables the condensation to take place in the form of a bank of mist. The famous "fogs" of Newfoundland are so produced. Fog differs from mist in not wetting solid objects with which it comes in contact. The light mists formed at night over low-lying meadows or valleys are usually very thin sheets, and as soon as the Sun appears the water particles are heated up and evaporated again, so that the mist clears quickly away. When a mass of warm air rises in the atmosphere its temperature falls (§ 163). On reaching a certain height the vapour becomes saturated, and as it still rises, and the temperature continues to fall, the vapour condenses upon the dust-motes forming a mist. *Clouds*, which are mists at high altitudes, often hang over a mountain or sail slowly through the air for hours. In such a case, though the form of the cloud does not change, the water globules composing it are always falling as fast as the friction of the air allows (§ 161); when they reach the warmer air below they are evaporated again and vanish, while new globules are as rapidly condensed on the dust above.[2]

168. **Classes of Clouds.**—The differences between clouds arise mainly from the height of the layer of mist composing them. Three types of cloud are distinguished by characteristic forms and by their usual elevation, and all other kinds may be classed as a mixture of two or more of them. The highest form of cloud is a mist of minute ice-specks, usually forming at a height of about $5\frac{1}{2}$ miles above sea-level. It appears like tufts or curls of snow-white hair, and is named *Cirrus*. In certain conditions this cloud

gives rise to halos, wide faintly-coloured rings which appear to surround the Sun or Moon. The name *Mare's tail* is sometimes applied to it. Little rounded tufts which often cover the whole sky in summer and are familiarly called *mackerel scales* belong to a class of cloud which floats about 3 miles above the Earth's surface, and may be looked upon as half-way between cirrus and the next type; they are termed cirro-cumulus. *Cumulus* is the cloud type which comprises the great white billowy clouds common in summer. They are usually flat on their under surface, and rise above into rounded forms often of wonderful beauty. The base of cumulus cloud is usually about $\frac{3}{4}$ of a mile above the Earth's surface, while the summits may rise as high as 2 or 3 miles. These clouds are formed by the condensation of vapour in ascending currents of air, and each mass of cumulus has been likened to a grandly carved capital topping the invisible column of rising heated air. The lowest clouds are sheets of fog floating within half a mile of the Earth's surface, and being so low they are usually seen edgeways when at any distance, and so appear as long layers parallel to the horizon. This arrangement gave rise to their name of *Stratus*. A cloud, presenting a dark gray or black colour and a ragged stormy appearance, from which rain usually falls is called *Nimbus*, or simply rain-cloud. It forms at the elevation of about a mile, and is described as a mixture of cumulus and stratus. The upper clouds act as floats, by the study of which much has been learned as to the movements of the upper atmosphere. The lower clouds are of great value as heat curtains, preventing the Sun's heat from being excessive by day, and almost entirely checking the loss of heat by radiation from the Earth at night.

169. **Rain.**—Sometimes the temperature of air remarkably free from dust falls below the dew-point, and a large quantity of water-vapour must condense, while there are very few solid motes to act as nuclei. Each mote consequently gets a very heavy coating of water, and drops are formed which are too large to be much checked by friction of the air as they fall. Thus a shower of rain may fall

from a cloudless sky. Rain more often originates in clouds. The upper part of a very deep layer of cloud is less dust-laden than the lower; the motes accordingly form larger water-drops, and these descend comparatively quickly, overtaking and embodying smaller globules as they fall, until they emerge from the cloud as large drops of water. If the cloud floats very high above warm air, the vapour of which is unsaturated, the raindrops will evaporate as they fall and may vanish before reaching the Earth. But if the cloud is low or the vapour in the air traversed by the rain-drops is nearly or quite saturated, there is so little evaporation that they reach the surface undiminished or even increased in size. When much water-vapour is rapidly condensed near the surface or over air which is fully charged with vapour, there must be a great fall of rain. Hence, when a hot vapour-laden sea-wind blows against the side of a mountain, the air rises, and growing cold in consequence (§ 163), the dew-point is reached and passed, and deluges of rain fall, while dark masses of clouds fill the sky. On the other hand, when wind blows over a mountain range and descends on the other side, it grows warmer as it sinks, evaporates all the cloud it carries, and becomes a drying wind upon the low ground. Rainfall is measured by the rain gauge, and its amount is stated in the number of inches of water which would accumulate on a level surface if the rain of a year were to rest where it fell.

170. **Snow** is produced when water-vapour condenses at a temperature below the freezing-point. The water forms small clear spicules of ice which always cross at an angle of 60°, so that snow-crystals usually have six rays uniformly arranged about a centre; but the variety of forms is very great. A number of crystals getting hooked or felted together form a snow-flake, and the fluttering showers of flakes rest lightly on the ground, sometimes covering it to the depth of several feet. One foot of snow is, roughly speaking, equivalent to one inch of rain. The whiteness of snow is produced by the reflection and refraction of light again and again amongst the numerous small crystals. The real colour is bluish or greenish like a block of ice.

A great quantity of air is entangled between the spicules of snowflakes, and this makes a covering of snow act as a non-conductor of heat—almost as perfectly as a covering of feathers—preventing radiation from the Earth at night, and so keeping the ground from freezing in cold weather. Under heavy pressure snow is compacted into solid ice.

171. **Hail.**—In winter there are often showers of tightly packed little snowballs about the size of small shot or rarely as large as peas. This is called soft hail, and it appears to be formed by the larger ice particles in a deep ice cloud overtaking and adhering to the smaller ones. True hail is a different thing, which only occurs in warm weather usually as an accompaniment of thunderstorms (§ 173) or tornadoes (§ 209). True hailstones are lumps of ice which sometimes weigh several ounces, and occasionally as much as 3 lbs. A shower of such masses is very destructive, breaking windows, cutting down standing crops, and often killing animals or even people. The hailstone when cut across usually shows alternate layers of clear ice and of compact snow. According to Ferrel such a hailstone is produced by an ordinary soft hailstone formed at a great height falling into a rain-cloud, where it gets a coating of water, and then being carried by an ascending current into a high cold region, where the water is frozen into clear ice and a deposit of snow takes place outside. The same hailstones may be caught in ascending and descending currents several times in succession, thus getting alternate coats of ice and snow. This theory accounts for true hailstones only occurring in summer, for it is only in hot weather that powerful ascending currents of air are formed.

172. **Electrification of the Atmosphere.**—Every change in the atmosphere, particularly evaporation, condensation and wind, gives rise to some disturbance in the distribution of electricity. As electricity resides on the surface of a body, it follows that when the minute particles of a cloud are uniting to form rain-drops, their electrical potential (§ 76) is rapidly rising, because the surface of a large rain-drop is smaller than the total surfaces of the

small water globules which combine to form it. A heavy shower of rain rapidly carries off the electricity, reducing the potential of a cloud to that of the Earth. In certain states of the atmosphere which are not yet thoroughly understood, silent electric discharge takes place between pointed bodies, such as flagstaffs or the masts of ships, and the air. This is accompanied by a pale brush-shaped light, which goes by the name of *St. Elmo's fire*. Air which is almost free from water-vapour is a nearly perfect non-conductor (§ 77), and in the dry climates of mountain observatories and high latitudes in winter, electricity produced by friction is not immediately conducted away to the Earth as it is in damp air. In Canada one can often light a gas-jet by an electric spark from the finger, produced by shuffling the feet on the carpet; and at Pike's Peak observatory in the United States the friction of opening a drawer or shutting a door often gave rise to electricity enough to give a severe shock.

173. Lightning and Thunder.—When the electric potential of a cloud becomes much higher than that of the Earth or another cloud, a disruptive discharge takes place between them through the air (§ 78). The electrical energy is mainly converted into heat by the resistance of the air, the particles of which become instantaneously white hot; but the passage of the electric current is so rapid that only a brilliant flash is visible. The intensely heated air expands suddenly, and then as suddenly contracts, setting up a succession of air waves (§ 58) all along the line of the flash. These reach the ear as a prolonged growl or roar, or as a sharp rattling explosion, according to the distance of the observer and to the direction of the flash. The sound is prolonged by echoes from the Earth's surface and hills, or from clouds. The electric discharge follows the path of least resistance, and as vegetable juices offer less resistance to it than air, trees are often traversed by the current. The sap between the wood and the bark is so heated by the discharge that steam is formed with explosive violence, splitting off the bark, tearing away branches, and ploughing deep furrows in the solid wood, as if the tree had been struck by a solid spear hurled with gigantic strength. An animal or a human body may form part of the path of least resistance and so be "struck," but this will never happen if there is a better conductor near. The impressiveness of a thunderstorm is largely due to the majestic roar of the thunder, the darkness of the sky, the lurid glare of the clouds, and the ominous stillness of the air; but apart from these the presence of highly electrified bodies produces an indescribable effect on the nerves of many people. Lightning-conductors attached to buildings serve to equalise the potential of the Earth and clouds, and thus tend to prevent a disruptive discharge from taking place. Thunderstorms occur most frequently in the tropics, and usually during the day; in polar regions they occur very rarely, and then only at night.

174. **The Aurora.**—In the north polar regions, where thunderstorms are practically unknown, beautiful luminous effects are produced at night by the *Aurora borealis* or Northern Lights (see small map on Plate XIV.) A similar appearance in the south polar regions is called *Aurora australis*. The Aurora forms an arch or ring of coloured light over the magnetic pole (§ 98) at a great height in the atmosphere, from 50 to 150 miles. Coloured fringes and streamers shoot from this arch in all directions, sometimes spreading over the whole sky, and again shrinking back with a pulsing motion. The Aurora appears to be caused by electrical discharges in rare air, as it very closely resembles the glow seen when a current traverses a "vacuum tube" containing a little highly rarefied air. This theory was recently confirmed by the Finnish physicist Prof. Lemström, who covered the top of Mount Oratunturi in the north of Finland with a network of wires and found a true Aurora produced when he sent a current of electricity from these wires to the Earth.[3]

175. **Wind.**—When air is heated at the Earth's surface it expands, and becoming less dense, rises and flows away in the upper regions of the atmosphere. The pressure of the air over the region where expansion has taken place thus becomes less than that of the surrounding atmosphere, and air is accordingly driven in from all sides until equili-

brium of pressure is restored. Moving air is known as wind, and always blows from regions where the pressure is higher to those where it is lower. The greater the difference of pressure, or rather the *gradient*, that is difference of pressure in a definite distance, the stronger is the wind. In English-speaking countries gradient is measured by the number of hundreths of an inch difference in the reading of two barometers at a distance of 15 nautical miles (17 miles). For example, if the barometer at one place read 29.14, and at another 34 miles away it read 29.00, the difference is 14 hundredths of an inch in 34 miles, or 7 in 17, and the gradient is spoken of as 7. The same gradient would result from a barometric difference of only 3.5 hundredths of an inch if the stations were only $8\frac{1}{2}$ miles apart. The strength of wind is proportional to the gradient as the following table shows:—

Gradient	0.5	3	7	15
Velocity of wind in miles per hour	7	25	50	80
Wind	Light breeze.	Fresh breeze.	Gale.	Hurricane.

Wind ceases to blow as soon as the difference of pressure ceases to exist. While blowing, currents of air move spirally from areas of high pressure to areas of low pressure, as is explained by Ferrel's law, deviating toward the right hand in the northern hemisphere and toward the left hand in the southern (§ 89). The strength of wind is measured by anemometers (§ 442), and is expressed either in terms of its velocity or of the pressure it exerts. Wind is named by the direction from which it blows, a wind blowing from east to west being called an East wind.

176. Circulation of the Atmosphere.—In order to understand the movements of the atmosphere as a whole, it is convenient first to consider the Earth as smooth and entirely surrounded by the hydrosphere. The air between the tropics, and especially over the equator, is always being heated by strong solar radiation, and it consequently expands and rises, through the rest of the air, as oil would rise through water. This region forms the furnace which furnishes motive power for the whole system of circulation. The cooler and denser air from the neighbouring temperate zones flows toward the equator along the surface to take the place of the ascending air, and is in turn heated and forced to rise. The polar regions receive little heat from the Sun at any time, and in the long dark winters radiate heat away into space. The air over them consequently becomes chilled, grows denser, and descends toward the surface. Thus by equatorial heating and polar cooling the air is constantly being raised at the equator, carried in the upper regions north and south to the poles, brought down there to the surface and drawn back toward the equator. The upper current blows spirally as a wind from the west-south-west in the northern hemisphere, and from the west-north-west in the southern hemisphere (as explained by Ferrel's Law), while the winds from the poles would blow from north-east in the northern hemisphere and from south-east in the southern.

177. **Ferrel's Theory of Circulation.**—The result of this arrangement, according to Professor Ferrel, is that in the upper layers of the atmosphere the pressure is highest above the equator and lowest over the poles. But the rush of air at a lower level from the poles toward the equator tends to carry the mass of the atmosphere in that direction, while the movement of the upper air toward the poles tends but more feebly to carry the mass of the atmosphere in the opposite direction. The two tendencies balance each other between latitudes 20° and 30° north and south, and the pressure of the lower strata of the atmosphere is thus greatly increased in the neighbourhood of the tropics. This is shown in Fig. 22 by the boundary line of the portion of the atmosphere shown being drawn nearest the surface at the equator and poles, farthest from it at the tropics. The arrangement of pressure at the surface is thus —Two belts of air at high pressure girdle the Earth a little poleward of the northern and southern tropics, a ring of air at lower pressure lies along the equator, and great regions where the atmospheric pressure is low surround the north pole and the south pole. The tropical zones of high pressure give rise to surface winds toward the equator, strengthening the north-east and south-east winds of the

lower atmosphere. They also produce air currents toward the poles in the opposite direction as southwest and north-west winds, which gradually die away about the polar circles, where the equator-seeking winds meet, check, and rise above them. Hence in the temperate zones the surface winds should be parallel to the

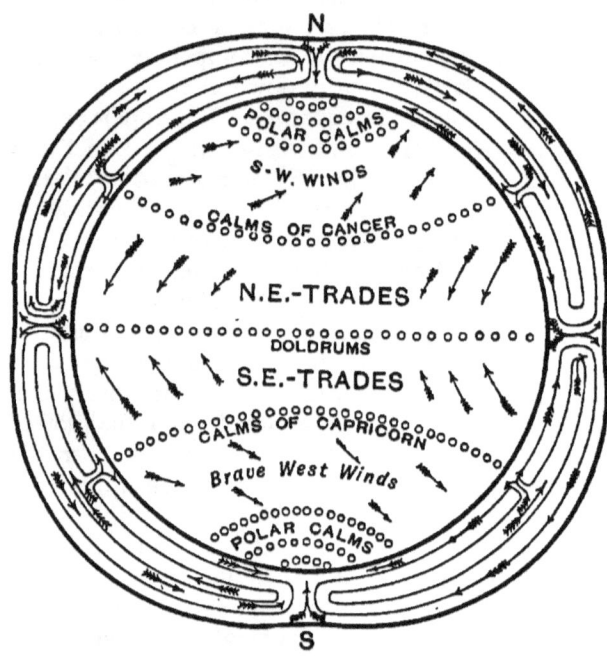

Fig. 22.—Theoretical Circulation of the Atmosphere, after Ferrel. The arrows show the directions of the winds over the surface and of the vertical movements of air.

pole-seeking upper winds, while between the two are the equator-seeking middle winds. In the tropics and the polar circles there are only the lower equator-seeking winds and the upper polar-seeking winds, as shown in the diagram.

178. **Zones of Winds and Calms.**—This theoretical circulation divides the Earth's surface into zones, which roughly correspond to those of solar climate (§ 125). In the tropical belts of high pressure, from which surface winds

blow poleward and equatorward, there is a calm. Since the upper air, which contains little vapour, is always descending, these regions are cloudless and the scene of enormous evaporation. The Temperate zones of poleward surface winds receive the hot vapour-laden tropical air and conduct it to colder regions, where much of its vapour is condensed. They are thus windy cool regions of moderate cloudiness and rainfall. The polar regions of low pressure are practically calm, and as most of the air descends from above they are relatively dry. The tropical regions swept by the equator-seeking winds are windy, hot, cloudless, but the scene of great evaporation from the hot sea surface. The narrow equatorial belt of low pressure into which the equator-seeking winds blow from north and south is also a region of calm. The air as it ascends here expands, cools, and the enormous supply of vapour swept in from the tropics condenses into the heaviest cloud, and falls as deluges of never-ceasing rain. The heat liberated by the condensation of so much vapour strengthens the equatorial up-draught. The equatorial belt of low pressure always lies nearly under the vertical Sun, consequently in the northern summer (§§ 122, 123) it swings to the north, and in the southern summer it swings to the south, displacing the belts of tropical high pressure northward and southward alternately. For reasons which cannot be explained here, this displacement is comparatively slight, extending over only five or six degrees of latitude. In the North Atlantic, for example, the equatorial low pressure belt never moves farther south than 5° N. All parts of the Earth's surface that the equatorial rain-belt traverses in its annual movement, experience a rainy season as it lies over them, and a dry season all the rest of the year, when swept by the equator-seeking winds. Near the equator, where the narrow rain-belt crosses a tract of the Earth both in its northward and in its southward swing, there are two wet and two dry seasons in the year. The theoretical circulation of the air and its resulting climates are affected by two causes, unequal heating of the air by land and sea surfaces (§ 164), and the deflection of the prevailing winds by plateau edges and mountain ranges. Regular zones of surface winds and climates consequently are found only in great expanses of ocean, and do not appear in narrow seas or on land (see Plates V. VI. VII.)

179. **Trade Winds and Doldrums.**—When the Spanish and Portuguese explorers of the 16th century found that north-easterly winds blew steadily all the year round on the Atlantic between 30° and 5° N. and enabled them to make quick voyages to the West Indies, they gave the name of Trade Winds to the favouring breezes. The name has since been extended to include all the permanent winds which blow from the tropical toward the equatorial calms. In the winter half of the year (November to April) the north-east trades of the Atlantic are felt as far north as 25° N. and reach southward to 5° N. ; and in the Pacific they sweep over the range of sea between 28° N. and 8° N., and the tropical calms reach as far north as 40°. The south-east trade winds at the same season are experienced in the Atlantic between a line drawn from the Cape of Good Hope to Rio de Janeiro, and the equator. In the eastern Pacific they reach farther north, crossing the equator to at least 5° N. The equatorial belt of calms and rains lies entirely to the north of the equator; its width varies from 120 to 200 miles in the Atlantic, and is about 300 miles in the Pacific. This calm belt, called by sailors the *Doldrums*, was greatly dreaded in the days of sailing ships, on account of the absence of wind, which often kept a vessel rolling helplessly for weeks, while the close damp air made the men dispirited and ill. Thunderstorms of terrific violence are very common in it. It was consequently of the greatest importance for a captain to know where the narrowest part of the belt could be found at each season, in order that he might pass quickly from the clear bright skies and fresh invigorating winds of the north-east trades to the equally pleasant and favourable region of the south-east trades. During the summer half-year (May to October) the rain-belt of the Doldrums with its calms moves farther north, and widens to from 300 to 500 miles. The north-east trades then begin in about 30° N. and die off about

12° N., while the south-east trades do not extend so far south, but cross the equator, blowing as far as 5° or even 8° N. The calm equatorial zone of rains always lies north of the equator, on account of the heating influence of the greater mass of land in the northern hemisphere. (Plate VII.)

180. **The Roaring Forties** is a name given by sailors to the belt of ocean between 40° and 50° S. in which the "Brave West Winds" blow all the year round, as regularly as the trades and more strongly. This belt is more nearly covered with a uniform stretch of ocean than any other part of the Earth, and exhibits the theoretical circulation of the atmosphere in great perfection. The prevailing wind is produced by the high pressure of the south tropical calm belt and the remarkably low pressure which surrounds the south pole. The strength and constancy of the brave west winds enable sailing vessels to compete with steamers in trading with New Zealand going by the Cape of Good Hope and returning by Cape Horn.

181. **The Northern Anti-trades.**—The south-west winds of the northern hemisphere, which blow from the northward edge of the north tropical zone of high pressure to the north polar region of low pressure, are sometimes called the Anti-trades; but they are much less constant and more variable in strength than the trade winds or the winds of the Roaring Forties. The trade winds blowing into the Gulf of Mexico in the summer months from the east or south-east are deflected by the edge of the great tablelands of Mexico into south-westerly winds, which blow up the Mississippi valley and sweep across the Atlantic, reinforcing the somewhat uncertain anti-trades.

182. **Daily Temperature Changes.**—The circulation of the atmosphere which has just been described was deduced by mathematical reasoning from a few simple data, and then proved by observation to be correct so far as disturbing causes allow. But the changes in the atmosphere which take place from hour to hour throughout the day were first observed in thousands of cases, and their cause has been subsequently ascertained by inductive reasoning (§ 17). Solar radiation goes on from sunrise to sunset, but the

temperature of the air reaches its maximum about 2 P.M. local time, or about 2 hours after the Sun has passed the meridian. Cooling then sets in, and the temperature reaches a minimum about 5 A.M., or shortly before sunrise. These hours apply to the tropics and vary slightly in different parts of the world, but the air is always coldest in the early morning and always warmest in the early afternoon. Sir David Brewster discovered, by comparing a long series of

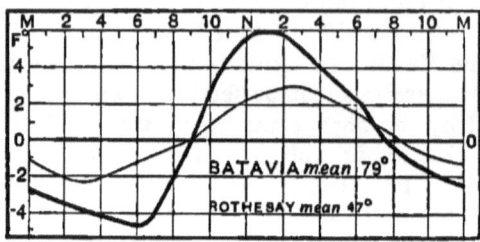

FIG. 23.—Daily Range of Atmospheric Temperature in temperate and tropical climates (after A. Buchan).

hourly observations, that the average temperature at any pair of hours of the same name (*e.g.* 9 A.M. and 9 P.M.) was almost exactly the average temperature for the whole day. Fig. 23 shows the range of temperature above and below the average for the day, the hours being marked along the top and the temperature in degrees above and below the average on the side (see § 444). The solid curve refers to a station in the tropics, the lighter curve to a temperate region.

183. **Daily Pressure Changes.**—The pressure of the atmosphere is least about 4 A.M. and 4 P.M. and greatest about 10 A.M. and 10 P.M. In Fig. 24 the diurnal range of the barometer above and below its mean value is given, the range in fractions of an inch being marked on the side, the hours from noon to noon along the top. The solid curve shows the typical range in the tropics, the lighter curve that in a temperate region. This regular increase and decrease of pressure twice daily, was for a long time supposed to be a tidal effect caused by the Moon's differential attraction, but Dr. A. Buchan in discussing the barometric observations made on the *Challenger* Expedi-

tion proved that it really depends on the changes of atmospheric temperature, and so is a result of the radiant energy of the Sun. The *Morning Minimum* of pressure about 4 A.M. results from the cooled dust-motes condensing upon themselves most of the water-vapour contained in the air, the vapour pressure is greatly reduced, and the total observed reduction of atmospheric pressure is thus accounted for. When the Sun appears, the dust-motes are warmed up, the vapour returns to the atmosphere, and the temperature of the air rapidly increasing, produces the *Forenoon Maximum* of pressure about 10 A.M. When the temperature of

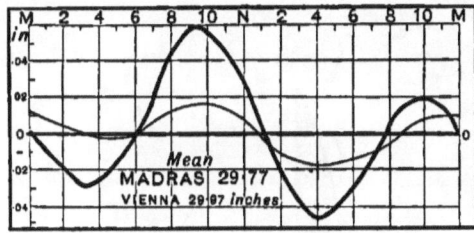

FIG. 24.—Daily Range of Atmospheric Pressure in temperate and tropical climates (after A. Buchan).

a gas is raised it must either expand or press more strongly on the vessel containing it; and in the forenoon the heated air is prevented from expanding for a time by the resistance of the cooler layers of atmosphere above, against which it presses with increasing force, and the barometer rises. After 10 o'clock the continued heating enables the air to overcome the resistance, and ascending currents set in, the air rises and, meeting the west winds of the upper atmosphere, is carried away to the eastward. The density of the whole column of atmosphere is diminished by the removal of the ascending air, and the *Afternoon Minimum* of pressure occurs about 4 P.M. As the surface air cools in the evening it grows denser and sinks, the upper air welling over from the heated regions to the west, where it is still only early afternoon, flows in above, cools and sinks, so raising the pressure to produce the *Evening Maximum* about 10 P.M. Thus the morning minimum and maximum are caused by the action of condensing and evaporating water in the atmosphere; the afternoon minimum is caused by a bodily removal of air upward and to the east, the evening maximum results from the sinking down and piling up of air from the west.

184. Land and Sea Breezes.—The different heating and cooling of land and sea (§ 164) produces a regular change in the daily winds of tropical coasts and islands, and in very calm clear weather similar effects may be observed in all latitudes. An island or strip of coast when heated by the Sun gives rise to ascending currents of air (Fig. 25). About 10 A.M. these ascending currents, having carried the

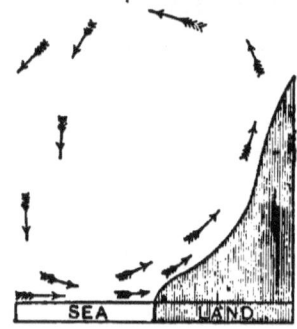 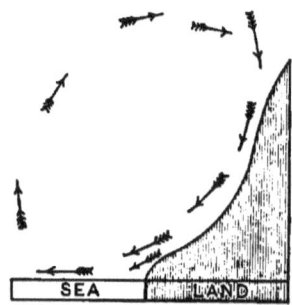

FIG. 25.—Sea-Breeze during sunshine.   FIG. 26.—Land-Breeze, at night.

air into the upper regions, produce a fall of pressure over the land compared with that over the cooler sea, and a sea-breeze sets in, at first as a very gentle air, but gradually increasing in force until about 3 P.M., when the land surface is most highly heated. After that hour the land cools down more quickly than the sea, and as the atmospheric pressure becomes equalised the sea-breeze dies away. The air over the land continues to cool down and to sink; more air consequently flows in above, and the pressure over the land thus becomes greater than that over the sea. A surface land-breeze (Fig. 26) sets in about 8 P.M., often with sudden squalls, which are dangerous to boats. It gradually increases in strength as the land grows cooler until it reaches a maximum about 3 A.M. In the trade-wind

ISOBARS AND W
After A

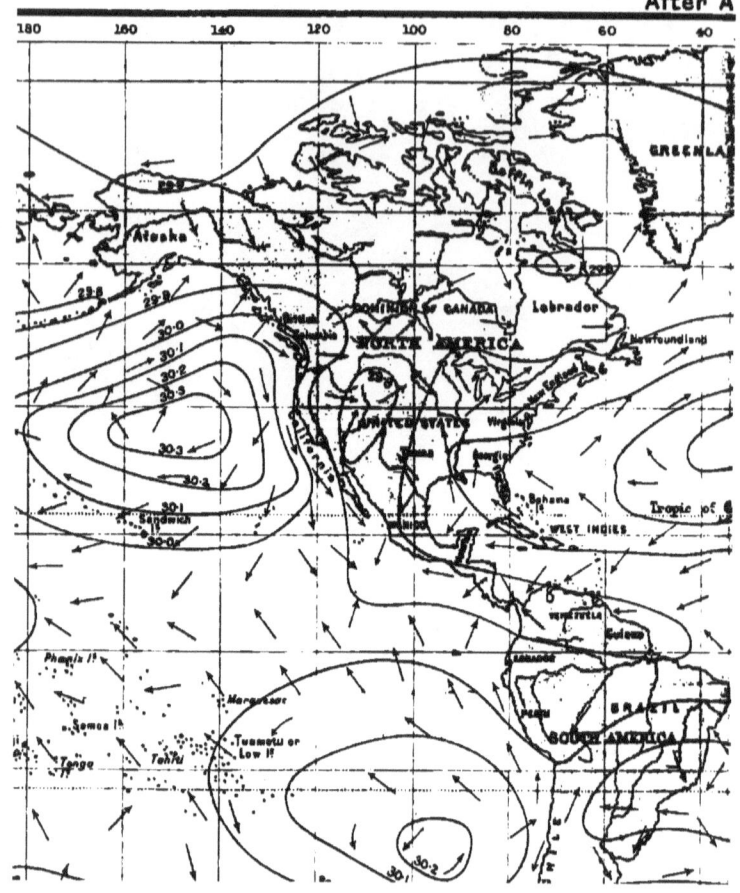

FOR JULY.
an.

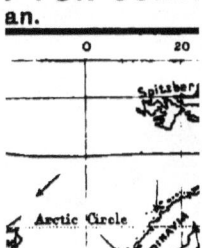

regions the land and sea-breezes are often not strong enough to reverse the direction of the prevailing winds, and merely alter the strength. On the south-east coasts of the Fiji Islands, for example, the prevailing south-east trade wind is intensified during the day and much reduced at night, while on the north-west coasts the wind is reduced through the day and strengthened at night. Land and sea-breezes are always light on a low flat island or coast; but when a range of mountains rises near the sea very strong winds are produced, the mountain slope acting like a flue, aiding the ascent of the hot air by day and the descent of cold air by night. On account of the lofty backbone of the Blue Mountains the sea-breeze in Jamaica is the strongest known.

185. **Monsoons.**—Over the centre of continents far removed from the ocean, the range of air-temperature is greatest, the great dryness (§ 164) favouring radiation and producing very high temperatures in summer and very low temperatures in winter. Over the sea the range of temperature is least. The continents by heating the air in summer set up ascending currents which last for months, so that the pressure of the air is greatly lowered, and surface winds blow in toward the continent from the surrounding seas. In winter the air being cooled by the continents produces descending currents; the pressure becomes much higher than that over the less chilled seas, and consequently surface winds blow outward from the continents during the winter months. These winds changing with the seasons are called Monsoons. They are produced exactly like land and sea-breezes, only with a period of a year instead of a day. Just as in the former case, monsoon winds may be too feeble to reverse the direction of the prevailing winds, and may succeed only in modifying their force (see Plate VII.) The monsoon effect of most continents is comparatively insignificant, and is confined to a small part of the coast. In the southern continents these winds are slightly developed, because in the widest part of South America and Central Africa the annual range of temperature is very small, and in the narrower part farther south the influence of the vast expanses of the neighbouring oceans predominates all the year round. In Australia the

K

monsoons are well-marked but not very strong, although the range of temperature is considerable; but with an equally great range the Sahara region of North Africa has a very much slighter monsoon-raising power. The flatness of these expanses of land and their low elevation partly account for this; the disturbing influence on atmospheric pressure of the expanses of sea to the north is also important. On the west coast of North America there are distinct monsoons, but it is in Asia with its steep mountain slopes rising from the sea that the monsoon blows with greatest power, and in India the name was first applied.

References

[1] J. Aitken, "On Dew," *Transactions R.S.E.* xxxiii. p. 9 (1885); also *Nature*, xxxiii. p. 256 (1885).

[2] J. Aitken, "Dust, Fogs, and Clouds," *Transactions R.S.E.* (1881); also *Nature*, xxiii. p. 195 (1881).

[3] See note on Lemström's Aurora Experiments in *Nature*, xxxv. p. 433 (1887).

Books of Reference

W. Ferrel, *Popular Treatise on the Winds*. Macmillan & Co. (An admirable discussion, but not easy reading.)

R. Abercromby, *Weather*. International Scientific Series.

CHAPTER IX

CLIMATES OF THE WORLD

186. **Configuration and Climate.**—In passing from the theoretical system of atmospheric circulation sketched in last chapter to the actual conditions of the atmosphere in different parts of the world, the disturbing influence of the land must be taken into account. The student should therefore read § 214 and Chapter XV. as far as it refers to the configuration of the continents, and study Plate XI., as well as the maps illustrating atmospheric conditions. Surface winds are altered in their direction in a very marked way by mountain ranges and the edges of plateaux. At the same time, sloping land differs from level ground by setting up a local vertical circulation, acting exactly as a chimney does in increasing a draught. In hot climates mountaineers find a strong wind sweeping up the slope by day helping their ascent, and on the summit the ascending air-current from opposite sides rises straight up, and is often strong enough to carry off hats and notebooks. At night the effect is reversed, and strong winds blow down the slopes. The same effects are produced in a more intense degree in narrow steep mountain valleys, the furious day and night winds of which make travelling difficult and dangerous in some of the Himalayan passes. Experienced hunters on the Rocky Mountains build their fires just below their tent, knowing that the night-wind will carry the smoke down the valley. In still winter weather the air, chilled as a thin layer on mountain sides, grows dense as its tempera-

ture falls, and flows gently down into the valleys, filling them to a certain level with intensely cold air. The peasants in many valleys of the Alps perch their wooden cottages on knolls or rocks, not so much for the picturesqueness of the site, but in order to stand above the surface of the flood of icy air which streams through the valley in winter. Rainfall is still more intimately connected with configuration. Meteorologists, in speaking of the *climate* of a place, mean the average state of the atmosphere with regard to warmth, wind, rain, and all other variable conditions.

187. **Atmospheric Temperature in different latitudes.**—The excess of land in the northern hemisphere, compared

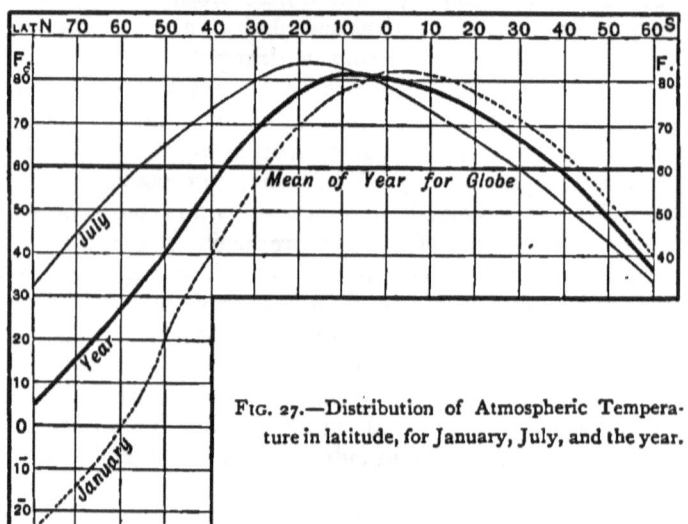

FIG. 27.—Distribution of Atmospheric Temperature in latitude, for January, July, and the year.

with the southern, alters the distribution of the solar energy shed equally on both, and prevents the simple astronomical climate zones (§ 125) from corresponding on the two sides of the equator. Fig. 27 shows by means of curves the mean temperature of the year, as calculated by Professor Ferrel for each 10° of latitude from 80° N. to 60° S. Latitude is marked along the top and temperature up the side of the diagram, the three curves of which correspond to the average temperatures of January, July, and of the whole year. The curve of average temperature for the year shows that the warmth of the air at 80° N. is only 5° F., that in 4° N. there is the maximum temperature of 81° F., and that in 60° S. it is 35° F., the northern hemisphere being as a whole a little colder than the southern. The mean surface air temperature of the whole Earth is about 59.5°. The curves for January and July show a great annual range of temperature in the northern hemisphere, increasing toward the north where the land preponderates, and a slight annual range in the southern hemisphere, decreasing toward the south where the sea influence prevails. The student should study this diagram, comparing the temperature at each season at various latitudes in the two hemispheres. To do this follow the vertical line of latitude until it cuts the curve; the point on the thermometer scale in a horizontal line with this intersection is the temperature at that particular latitude.

188. **Isotherms.**—If the temperature of every place in the world at some one instant were marked in figures on a large map the result would be very confusing to look at. But if all the figures except those showing a difference of 10 degrees were blotted out the map would be much simpler. Near the equator the number 80 would occur frequently, farther north and south there would be rows of 70, still farther strings of 50, and so on. A line might be drawn through all figures 80, and the figures themselves might then be blotted out, except one left to mark the line, and the same might be done for 70, 60, and the rest, greatly simplifying the map. Such lines are termed isotherms, or lines of equal warmth, as they pass through all the places where the air temperature is the same. In interpreting the maps (Plates III. and IV.) it is usually assumed that the temperature between two isotherms is proportional to the distance. For example, in the January map (Plate III.) the line of 70° temperature in Central America is one inch from the line of 80° in South America, so that between them one-tenth of an inch on the map corresponds to a change of 1° of temperature. The lines of 40° and 50° in

North America approach at one place on the same map to within one-tenth of an inch of each other, so that between them one-hundredth of an inch corresponds to a change of 1°. The space between the isotherms is coloured to bring out the difference of temperature, the hottest regions being shown in deepest red, the coldest in deepest blue. Isotherms are constructed to refer to sea-level, so in order to find from the maps the actual temperature at any place a deduction of 1° for every 270 (or for convenience say 300) feet must be made (§ 163). For every place on the contour-line of 600 feet of elevation 2° must be deducted, and for every place on the 6000 foot contour-lines 20° must be deducted from the isothermal temperatures. Those two contour lines are marked on the maps as a guide to the interpretation of the results.

The maps in this volume are reduced from the most recent set of isotherms, compiled in connection with the scientific reports of the Voyage of the *Challenger*, by Dr. A. Buchan, and they give the average temperatures for the fifteen years from 1870 to 1884.

189. **Air Temperature in January.**—January is the midsummer of the southern hemisphere. The map (Plate III.) shows that the region with a temperature over 70° lies south of the Tropic of Cancer on land, and the only places warmer than 90° are under the Tropic of Capricorn in Africa and Australia, the land being more heated than the water by the nearly vertical Sun. The eastern sides of the southern continents are warmer than the western; thus on the Tropic of Capricorn, the east coasts of Africa and South America have a temperature of 80°, and the west coasts less than 70°. This is explained by the prevailing winds and ocean currents (§ 243). The isotherm of 32° in the southern hemisphere occurs about 64° S., and its direction is nearly east and west, being uninfluenced by any land. Farther north, the direction of the isotherms becomes more irregular on account of the increasing interference of land in altering the temperature of the air. In the northern hemisphere, where January is midwinter, the sea as a rule is warmer than the land in the same latitude, and the

coldest regions are the centres of the great continents. The coldest place where observations have ever been made, is the Siberian village of Verkhoyansk just within the Arctic Circle (see Plate VII.) On account of the Arctic Sea being frozen across in winter, this village lies close to the centre of the northern continental mass. The mean January temperature at this station is 61° below zero, Fahrenheit; and the absolutely lowest temperature ever experienced by human beings occurred there in January 1886, which was − 89° F. The powerful influence of the warm surface-water of the Gulf Stream (§ 244) on the air is shown by the temperature of the Lofoten Islands, in the same latitude as Verkhoyansk, being above 32°, the difference between the two being more than 93°. The coldest point in the American continent lies a little north of the magnetic pole (§ 98), and has a temperature of − 40°. In order to appreciate the effect of land and sea in modifying climate the student should carefully follow the isotherms of 30° and 40°, noting carefully the latitude at which these temperatures prevail near the coast and in the heart of continents. To make this exercise still more instructive, the lines might be traced on the contoured map (Plate XI.), and the actual surface temperatures calculated.

190. **Air Temperature in July.**—The lapse of six months brings round the northern summer and southern winter. The Sun now vertical near the Tropic of Cancer beats down upon a far greater breadth of land surface than in January, and so the area with a temperature exceeding 90° in North America, North Africa, and Asia extends far to northward of the tropic. The sea now exercises a cooling influence on the air of the middle latitudes in the northern hemisphere. The isotherm of 70° F., for example, runs far to the north over the continents, reaching 55° N. in North America, and 58° N. in Eastern Asia, but it scarcely gets north of 40° N. in the Atlantic, and is carried south to 25° N. by the Pacific. In higher north latitudes the slight north-eastward trend of the isotherms shows that some warming effect is still due to south-west winds and currents. In July the Lofoten Islands, having warmed up

only by 20° on account of the sluggish heat transactions of water, are at the same temperature, 55° F., as Verkhoyansk, where, however, the air has been heated no less than 116° since January, this being the greatest annual range known. The purely continental character of Verkhoyansk is modified by the fact that in summer it is not far from the shore of the cold Arctic Sea, whence cool monsoon winds blow. In the southern hemisphere the temperature of the land has fallen by radiation a little below that of the sea. The prevailing winds, however, are so powerful, and the oceanic influence predominates so greatly, that temperatures below 70° are found north of 20° S. only on the west sides of the continents which are cooled by the increasing upwelling of ocean water due to the trade winds. The extreme tip of South America is the only southern continental land which has a winter temperature below 40° in July, and the isotherm of 32° encircles the globe in 55° S., not touching any land at all, and showing but a slight range from its winter position.

191. **Land and Sea Climates.**—The comparatively cool summers and mild winters of the extremities of the southern continents compared with those of the same latitudes in the northern hemisphere is a direct result of the arrangement of land and sea on the globe. A land climate is everywhere extreme, a sea climate is always mild; but an examination of Fig. 27 will show that the average temperature for the year is nearly the same in both hemispheres. Fig. 28 shows the annual range of temperature in degrees above and below the mean for the year in typical continental and oceanic climates. The

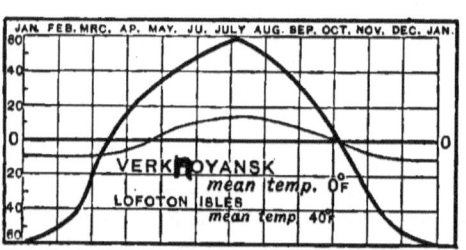

FIG. 28.—Curve of monthly mean temperature for a typical land-climate (dark) and sea-climate (light). The dark horizontal line marked 0 represents mean annual temperature; the figures show number of degrees above and below the mean.

The LIGHT BLUE TINT indicates the PAST Distribution of Glaciers

solid curve shows the range at Verkhoyansk, the finer line that at the Lofoten Islands. In continental or land climates the range of temperature is great and the rainfall very small, in oceanic or sea climates the range of temperature is small and the rainfall great. Prevailing winds carry an oceanic climate for a considerable distance inland on the west coasts of northern continents, and they carry a continental climate a considerable distance seaward on the west shores of northern oceans. The student should verify these statements by making a detailed comparison of the two isothermal maps, and the map of rainfall (Plate VIII.)

192. **Isobars.**—The invisible differences of atmospheric pressure may be laid down on a map in the same way as the invisible differences of temperature. Lines running through places over which the atmospheric pressure, as measured by the barometer and reduced to sea-level, is the same, are called isobaric lines, or shortly *isobars*. Those shown on the maps (Plates V. and VI.) express the pressure in inches of mercury at intervals of every tenth of an inch, and the spaces between them are coloured, so that the regions of highest pressure are deep red, and those of lowest pressure deep blue. When adjacent isobars are drawn far apart on the map the barometric gradient (§ 175) between them is slight, and the wind set up is consequently gentle; but when the isobars are crowded closely together, a steep gradient is indicated, giving rise to furious wind. A gradient of 0.5 corresponds to a difference of pressure of 0.20 inches in one degree of latitude; a gradient of 15, giving rise to a hurricane, corresponds to a difference of pressure equal to 0.60 inches in one degree of latitude. In the maps the arrows are represented as flying with the wind. The shortest path from a region of high to one of low pressure is at right angles to the isobars, but in consequence of the rotation of the Earth the actual path of the wind is that stated in Ferrel's Law (§ 89). The deviation is proportional to the latitude, so that in the far north and south, wind blows nearly parallel to the isobars. Dr. Buys Ballot, the late eminent Dutch meteorologist, independently discovered the *Law of the Winds*, which Dr. Buchan has

put in this form :—" Stand with the Low pressure on your Left hand and the high pressure on your right; then in the Boreal (northern) hemisphere the wind is Blowing on your Back; but in the southern hemisphere in your face." The student should impress these statements by studying Plates V. and VI.

193. **Winds of the Southern Hemisphere in January.**—The theoretical arrangement of atmospheric pressure and winds (§§ 178-181) is changing from hour to hour in response to the changes of day and night (§ 184) and summer and winter (§ 185). The two maps (Plates V. and VI.—which should be referred to continually in reading what follows) are reduced from Buchan's *Challenger* maps, and give the average conditions of the atmosphere in January and July for the fifteen years 1870 to 1884. In the map for January the equatorial zone of low pressure, as limited by the isobars of 29.90 inches, is narrow over the ocean but widens greatly over the three southern continents, where the heat of summer causes the air to ascend, flow away, and reduce the pressure over the land. Another consequence of high temperature over the continents is, that the south tropical belt of high pressure is broken into three isolated portions lying altogether south of the tropic in the three oceans. The southern area of low pressure and of steep gradient, as limited by the isobar of 29.90 inches, occupies the whole surface south of 40° S., the isobars running nearly straight east and west. From the three south oceanic regions of high pressure, surface winds blow outward, forming the south-east trades on the northern margin toward the equatorial low pressure, and the brave west winds on the southern toward the great south polar low pressure. The portions of the equatorial low-pressure zone extended to the south by the continents produce monsoons, or an indraught of surface air toward the land. On the north-west coast of Australia the Australian low-pressure area draws the trade wind round to form a north-west monsoon. On the west coast of Africa the south-east trade is drawn in to form a light south-west monsoon, and in the Gulf of Guinea it is drawn in strongly

from the west. The same action is seen on the west coast of South America, but there the uniform face of the Andes deflects the wind back again. On the eastern shores of the southern continents the monsoon effects strengthen the prevailing trade winds.

194. **Winds of the Northern Hemisphere in January.**—The north tropical zone of high pressure forms a continuous belt round the world, narrow over the oceans, but extending right up to the polar regions over the two northern continents. Over these continents, air is continually descending from the upper atmosphere, where its place is being taken by the air driven up from over the northern oceans and southern continents. The Arctic low-pressure area is cut up into two comparatively small depressions, one with its centre between Iceland and Greenland, the other in the North Pacific Ocean. In consequence of this arrangement the north-east trade winds of the Atlantic blow into the Carribean Sea and against the coast of South America beyond the equator, and under the influence of the South American low-pressure area, unite with the south-east trades, blowing up the valley of the Amazon, and obliterate the belt of calms along the coast. The North Atlantic low-pressure area, while maintaining the south-west winds of western Europe, draws in cold north-east winds on the east coast of North America. The high pressure over North America gives rise to monsoon winds which attain considerable force as north-westers along the west coast of Central America, and also sets up prevailing northerly winds down the Mississippi valley.

195. **Winter Monsoon of Indian Ocean.**—About the month of October, when the pressure over the great Asiatic continent becomes higher than that over the ocean, light northerly winds set in in the Bay of Bengal and the Arabian Sea, gradually changing to north-east winds at sea, where they represent the trade winds, but rarely attaining great force, and often broken by calms. Along the base of the Himalayas in the plain of the Ganges the wind is north-westerly. This state of matters lasts for several months, coming to a climax in January. Over

most of India it is a dry season, as the air of the North-East or Winter Monsoon has descended from the upper region of the atmosphere, and contains little water-vapour. On the east coast of the Indian peninsula and of Ceylon, the north-east wind having traversed the Bay of Bengal, sweeps along a considerable amount of vapour, which is precipitated on the Eastern Ghats and the eastern side of the Ceylon hills, winter being their rainy season.

196. **Winds of the Southern Hemisphere in July.**—Notwithstanding the change from summer to winter in the southern hemisphere, the southern region of low pressure is practically unaltered in position, but the gradient southward is reduced, and the winds of the Roaring Forties blow with slightly diminished strength. The south tropical belt of high pressure has reunited in consequence of the cooling down of the southern continents, and it now stretches far north of the tropic. In consequence of the small range of temperature and slight winter cooling of the southern continents, the highest pressure in the southern hemisphere is over the oceans even in winter, and this fact accounts for the permanence and steadiness of the brave west winds. The south-east trades blow across the equator far to the north in all the oceans. In the Indian Ocean the calm belt is completely obliterated on account of the great suction over Asia which draws supplies from the southern hemisphere and turns the south-east trade winds to feed the south-west monsoon. At the same time pressure is high over the continent of Australia, from which monsoon winds blow outward.

197. **Winds of the Northern Hemisphere in July.**—The equatorial belt of low pressure extends over the whole land surface of the northern hemisphere and unites with the North Polar region of low pressure, centres of lowest pressure lying in the south-west of Asia and in the west of North America. The north tropical belt of high pressure is broken into two great isolated high-pressure areas, which occupy the North Atlantic and the North-West Pacific, keeping up the north-east trade winds in those oceans; and giving rise to south-westerly winds over eastern North-

America and North-Western Europe. The monsoon influence of North America is very slight, on account of the position of these two high-pressure areas, and that of North Africa is also remarkably feeble (§ 185). The winds of the Indian Ocean and Western Pacific are completely dominated by the vast furnace flue of Western Asia, which attains its maximum effect in July, and destroys the theoretical atmospheric circulation of the northern hemisphere.

198. **Summer Monsoon of the Indian Ocean.**—Round the coast of Asia the north-east wind falls off in February, and gradually shifts to the south as the winter high pressure over the continent is reduced, and gives place to the summer low pressure. March and April are characterised by variable winds and frequent storms. By May the northeast wind has died away, and in its place south-west winds, usually spoken of in India as *The Monsoon*, blow strongly across the Arabian Sea and Bay of Bengal, and wheel round along the foot of the Himalayas, blowing up the Ganges valley as south-east winds. This state of matters lasts until August or later. As the wind blows for a long distance over the heated surface of the ocean it reaches land laden with vapour, and, rising up the steep and almost unbroken slopes of the Western Ghats, condenses in tremendous showers. The first deluges of rain are known as the bursting of the monsoon. A heavier rainfall reaches the western edge of the Indo-Chinese peninsula, and the heaviest of all is found in the converging valley of Assam, at one place over 500 inches a year. After August the south-west monsoon diminishes in force and gradually dies away as the pressure over the land increases. The monsoon owes much of its strength to the energy set free by the condensation of the vapour it carries. On the coast of China the summer monsoon blows from the south-east, and the winter monsoon from the north-west.

199. **Yearly Swing of the Atmosphere.**—The disturbing effect of land and sea on the normal arrangement and movements of the atmosphere may be put briefly thus. In winter the chilled land draws down the blanket of air which the less-cooled sea is tossing off upward. In summer the

heated land throws off as much of its air-covering as possible, piling it upon the colder sea which eagerly draws it down. While the land is throwing off the air above, which descends upon the sea, the sea commences to return it to the land along the surface more slowly than it receives it at first, more rapidly afterwards. When the land has drawn down on itself from above a greater supply of air at the opposite season the sea gradually draws it off along the surface. There is thus a constant effort to restore the equilibrium of atmospheric covering between land and sea, disturbed by the rapid radiation of the land. The prevailing winds of the year, disregarding minor seasonal changes, are shown in Plate VII.

200. **Rainfall and Evaporation.**—A continual circulation of water takes place between the hydrosphere and atmosphere. Sea winds blow water-vapour against the land and ascending currents carry it into the upper atmosphere, where it condenses and returns either directly as rain or through springs and rivers to the sea. The amount of evaporation at sea, and of rain falling on land depend mainly on temperature and winds. Dr. John Murray has calculated that nearly 130 million million tons of water, or about $\frac{1}{40}$ of the whole mass of the atmosphere, are transferred from the sea surface to the land, and find their way back again in rivers every year. More than half of the rain falls between the tropics, and probably not more than $\frac{1}{10}$ of it reaches ground as snow beyond the polar circles.[1] The average rainfall of the globe is about thirty-three inches. A calculation has been made that one quarter of the land surface has a rainfall less than one foot in a year, one quarter has a rainfall between one and two feet, one quarter, of which the British Islands form part, has a rainfall of between two and four feet, and over the remaining quarter the rainfall exceeds four feet in a year. In all regions not reached by sea winds the rainfall is very slight, and evaporation preponderates, a nearly rainless area containing dwindling salt lakes occupying part of the interior of each continent (§ 356).

201. **Distribution of Rainfall.**—Plate VIII., the data

of which were mainly compiled by the American meteorologist Professor Loomis, shows the rainfall on the land by deepening blue tints according to the number of inches which fall at each place in a year. It also shows, mainly from the data of Mr. Buchanan of the *Challenger*, the salinity of the ocean by deepening red tints; salt areas in the sea are produced by evaporation of the water which supplies the rainfall of the land, and they may be termed the comparatively dry regions of the sea. They correspond very closely with the centres from which the trade winds blow. The lightest blue colour on the map denotes regions where the rainfall is under ten inches per annum. These correspond exactly with the regions of extreme range of temperature, lying as a rule in the interior of continents. The regions of greatest rainfall coloured in deepest blue are all steep land slopes exposed to a sea wind. In North America, for example, the trade winds blowing round the Gulf of Mexico, and the south-west winds beating on the coast of Oregon and British Columbia, ensure heavy rainfall. South America shows a very interesting relation. In the trade-wind region vapour is carried up the flat valley of the Amazon and condensed on the eastern slope of the Andes, the western slope of which is rainless. In the south of the continent the west winds of the Roaring Forties dash perpetual showers against the western face of the Andes, and descending sweep as drying winds across Patagonia. In India and the Malay Archipelago the heavy rains are produced entirely by the summer monsoons. Attentive study of the rainfall map, along with those of winds and configuration, will bring out similar reasons for the local distribution of rainfall in all parts of the Earth.

202. **Winds of the British Islands.**—The British Islands are usually covered by the edge of the North Atlantic area of low pressure. The pressure being lowest in the north-west, and highest in the south-east, corresponds to prevailing south-westerly winds. In January the isobars are closely crowded together; in that month the average gradient over the British Islands is steeper and the winds are consequently stronger than in any other part of the

world. From January onward the atmospheric pressure increases rapidly in the north and much more slowly in the south, so that in the month of April the gradient, though still for westerly winds, is very slight. A small temporary rise of pressure in the north may thus reverse the gradient, and as soon as the pressure in the north becomes higher than that in the south, east wind sets in. A similar state of matters occurs again in November, on account of the pressure in the south falling more rapidly than that in the north, and the months of April and November are famed for bitter east winds in all parts of Britain.[2]

203. **Temperature of the British Islands.**—The temperature of the British Islands on the average for the year is about 48°, increasing from 45° in Shetland to 53° in Scilly, or an average rise of temperature of 1° for every 100 miles toward the south. In winter the temperature has no relation to the latitude, the islands grow colder from west to east. The isotherms of January (Plate IX.) run from north-west to south-east. A broad strip of country from Caithness to Lincoln has an air temperature of 38° or less. Shetland, Orkney, Ayr, Liverpool, Oxford, and London are traversed by the isotherm of 39°. The points of Kerry, Cork, and Scilly are at 45°. The south of England is mild in winter, not because it is the south, but because it runs so far to the west. By the month of April the isotherms run nearly east and west; the temperature is 42° in Shetland, 45° from Skye to Aberdeen, and 48° from Erris Head through Dublin and Liverpool to Harwich. In this month land and sea have practically the same temperature. In July the land has heated up more than the sea, so that the south-west wind now has a cooling effect, and the isotherms (Plate X.) run roughly from S.W. to N.E. Shetland is at 54°, the line of 58° runs from Malin Head, near Rothesay and Inverness, to Peterhead, and that of 60° from Killarney across Ireland, through Lough Neagh southward, north through Whitehaven to Selkirk, and then south to Newcastle. The hottest region is round London, where the temperature averages 64°. As autumn advances the air cools down most rapidly on the east coast,

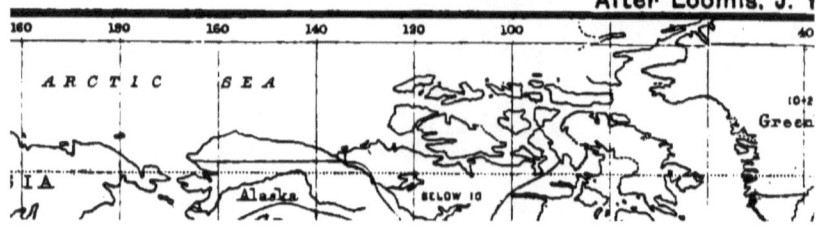

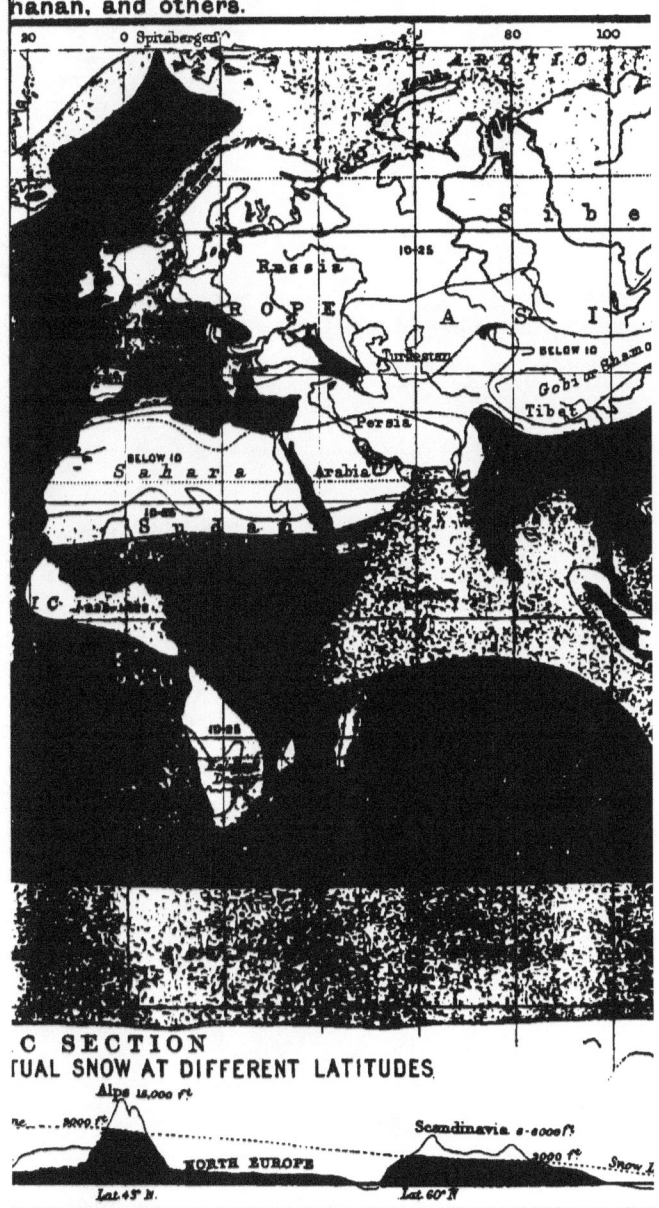

and in September the isotherms run west and east once more, and the temperature varies from 52° in Orkney to 58° along a line from Pembroke through Bristol and Reading to Lowestoft. The way in which proximity to the sea and exposure to the prevailing winds influences the range of temperature is shown in the following tables :—

	Shetland	Rothesay	Plymouth	Inverness	Edinburgh	London
Jan. temp.	39.5	39.5	43.0	37.5	38.0	39.0
July temp.	53.5	58.0	62.5	58.0	59.5	64.0
Annual range	14.0	18.5	19.5	20.5	21.5	25.0

If it were not for the warm south-westerly winds the January temperature would be 7.5° in Shetland, 12.5° at Edinburgh, and 22° at London, and the sea all round the islands would be frozen.

204. **The Rainfall of British Islands** has been studied by Mr. G. J. Symons, who publishes an annual volume on *British Rainfall*. The rainfall is greatest on the west and least on the east coast, warmth always going with wetness (Plate XVII., compare also Plate XVI.) In Ireland, on account of the mountains forming irregular isolated groups, the rainfall is remarkably uniform over the whole island, averaging about forty inches in the year. In Great Britain the low outer Hebrides have a rainfall of about fifty inches, but the high mountains of Skye and the Western Highlands condense more than eighty inches in the year over an area stretching from Skye to Loch Lomond. The mountains of Cumberland and Wales and the high land of Cornwall have also a large rainfall; but the whole east of Britain has less than forty inches. Most of the district between the Humber and the Thames, the driest part of the British Islands, receives less than twenty-five inches of rain per annum. Contrary to the usual opinion, November is nowhere the rainiest month in the British Islands. The heaviest rainfall in the west and north of Ireland and the west of Scotland takes place in December and January. In England and the east of Scotland it occurs in October, except in the very dry region between the Thames and Humber, where most rain falls in August. In the south of England the least rainy month is March, in

the north of England and south of Scotland it is April, in the Scottish highlands it is May, and in Orkney it is June. The average distribution of climate shown in the maps, although correct on the whole, cannot be depended upon to hold good at any special place for any particular month. Such maps are of great value in choosing a place to reside in, but of very little use for planning a pleasure trip. The conditions of weather are somewhat complicated, but appear to depend mainly on the distribution of atmospheric pressure, which may be classified into certain well-marked types.

205. **Anticyclones.**—An anticyclone is a portion of the atmosphere in which the pressure is highest at the centre, and diminishes nearly uniformly in all directions. The wind in an anticyclone blows spirally outward, as is illustrated in the high-pressure regions shown in the Isobaric maps. In the northern hemisphere the circulation of surface wind round the edge of an anticyclone is in the same direction as the hands of a watch move, in the southern hemisphere in the opposite direction, as explained by Ferrel's or Buys Ballot's Law. An anticyclone when once formed is a very steady arrangement of pressure, and usually lasts for many days or even weeks at a time. This being so, it is evident that a supply of air must be continuously renewed from above to take the place of that passing out as surface winds. Air in fact passes through an anticyclone much as grain does through a pair of mill-stones, though of course without suffering any physical change. In the upper regions of the atmosphere air must be moving inward and sinking downward to maintain the anticyclone, and the pressure in the upper region of the atmosphere must thus be least above the spot where it is greatest on the Earth's surface. This deduction has been proved to be true by observations at mountain meteorological stations. The surface winds of an anticyclone are usually light and variable. As the air is descending from above, it contains very little water-vapour, and no clouds are formed. Hence in summer, anticyclonic weather is brilliant, hot, and calm, with haze at night or heavy deposits of dew, on account of great cooling by radiation. In winter an anticyclone is

calm and clear, and by intense radiation the land cools down greatly at night, and the temperature of the air falls. This is the condition required for long spells of frost, and in large towns and over lakes and estuaries it produces dense, low-lying fogs. The low temperature tends to increase the density of the lower air in an anticyclone, and until very recently was viewed as the main cause of the formation of this arrangement of pressure. Fig. 29

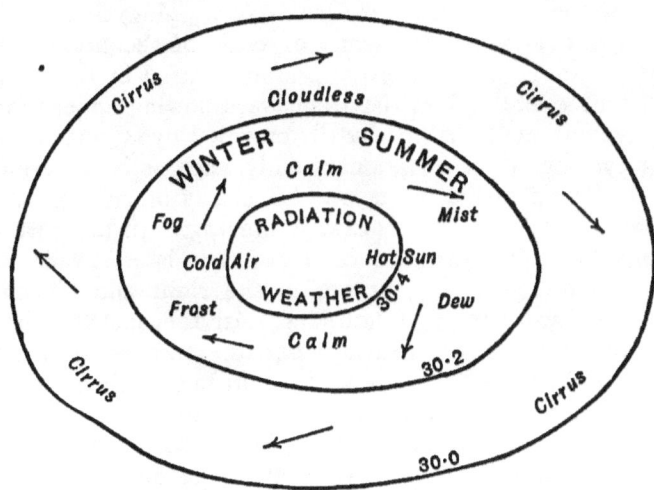

FIG. 29.—Isobars of an Anticyclone. (After the Hon. Ralph Abercromby.) Direction of wind shown for the northern hemisphere. The prevailing weather in winter is shown on the left side, that in summer on the right side of the diagram.

shows the form of isobars and the kind of weather in a typical anticyclone, which may be summarised as a very steady and nearly stationary descending eddy or whirl of banked-up air crowded into one place by the converging currents of the upper atmosphere as they flow toward the poles.

206. Cyclones.—An area of low atmospheric pressure which has the lowest pressure in the centre was called a *Cyclone*, because the early observers believed that the wind blew round it in circles. We now know that wind blows in toward the centre of low pressure in a spiral curve with a strength proportional to the gradient. The circulation of winds in a cyclone is—following Ferrel's or Buys Ballot's law—in the direction opposite to the movement of watch-hands in the northern hemisphere, and in the same direction as the hands of a watch in the southern. As the centre of a cyclone remains at the lowest pressure in spite of surface winds pouring in from every side, the air must rise in the centre and flow out above. A cyclone is thus an inward and upward whirl or eddy of the atmosphere. The air above has its normal pressure raised by the upflow, and consequently gives rise to outward-flowing upper winds; the cyclone in fact is necessarily crowned by an anticyclone. The cyclone is not a stationary eddy, such as is represented by the low-pressure areas on the charts of *average* atmospheric conditions, but is always moving. In the northern hemisphere the direction of its motion is westward near the equator, gradually turning to the right and becoming north-eastward in high latitudes. In the southern hemisphere it is westward near the equator, turning toward the left and becoming south-eastward in high latitudes. The rate of motion of the centre is from 15 to 30 miles an hour in the temperate zones, but only from 3 to 8 miles an hour in the tropics. The actual particles of air do not move forward, but the diminution of pressure is passed on like a wave (§ 57) through the air. Cyclones usually form on the edge of the permanent regions of high pressure, and travel away along their margins. Professor Hann, the great German meteorologist, has concluded that the cyclones and anticyclones of the temperate zone are true eddies and bankings-up formed in the great streams of air which set poleward from the equator.[3]

207. **Cyclonic Weather.**—There are certain changes of weather associated with a cyclone which result from the fact that it is an eddy of ascending surface air. The air on rising near the centre is cooled by expansion, and the vapour condenses into cloud, and ultimately falls as rain. Hence, when the cyclone is approaching an observer and condensation has just begun to take place in the upper regions, a halo produced by reflection from the condensed

particles of ice is commonly seen round the Sun or Moon. Later the sky becomes gloomy, the air feels warm and oppressive even in winter, thick clouds form, and there is heavy rain, while the barometer is all the time falling, and the wind shifting its direction. As soon as the barometer begins to rise, the centre of the cyclone has passed; and as the atmospheric pressure increases in the rear of the depression the sky clears, the wind freshens, and the air feels peculiarly exhilarating. Fig. 30 shows the form

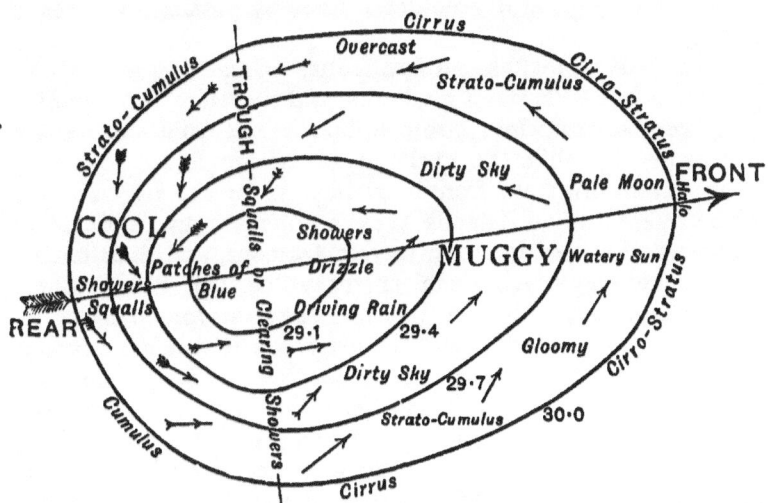

FIG. 30.—Isobars of a Cyclone. (After the Hon. Ralph Abercromby.) Direction of wind and distribution for weather shown for the north temperate zone.

of the isobars, direction of wind, and the different conditions of weather in various parts of a cyclone. It shows what changes an observer would notice according as he was to the north or south of the cyclone as it passed. The long arrow shows the direction in which it moves. In a typical cyclone, such as that represented, the gradients are always steeper in the rear than in the front, so that the strongest winds are experienced after the barometer has begun to rise. The succession of weather is the same in every cyclone; but the intensity of it depends on the gradient of pressure. If the cyclone is of great size and the diminution of pressure in the centre very slight, gentle winds and light showers only will be produced. But if, on the other hand, the cyclone is of small dimensions, less than a hundred miles across, and the diminution of pressure in the centre is great, terrific winds and deluges of rain result. The centre of a cyclone is always calm, although a gale may be blowing round it in every direction a few miles distant. The weather of the northern hemisphere is to a very great degree determined by passing cyclones of large size.

208. **Hurricanes and Typhoons.**—Small cyclones of slow motion but with steep gradients, and therefore accompanied by very severe winds, are common in the tropics at certain seasons. Unlike the cyclones of the temperate zone they appear to originate from local heating of the air. Among the West Indian Islands such storms are liable to occur during the months from July to October, and their terrific violence has given wide currency to their native name of *Hurricanes*. In the Bay of Bengal at the changing of the monsoons, and along the east coast of Africa, similar storms, to which the name *Cyclones* was first applied, are experienced. In the China Sea they are common from July to November, and are known as *Typhoons*. These tropical storms differ from the less violent cyclones of the temperate zone in always having a patch of clear blue sky over the central calm where the barometer is lowest; this is called the eye of the storm. Although the calm centre of a cyclone is referred to poetically as "the whirlwind's heart of peace," it is the part most dreaded by sailors. There is no wind to move a sailing ship, and a terrible chaos of heavy waves is driven in by the winds raging on every side. A ship-captain in the season when these storms are prevalent is always on the watch for them, and as their approach is heralded by a fall in the barometer and the way in which the wind changes, he can find the direction of the centre. A steamer in many cases can then by changing its course let the storm pass harmlessly.

The commonest Cyclone tracks of the tropics and the

usual direction of motion of the storm-centre are represented on Plate VII.

209. **Whirlwinds.**—Eddies of ascending air which are of small diameter compared with their height, and move rapidly forward over the Earth's surface, are called *Whirlwinds*. They are often set up by the sudden heating of the lower layers of the atmosphere. The dreaded *Simoom* of the Sahara is a whirlwind which raises great gyrating clouds of sand, and sweeps forward with irresistible force, darkening the sky at midday. The *Tornado* of North America is even more destructive. It is most often formed in the southeast side of a slowly moving cyclone, and usually acquires its full force suddenly in sultry summer afternoons. The origin of a tornado has given rise to much controversy, but is usually believed to be the rapid heating by the Sun of a lofty column of air fully charged with water-vapour. The heated air expands upward and rotates as it rises, drawing the surface air in from all sides. The water-vapour, condensing as the air cools in ascending, adds to the heat-energy (§ 159) of the whirl, and helps to produce a tremendous reduction of pressure in the centre. Surface winds rush into this partial vacuum, and whirl with terrific violence up the central hollow as if it were a furnace chimney. In consequence of their force the tornado cuts a clean path through forests or towns that lie in its path. The breadth of the zone of destruction is seldom more than a quarter of a mile. Houses are not simply knocked down but burst up when a tornado passes over them. The low pressure of the centre creates a partial vacuum, and the air inside a house consequently expands so rapidly that the roof is blown off and the walls thrown outward. Sheep and fowls when caught up are completely plucked of wool or feathers by the fierce whirls of wind before they are dropped. After about an hour the heated vapour-laden air that originates the tornado is dispersed, and as the whirl travels at the rate of 30 miles an hour the track of destruction is usually 30 miles long, although instances of papers being carried 45 miles are on record. Tornadoes are most common in the United States east of 100° W.; but it is only in a

small district of Kansas on the Missouri River, and in the south-west of Illinois, near the Mississippi and Ohio, that more than 50 have been recorded in the last hundred years.

210. Waterspouts and Cloudbursts.—The rapid condensation of water-vapour in the axis of a tornado, or in the comparatively harmless whirlwinds that sometimes occur in all parts of the world, produces a dark funnel-shaped cloud tapering downward to the Earth. Such a cloud occupying the centre of an ascending eddy of air is called a waterspout. When it strikes the ground the heavy fall of rain on a very small area sometimes produces great destruction. At sea, or in passing over a lake or river, the low pressure of the whirling air of a waterspout often sucks up a column of water and whirls it on for considerable distances. In this way shoals of fish or swarms of frogs are sometimes raised high in the air, carried for miles inland, and dropped as showers of fish or frogs to the wonder of country people. It often happens that the upward rush in a tornado is strong enough to prevent the condensed water from falling until a great quantity has accumulated; then it descends not as rain but like a river, and the phenomenon is spoken of as a *Cloudburst*. On mountain slopes cloudbursts have been known to hollow out deep ravines in a few minutes. Hail as well as rain may be similarly accumulated, and the worst hailstorms occur during the passage of a tornado.

211. Weather-charts.—The gradual growth of knowledge about the atmosphere showed that the barometer could be used for predicting changes of weather in certain cases. Most barometers have a series of words from "Set fair" to "Stormy" engraved on the scale, as if high or rising pressure always means calm and fine weather, and low or falling pressure always foretells wind and rain. A few weeks' observation will in most cases convince any one that this is a mistake, and that a single barometer is of little value for forecasting the weather. Fig. 30 shows that it is not the actual height of the barometer at one place, but the difference in the height of many barometers at con-

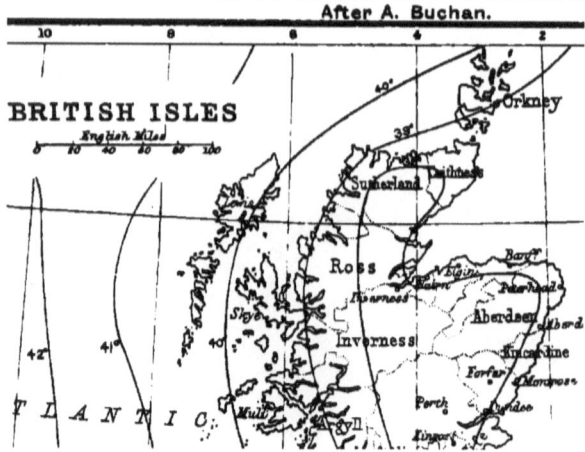

siderable distances apart, that can throw light on the state of the atmosphere and the associated weather. About thirty years ago the first synoptic charts showing the isobars of a country were introduced as an aid to the study of weather, and such weather-charts enable storms to be foreseen in many cases. In nearly every country there is now a number of meteorological stations where observations of barometer, thermometer, wind, etc., are made at the same hour morning and evening, and telegraphed to a Central Meteorological Office maintained by Government. Here charts are prepared showing at one glance the state of the atmosphere both as regards pressure and temperature (corrected to their value at sea-level) over the whole country and surrounding districts. If the student will take the trouble of tracing in red ink on thin paper the figures of a cyclone and anticyclone given above (Figs. 29 and 30), and will then lay this tracing over the map of the British Islands (Plate IX.), he will see exactly how the weather varies in different parts of the country according to the distribution of these types of atmospheric pressure.

212. **Weather Forecasts.**—Several arrangements of isobars besides those into cyclones and anticyclones may occur. Isobars drawn from actual observations may be straight, showing that they form part of neither cyclone nor anticyclone; sometimes they are sharply curved, forming V-shaped areas of low pressure or wedge-like areas of high pressure lying between adjacent anticyclones or cyclones; and they very often form loops, showing the existence of a small secondary cyclone inside a larger. Each type of pressure-distribution corresponds to a special kind of weather, and the relation between isobars and weather has been carefully studied and is well known to practical meteorologists. The commonest weather in the British Islands is that produced by the passage of cyclones eastward from the Atlantic, and this may be taken as a characteristic example to illustrate weather forecasts. If the student places a tracing of Fig. 30 on Plate X. so that the large arrow points north-east and its head is on the south-west of Ireland, and then moves the tracing gradually north-east-

ward, he will see how the weather varies in all parts of the islands as the cyclone passes along its path. By shifting the centre to north or south, and changing the direction of passing (but always moving the tracing as the arrow flies), the effect on the student's own locality of cyclones passing in any direction may be followed. Remembering that as isobars of successively lower value are passing the barometer is falling, and that as isobars of higher value are passing the barometer is rising, it will be found possible to identify the actual movements of a cyclone by watching the barometer and the changes of wind and weather. In order to predict on Monday the kind of weather and direction of wind on Tuesday when a cyclone is passing, it is necessary to know where the centre is, at what speed, and in what direction it is moving, so that a map of the conditions expected on Tuesday can be drawn up from the data supplied by Monday's observations. But in order to predict the intensity of the weather and the force of the wind, it is necessary to know whether the cyclone is "deepening" or "filling up," that is, whether the gradient of pressure from circumference to centre is growing greater or less. Only experience and practice can guide a forecaster in these matters, and the success of the predictions issued daily by all civilised governments depends on the knowledge and skill of the men who make them. It often happens that a cyclone does not follow the usual path, or that the pressure at the centre increases when the forecaster thought it would diminish, or that a secondary depression suddenly forms in an unexpected place, and of course in all such cases the forecast made is a failure. Yet on the whole more than 80 per cent of the predictions issued in Britain and America are successful. The British Islands are divided for purposes of forecasting the weather into eleven districts. At 10 A.M. and 8.30 P.M. forecasts for the next 24 hours of the weather in each of these districts are published at the Meteorological Office in London from observations made all over the country at 8 A.M. and 6 P.M. The weather-charts, reports, and forecasts in a daily and weekly form are sent out to subscribers by the Meteorological Office. The Reports and Forecasts are published in all the morning and evening newspapers, but only the *Times* prints a daily weather-chart of the British Islands.

213. **Storm Warnings.**—A sudden fall of the barometer at any of the special British meteorological stations is at once telegraphed to London, and if it is found to indicate the discovery or sudden deepening of a ' cyclone crossing the islands which is likely to cause a dangerous storm at sea, warnings are telegraphed to all the important harbours and fishing stations on the coast, where signals are immediately hoisted to give notice to fishermen and sailors. Such signals are most valuable on the east coast because the disturbances usually come from the west. Farmers profit by weather predictions as well as sailors, particularly in the hay and harvest seasons. The escape of gas in coal-mines and consequent risk of explosions has been found to depend largely on variations of atmospheric pressure, and miners' warnings are now regularly issued when any serious change of pressure over the coal-mining regions is anticipated. In many ways the British Islands are in the worst position for forecasting the weather as they lie in the most disturbed region of the atmosphere. The most westerly observing station is on Valentia Island in the south-west of Ireland, which often does not give time to warn the country before a storm appears, and affords very little opportunity of tracing the probable path in which it will travel. A floating station in the Atlantic, west of Ireland, would be an enormous help in framing British forecasts, and would undoubtedly save many lives and much money. On the continent of Europe forecasting is comparatively easy, as the British stations give early notice of all changes. Similarly, in a broad stretch of land like North America, Australia, or India, where the stations are widely distributed and well equipped, there are great advantages for the prediction of weather. In the United States the Weather Bureau of the Agricultural Department has charge of meteorological observations, and the forecasts are not only distributed as in Britain, but in the thinly peopled districts the trains are fitted with special signals so that the farmers along the railway have only to look out as the train

passes in order to know what weather to expect for the day. The attempt to time the arrival on the coast of Europe of cyclones whose path across America has been tracked out is rarely successful, as most depressions either fill up or change their path or rate of moving on the way across the Atlantic. There are many prognostics or signs, such as the appearance of halos, of mist on hill-tops, great clearness of the atmosphere, exceptionally bright reflections in water, the movements of animals, by which experienced people can foretell the weather of their own district with marvellous correctness. Indeed, for any mountain valley or seaside town the opinion of an observant old shepherd or fisherman on the approaching weather is likely to be more correct than the somewhat general Meteorological Office forecast.

REFERENCES

[1] J. Murray, "On the Total Annual Rainfall," etc., *Scot. Geog. Mag.* iii. 65 (1887).

[2] A. Buchan, "Climate of British Islands, Pressure and Temperature," *Journ. Scot. Met. Soc.* for 1882. "Rainfall," *ibid.* for 1885.

[3] H. F. Blanford, "Cause of Anticyclones and Cyclones," *Nature*, xliii. 15 (1890). "The Genesis of Tropical Cyclones," *Nature*, xliii. 81 (1890).

BOOKS OF REFERENCE

Challenger Reports, Physics and Chemistry, Circulation of the Atmosphere, by A. Buchan. (A unique collection of isobaric and isothermal maps for every month of the year.)

A. Buchan, Art. "Meteorology," *Encyclopædia Britannica.*

R. H. Scott, *Elementary Meteorology.* International Scientific Series.

H. F. Blanford, *Climates and Weather of India.* Macmillan and Co.

N. S. Shaler, *Aspects of the Earth*, pp. 197-257. Smith, Elder, and Co.

Consult also the publications of the Royal Meteorological Society, the Scottish Meteorological Society, and of the Meteorological Office, 116 Victoria Street, London, S.W.

CHAPTER X

THE HYDROSPHERE

214. Land and Water.—The hydrosphere does not completely cover the globe, because the lithosphere which supports it is diversified by great heights and hollows. The portion of the heights projecting above the water surface forms land, which is estimated at the present time to cover 28 per cent or a little more than one quarter of the globe. Most of the hydrosphere is retained in the great world-hollows forming the ocean, which covers about 72 per cent of the surface; but on account of evaporation and condensation a small part is always present as vapour in the air, and a larger amount rests as lakes in hollows of the land or flows across the surface in rivers. The proportion of land and water in different latitudes is represented in Fig. 31, where the land area is indicated by shading. The largest proportion of land is in the northern hemisphere, where

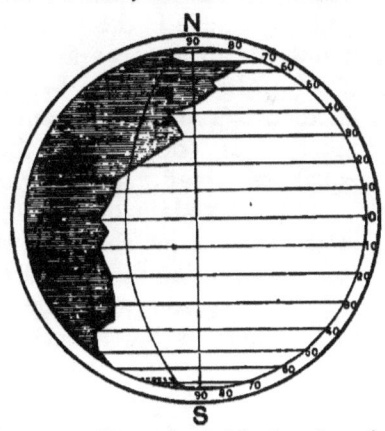

FIG. 31.—Proportion of land and sea in different latitudes. Land area shaded (after Krummel).

it occupies about 42 per cent of the surface, while water largely predominates in the southern hemisphere, where about 17 per cent of the surface is dry land. The fine curve in the figure shows the average distribution of 28 per cent of land in all latitudes. All the great land masses of the globe are widest in the north, and taper to a point toward the south. Only a few small islands lie beyond 56° S. if the unexplored Antarctic region is excepted. The inequality of the distribution of land and water appears greatest in the hemisphere having its centre near New Zealand, which comprises two-thirds of the entire ocean surface and only one-eighth of the land; and in the opposite hemisphere (with its centre in the English Channel) which contains only one-third of the ocean and seven-eighths of the land of the Earth. In the water hemisphere the proportion of land is about $\frac{1}{13}$ or 8 per cent; in the land hemisphere it is about $\frac{1}{2}$ or 50 per cent, the areas of land and sea being equal (see small maps on Plate XI.)

215. **Divisions of the Hydrosphere.**—The Caspian Sea is the only large sheet of water which is cut off by land from the rest of the hydrosphere, and its separation from the ocean is comparatively recent (§ 335). Otherwise the hydrosphere is a connected whole, made up of four wide open expanses called *Oceans*, from which smaller portions called *Seas* are more or less distinctly marked off by the land. It is a matter of opinion where to draw the line between oceans and seas; the expanse of water within the Arctic Circle, for example, is by some authorities considered the smallest ocean, and by others with more show of reason it is held to be the largest sea. Seas may be classed in three groups—(*a*) *Inland Seas*, entirely surrounded by land, of which the Caspian is the only example; (*b*) *Enclosed Seas*, nearly surrounded by land but connected with the ocean or with another sea by one channel, which is narrow and shallow compared with the general breadth and depth; (*c*) *Partially Enclosed Seas*, which (a) have two or more entrances, or (β) are marked off from the ocean by a line of islands, or (γ) by an entirely submerged barrier.

216. **The Oceans.**—No natural boundaries mark off the hydrosphere sharply into separate parts, but it is convenient to distinguish four divisions called oceans, the positions

of which are shown on Plate XIII. The *Southern Ocean* may be characterised as the shoreless ocean, for it extends round the Earth from 40° S. to the Antarctic ice, only a portion of South America, the islands of Tasmania and South New Zealand, and some smaller ones projecting into it. Its area is about 30,000,000 square miles. The *Pacific Ocean*, with an area of 55,000,000 square miles, as large as all the land of the globe, is well called the Great Ocean by the Germans. It contains many islands and partially enclosed seas, the names of which are given in the following table. The Pacific is the only ocean parts of which lie more than 2500 miles from the nearest continent (see Plate XII.) The *Indian Ocean* is entirely enclosed by land on the north, and has an area of 17,000,000 square miles. The *Atlantic*, with an area of 33,000,000 square miles, has a more indented shore than any other, and may be called the ocean of enclosed seas. The largest of these, often itself termed an ocean, is the Arctic. More than half the land of the globe sends rivers into the Atlantic and its associated seas.

OCEANS AND SEAS

ATLANTIC.		PACIFIC.		INDIAN.	
Enclosed.	Partially Enclosed.	Enclosed.	Partially Enclosed.	Enclosed.	Partially Enclosed.
Mediterranean	Arctic	Yellow	Bering	Red	Andaman
Black	Kara		Okhotsk		
Adriatic	Norwegian	Gulf of	Japan	Persian	
Baltic	North	California	China	Gulf	
White	Caribbean		Celebes		
Hudson Bay			Banda		
	Gulf of Mexico		Java		
			Sulu		
			Arafura.		

217. Ocean Tides.—If the hydrosphere were continuous, or if the land were arranged in narrow strips from east to west, a double tidal wave (§§ 103, 114) would travel round the globe every day, the velocity of this free wave form being thus about 1000 miles an hour at the equator, and its length half the circumference of the Earth. If the land of the globe were arranged in strips from north to south, cutting up the hydrosphere into a series of narrow compartments, there would be no appreciable tidal effect. By the actual arrangement of land there is a free water ring in the Southern Ocean only; there is one long comparatively narrow compartment, the Atlantic Ocean; another wider and shorter, the Indian Ocean; while the rest of the hydrosphere forms the wide open surface of the Pacific extending half-way round the globe at the equator. In the Pacific and the adjacent Southern Ocean alone the tidal wave has full room to form, and from them the wave passes westward, being deflected northward into the other oceans. Co-tidal lines on a map (Plate XIII.) show the places which the same phase of the tidal wave reaches at the same hour. Starting from 12 the position of the crest of the wave at each successive hour is marked by 1, 2, 3, up to 12. The tidal wave travels most rapidly, and is longest and of least amplitude in deep water; in the central Pacific the range between High Water and Low Water (the amplitude of the tidal wave) is less than 2 feet, and no current is produced.

218. **Tidal Currents.**—When the tidal wave enters shallow water it becomes shorter and moves more slowly. The under side of the wave becoming more retarded than the top, the surface water is carried forward as a true current, the energy of which is derived from the Earth's rotation. In this way shoals or submarine peaks convert the simple up and down movement of the tide in the open ocean into rapid currents, usually for a very short distance but sometimes extending to a great depth. These are more definite along the shores. The usual tidal effects observed on a broad gently-shelving shore are the gradual rise of the level of the water, the submergence of the beach and advance of the sea on the land; then after the highest point has been attained, the gradual lowering of level with corresponding uncovering of the beach and retreat seaward of the sea-margin. At New and Full Moon, when spring-tides (§ 114) occur, the rise and fall is at the greatest, and then, at any one place, high water occurs at the same

VEGETATION ZONES OF CONTI
After Engle

Note to Monsoon Drifts &c.
In the Indian Ocean, the China Sea, and the West Coast of Mexico and Central America, the Currents change with the Monsoons. The simple arrows ⟶ show the S.W. and S.E. Monsoon Drift during the Northern Summer. The arrows marked thus ←●— show the N.W. and N.E. Monsoon Drift during the Northern Winter.

Sea-Weed   Warm Currents coloured Red
                                         The direction of the C

ITS AND OCEANIC CURRENTS.
d others.

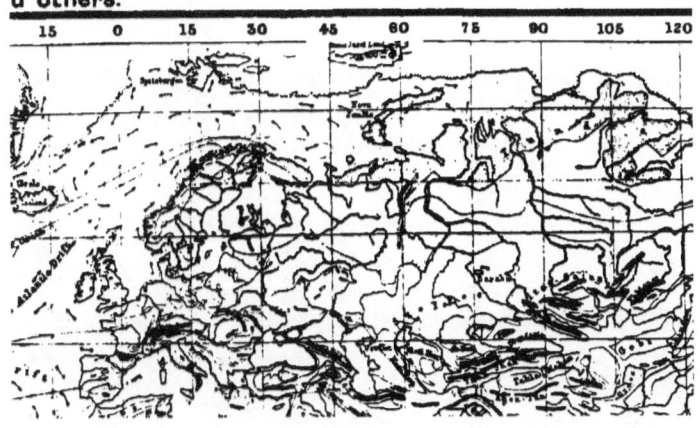

hour. Admiralty charts show the tidal data for each seaport, thus, *e.g.*, "High Water, Full and Change, X. rise 10 feet." This means that on the day of Full Moon and of Change or New Moon high water occurs at 10 A.M., and the rise of the sea between low water and high water is 10 feet. Each successive high tide after Full Moon occurs at an interval of about $12\frac{1}{2}$ hours, rises to a somewhat less height and falls to a somewhat less depth, thus covering and laying bare a narrower strip of the beach until the Moon's phase is the third quarter, when the time of morning high water is 4 A.M. and neap-tide occurs. After this the tides increase in amplitude again until the period of Change or New Moon, when the time of morning high water is once more 10 o'clock. The time during which tidal currents run in one direction and in the opposite bears little relation to the hours of high water and low water, depending largely on the form of the coast. In partially enclosed seas a branch of the tidal wave usually enters by each channel, as shown in the co-tidal map of the British Islands (Plate XVII.).

219. **Tides in Bays and Estuaries.**—When the tidal wave of the ocean enters a narrowing bay or sea inlet, the depth of which diminishes rapidly, the tidal currents become rapid and tumultuous and the water is heaped up to a great depth against the land. At the entrance of the Bay of Fundy the tide rises 8 or 9 feet, but at the head the rise at spring-tides is more than 70 feet, the greatest tidal range known. The highest spring-tide at Cardiff docks rises 42 feet, and the lowest neap-tides 20 feet, while at the mouth of the Bristol Channel the rise of spring-tide is only about 10 feet. The tidal wave rushes up some rivers with great violence, forming a *bore* or wall of foaming water stretching right across the stream, and often producing much destruction to shipping in the Amazon and Yang-tse-kiang. A tidal current sweeping through a narrow irregular channel gives rise to eddies or whirlpools sometimes of great size, like that of the Maelström in the Lofoten Island group.

220. **Properties of Water.**—In order to understand the action of solar energy on the hydrosphere, we must

know something of its composition and physical properties. The hydrosphere is composed almost entirely (about 96.5 per cent) of water, and the total amount of this substance which exists upon the Earth is estimated at about 335 million cubic miles or 1,500,000 million million tons. The mass of the hydrosphere is thus about 300 times as great as that of the atmosphere, but its volume is at least 100 times less. Pure water is a chemical compound of oxygen and hydrogen united together in the proportion of one-ninth hydrogen and eight-ninths oxygen by mass. Intense heat (§ 71), the action of some heated metals, or the passage of an electric current, separate these constituents, giving to water in some rare circumstances the character of an explosive (§ 294). The student should read again §§ 66-73. Water, on account of its singularly high specific heat and latent heat, is better fitted than any other fluid for the part it plays in transmitting and regulating energy in Nature. Water is capable of dissolving all natural substances, although some, such as glass or silica, are taken up in minute proportions. Natural water is consequently never pure; however clear it appears, it contains various gases and solids in solution.

SALTS OF RIVER-WATER

Calcium Carbonate .	42.90	Carbonates
Magnesium Carbonate	14.80	= 57.70
Silica . . .	9.90	
Calcium Sulphate .	4.50	Sulphates
Sodium Sulphate .	4.20	= 11.40
Potassium Sulphate .	2.70	
Sodium Nitrate .	3.50	
Sodium Chloride .	2.20	
Iron Oxide and Alumina	3.60	
Other Salts . .	1.30	
Organic Substances .	10.40	
Total	100.00	

221. **River-water** contains salts of many kinds in solution derived from the surface over which it flows. The amount

of dissolved solids in river-water may vary from about 2 grains in the gallon where a river flows over granite rocks, to more than 50 grains per gallon where the streams traverse a limestone country; the average salinity of river-water is about 12 grains per gallon or 0·018 parts in 100. The composition of the dissolved solids is different for each river on account of the different rocks traversed, but the accompanying table gives the composition of 100 parts by weight of the dissolved salts of an average sample of river or lake water. The large proportion of carbonates and of silica and the small proportion of common salt (sodium chloride) present are characteristic.

222. **Sea-water.**—The water of the ocean contains nearly 200 times as much dissolved solids as the water of the land. Sea-water, indeed, is at once recognised by taste as *salt*, while rivers are pronounced *fresh*. Although the salinity of sea-water varies from place to place and from time to time within certain narrow limits, the composition of the dissolved solids remains almost the same everywhere. In other words, water collected in any part of the great oceans, and boiled down with suitable precautions so as to leave the solids behind, yields "salt" of almost exactly the

SALTS OF SEA-WATER

Sodium Chloride	77·70
Magnesium Chloride	10·80
Magnesium Sulphate	4·70 ⎫ Sulphates
Calcium Sulphate	3·60 ⎬ =10·80
Potassium Sulphate	2·50 ⎭
Calcium and Magnesium Carbonate	0·30
Magnesium Bromide	0·20
Other Salts	0·20
Total	100·00

same composition which is shown in the accompanying table. The only exception which has been proved to this statement is that at great depths there is a slightly greater proportion of calcium or magnesium carbonate than near the surface. It is remarkable that more than three-quarters of the whole is made up of common salt, while the proportions of carbonates and of silica are very minute. Silica in carefully filtered sea-water never appears to exceed 1 part in 250,000 or 0·0004 per cent. The proportion of sulphates is nearly the same as in the salts of river-water. Some geologists suppose that the sea consists merely of concentrated river-water; and even on the more probable assumption that sea-water contained salts in solution derived from the primeval atmosphere, it is evident that some agent must be at work withdrawing silica and carbonates from river-water as it enters the sea. That agency is known to be the power of living creatures—plants and animals—to make themselves shells or skeletons of silica or of calcium carbonate secreted from the water (§ 273). Sea-water is slightly alkaline, probably on account of its containing bicarbonates in solution. It dissolves carbonate of lime, especially when subjected to great pressure.

223. **Salinity.**—The salinity of sea-water is the amount of dissolved salts contained in 100 parts. One hundred pounds of average sea-water contain about 3·5 pounds of dissolved salts, and thus the average salinity is said to be 3·5 per cent. It is difficult to measure salinity directly, as some of the salts decompose when the water is boiled down. The density of sea-water, however, depends on its temperature and on the salinity, so that if the density is always measured at the same standard temperature, or corrected to it, the differences of density are due to differences of salinity alone. For example, if a bottle contains exactly 1000 grains of pure water at the temperature of 60° F. it would contain 1013 grains of sea-water which held 1·75 per cent of salts in solution, and 1026 grains of water holding 3·5 per cent of salts. Density (specific gravity) is measured most easily by means of a delicate hydrometer, but most accurately by weighing a carefully measured portion of the water. The standard temperature to which density of sea-water is calculated is usually 32° F. or 60° F. in English-speaking countries, and 0° C. or 17·5° C. on

the continent of Europe. The density at 60° F. corresponding to various degrees of salinity is as follows :—

Salinity	0·00	1·00	2·00	3·00	3·25	3·50	3·75	4·00
Density	1·0000	1·0058	1·0138	1·0220	1·0240	1·0260	1·0280	1·0300

224. Salinity of the Ocean.—As a rule the surface water of the ocean is salter than that lying beneath, the fresher water below being denser in its position, because its temperature is much lower and the pressure upon it greater. In those parts of the ocean where the rainfall is heavy the surface water is always being freshened, and its salinity is consequently lowered. The map (Plate VIII.) shows the freshened regions by a lighter tint of pink, the figures referring to the density. There is one band of comparatively fresh water in the rainy equatorial region of each ocean, and fresh zones around the melting ice of the Arctic and Antarctic coasts. Seas and ocean shores situated in regions of great rainfall, or receiving large rivers, are also usually fresher than the average. The saltest water occurs in the regions of greatest evaporation and least rainfall, pre-eminently the Mediterranean and Red Sea, and in the trade-wind regions of the open oceans. The track of fresher water along the west coast of Africa and of South America is probably produced by upwelling in consequence of off-shore winds (§ 240, 241). The way in which the very salt water extends close to shore along the coast of South America, between the mouths of the rivers Amazon and La Plata, is accounted for by the westward trade-wind drift of surface water. All the salts dissolved and invisible in the whole ocean would suffice to form a solid crust 170 feet thick over the entire sea surface.

225. Absorbed Gases in Sea-water.—All atmospheric gases are to some extent dissolved by sea-water. The amount absorbed depends conjointly on the pressure of the gas (being greater as the pressure is greater), the temperature of the water (being greater as the temperature is lower), and the nature of the gas itself. Under the same pressure oxygen is nearly twice as soluble in water as nitrogen ; but nitrogen exerts on the sea surface four-fifths, and oxygen only one-fifth, of the whole atmospheric pressure; thus sea-water in contact with air absorbs twice as much nitrogen as oxygen. Still the proportion of oxygen in the air which is breathed in the water by sea creatures is twice as great as that in the atmosphere. At the average pressure and 32° F., 100 parts of water by volume absorb from air 1·56 parts of nitrogen and 0·82 of oxygen; at 70° F. the quantities absorbed are 1·00 part of nitrogen and 0·52 of oxygen, and so on in inverse proportion to the temperature. The amount of absorbed nitrogen in sea-water does not change after it has sunk below the surface; thus by finding how much nitrogen is dissolved in any part of the ocean one can calculate the temperature the water originally had at the surface, and also the amount of oxygen which must have been absorbed at the same time. The creatures living in the sea, and dead animals and plants decaying, diminish the amount of oxygen, so that the full quantity which was absorbed by the sea-water is hardly ever found in samples taken from a considerable depth. If any part of the ocean were quite stagnant, and never renewed from the surface, the dissolved oxygen would in time become exhausted. The chemists of the *Challenger* and of other deep-sea expeditions have never found a sample of sea-water free from oxygen, and this is a sure indication that all parts of the ocean are moving, however slowly. Very little carbonic acid is absorbed from the air, on account of the small proportion of that gas in the atmosphere; but the oxygen, when used up as described above, is changed in great part into carbonic acid, which remains in the sea-water chemically combined with the carbonates.

226. **Pressure and Sea-water.**—Professor Tait has found by experiment that sea-water is very slightly compressed by its own weight. Under the surface the pressure increases about 1 ton per square inch for every mile of depth. At the bottom of the deepest part of the ocean the vast pile of water exerts a pressure more than 500 times that of the atmosphere on the surface, or about 4 tons to the square inch. At this depth 11,000 cubic feet of sea-level air would be squeezed into 22 cubic feet; but 11,000

cubic feet of sea-water would only be reduced to about 10,000 cubic feet, the density being only slightly increased. If sea-water were absolutely incompressible the oceans would be about 200 feet deeper than they actually are. Sea-water is perfectly elastic. When pressure is removed from a portion it returns at once to its original volume.

227. Heat and Sea-water.—When sea-water is warmed it expands, steadily diminishing in density as the temperature rises. The specific heat is less than that of fresh water, for while 100 units of heat (§ 65) are needed to raise 100 lbs. of pure water from $32°$ to $33°$, 93.5 units can raise the temperature of 100 lbs. of sea-water (density 1·0260) through the same range. Sea-water conducts heat better than fresh water, so that the heat of the surface penetrates to a greater depth in the sea than in a deep lake in the same time. When heat is removed from sea-water, *i.e.* when it is cooled down, its density increases steadily, for its maximum density occurs below the freezing point. The chilled surface layer in contact with a very cold atmosphere always sinks, unless it is much fresher than the lower layers, which only happens in polar regions or near shore. Sea-water freezes about $28°$ F., or at a temperature $4°$ lower than fresh water, and in the process of freezing most of the salts separate out, so that the ice formed is nearly fresh, while the water yielding it is left much salter. All the salts are not excluded equally, the ice retaining a larger proportion of sulphates than of chlorides.[1] Sea-water ice has a soft and spongy texture, full of cavities containing residues of unfrozen brine, and the water produced by melting it is consequently bitter and unwholesome.

228. Circulation of Deep Fresh Lakes.—When the Sun shines on a deep lake in summer the upper layer of water is warmed, and expanding maintains its position, heat being passed on to the lower layers by the slow process of conduction. There is no tendency to transmit the heat by descending hot currents as in the sea. When winter sets in, the surface water cools rapidly by radiation, and contracting, it becomes denser and sinks allowing warmer water from beneath to take its place. This process goes on just as in the sea, until the lake cools down to 39° F., but at that temperature fresh water attains its maximum density, and the similarity to the cooling of the sea ceases. Further cooling of the upper layer makes the water expand, and therefore it remains at the surface until the temperature falls to 32°, when it solidifies to form a sheet of ice. Ice is not formed as long as any of the water in the lake is warmer than 39°. The heat from the water under the ice is conducted upward very slowly, so that the whole mass of water can only become solid in very shallow lakes when the winter is long and severe. A deep fresh-water lake in a region where the summers are warm is rarely altogether cooled down below 39° during winter, unless the season is very severe, hence the common observation that deep lakes do not freeze. In calm weather the study of the Swiss lakes, carried on by Professor Forel and others, shows that the upper 5 fathoms of water may be affected by the diurnal range of air temperature between day and night, but the annual change of temperature between summer and winter exerts some influence to a depth of from 50 to 80 fathoms. Beneath that depth the temperature remains unchanged all the year round at 39°. A steady wind blowing in the direction of the length of a long narrow lake (§ 240) may, however, mix the water so thoroughly that the temperature is made practically uniform from surface to bottom at any season of the year.

229. **Phenomena of Sea-lochs.**—Fjords or sea-lochs are miniature enclosed seas of great depth, surrounded on all sides but one by lofty mountains, and barred off from the deep sea outside by a sill rising to within a few fathoms of the surface, as shown in Fig. 55, § 339. The sea-lochs of Scotland have been studied in some detail by Dr. John Murray and the author of this book.[2] The lochs are filled with sea-water much freshened on the surface by numerous small mountain torrents, but scarcely less salt at the bottom than the open sea. In summer the surface temperature is greatly raised, but at the bottom, which is cut off from tidal influence, the temperature falls steadily, and comes to a minimum when the surface is warmest. As winter

advances the surface cools rapidly, and since the water is comparatively fresh it continues, in spite of its increasing density, to float on the warmer sea-water below, and sometimes freezes, while at a depth of a few fathoms the temperature of the salt water may be more than 45°. The heat of summer is conducted downward so slowly that the highest temperature of the year is reached at the bottom when the surface is at its coldest in January or February; the seasons at the bottom of Upper Loch Fyne or Loch Goil, for example, being six months behind those at the surface. In the far deeper basins of the fjords of Norway seasonal changes of temperature penetrate to about 200 fathoms, but no farther.

230. **River and Sea-water.**—When a large swift river flows directly into the sea it spreads out over the surface for many miles, floating on the salt water, which it freshens superficially. The form of the fresh stream may often be traced by the contrast of its colour with the clear blue of the ocean. Off the mouths of the Amazon and the Orinoco, for example, muddy fresh water is found floating on the surface of the sea several hundred miles from land. The Sun's heat rapidly evaporates the floating fresh water, and salt from below diffuses up and increases its density, thus enabling it to mix with the mass of the ocean, a process assisted by wind and waves. When rivers pour directly into a sea affected by tides it may happen that the current of fresh water is only slackened, but not reversed, by the rising tide. In the Spey, which is the swiftest river in Britain, salt seawater is forced, like a dense fluid wedge, for a considerable distance up the bed of the river by the rising tide, and lifts the fresh stream to a higher level, so that perfectly fresh water is found on the surface, separated by a brackish layer a foot or two thick from the salt water below. The salt wedge is withdrawn by the ebb-tide, and the river current resumes its rapid flow to the sea.[3] Rivers which enter the sea directly have little influence on the salinity and temperature of the deeper layers of sea-water.

231. **Estuaries and Firths.**— In the La Plata, the Thames, the Severn, the Forth, the Tay, the Garonne, and

other rivers where the fresh water meets the sea gradually in a narrow inlet, the wedge-like action of the salt water at high tide is scarcely perceptible. The effect of the tidal currents sweeping to and fro in the funnel-shaped channel is to mix the river and sea-water together as if they were being shaken in a bottle. In such an inlet as that of the Thames or the Firth of Tay, where the river is large, the water is found to grow rapidly salter from river to sea, the surface is much fresher than the lower layers, and the change of salinity between high and low tide is very marked. This form of river entrance is appropriately called an *Estuary*. When, however, the inlet is very large compared with the river, and when there is no bar at the opening, the estuarine character is only shown at the upper end. In the Firth of Forth, for example, the landward half is an estuary, but in the seaward half the water has become more thoroughly mixed, the salinity is almost uniform from surface to bottom, and increases very gradually toward the sea. The result is that the river-water meets the sea diffused uniformly through a deep mass of water scarcely fresher than the sea itself, so that the two mix uniformly, and the sea becomes slightly freshened throughout its whole depth for many miles from land.

232. **Temperature in River Entrances.**—The temperature of a river in the temperate zone follows that of the land over which it flows, and is thus subject to considerable variations between day and night. River-water, unless it flows very rapidly, can never become colder than 32°; but in summer its temperature may be raised to a very high degree if there is little rain and strong sunshine. Rain lowers the temperature of rivers in summer, especially when it floods torrents descending from cold mountains. Such rivers are warmer than the sea in summer and cooler than the sea in winter. In an estuary or firth in summer the temperature is highest on the surface and in the river, diminishing at first very rapidly, but afterwards more slowly as the sea is approached. In autumn, on account of the more rapid chilling of the land, the temperature becomes nearly uniform in river, estuary, and sea, and from surface

to bottom. In winter the water is coldest on the surface and in the river, growing warmer, at first rapidly, and then more gradually, toward the sea. In spring, on account of the land heating up more rapidly, the temperature becomes once more uniform throughout.[4] Fig. 32 shows the actual

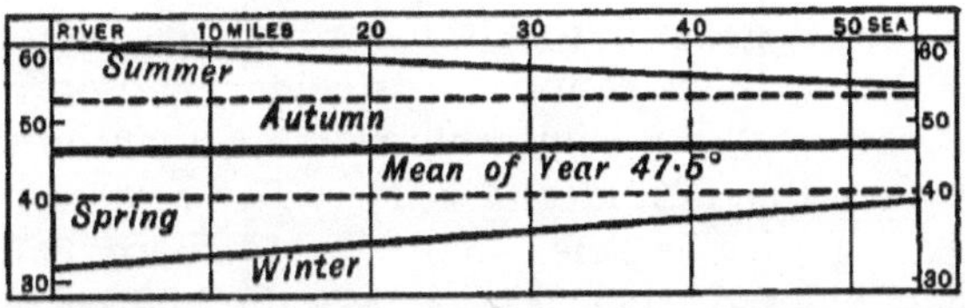

```
  RIVER    10 MILES   20        30        40      50 SEA
60                                                     60
     Summer
50                                                     50
              Autumn
              Mean of Year 47·5°
40                                                     40
     Spring
30                                                     30
              Winter
```

FIG. 32.—Temperature of surface water at different seasons along the middle line of the Firth of Forth. Distances from Alloa are shown in miles horizontally from left to right; temperature in degrees Fahrenheit is shown vertically.

distribution of temperature along the Firth of Forth, from Alloa to the sea, at four typical seasons, on the surface.

233. Surface Temperature of the Ocean.—The isothermal lines on the ocean in Plate XV. represent the average temperature of the surface water for the year. Although more easily heated than fresh water (§ 227), the sea surface has a less range of temperature than that of fresh lakes. This results in part from the greater clearness of sea-water, in part from its distance from heated land. The average temperature of the surface of the open ocean varies less than 1° between day and night, but between summer and winter there is a range of from 5° to 10°. Along a line, drawn from Newfoundland to Iceland, the annual change of temperature between the coldest month, February, and the hottest month, August, is as much as 20°; but this is due less to the heating and cooling of water than to a seasonal change in direction of warm and cold currents (§ 242). In the tropical zone the sea surface has a temperature higher than 80° for the whole year. This zone of very hot water is widest in the Indian Ocean and narrowest in the Atlantic; and in all three oceans it is wider on the western than on the eastern shores. The temperature falls very uniformly toward the south, reaching 40° F. about latitude 48° S. south of Africa, but not until latitude 58° S. south of

New Zealand. In the Southern Ocean there is practically no annual change of temperature, the water growing steadily colder toward the Antarctic ice at all seasons. Toward the north the ocean grows cooler more gradually, 40° being found in summer only in the Arctic Sea, but in winter between New York and the Lofoten Islands, and between Japan and Alaska. As a general rule the sea surface on the west coasts of the southern continents is colder, and on the west coasts of the northern continents warmer, than on the east coasts in the same latitudes (§ 241). The northern half of each ocean is also warmer than the southern half at all seasons. Enclosed tropical seas have the highest temperature of any water surfaces in the world. In the Red Sea readings of from 90° to 100° F. have been reported.

234. **Polar Seas.**—The Arctic Sea, lying in the coldest region of the globe, appears to be frozen over every winter, and the ice, measuring from 2 to 10 feet in thickness, is only partially dissipated in summer. Ice first forms along the shore-line, remaining attached to the land as a flat shelf, termed the *ice-foot*, which is often strewn with boulders and shattered rocks from the cliffs that tower above it. Thence the surface gradually freezes across. When the winter covering of the ocean breaks up, ice-islands, or *floes*, some of which have been seen 60 miles long, drift away with the wind. Open lanes and wide expanses of water thus appear in summer across the Arctic Sea, but these are liable to be closed at any time by a change of wind driving the floes together. Two floes in collision present a grand and terrifying scene, the ice cracking and rending with a noise louder than thunder, while the shattered sheets are piled up one above another to a great height, forming irregular hummocks or ice-hills. Sir George Nares, in the last great North Polar expedition, found the ice-floes in what he called the Palæocrystic Sea more than 150 feet thick, and he estimated that some of them were 500 years old. The water in which the floes float has the temperature of melting sea ice (about 28°), and the lower layers are usually considerably warmer. Indeed, in Polar regions there are often alternate layers of cold and warm

water, one above another, the greater salinity of the warmer water making its density greater than the colder but fresher water above. A temperature curve of such a region ("Atlantic 71° N. lat.") is shown in Fig. 33. The ends of great glaciers reaching to the sea break off in the water and float away in summer as *icebergs* (§ 338). In the Arctic regions the icebergs are lofty pinnacled masses, often resembling cathedrals or castles several hundred feet in height, with a covering of dazzlingly white snow, but showing the true ice-colour of intense blue in their cracks and caves. Lofty as these icebergs are, we know that as ice has a density of about 0.9, only one-ninth of its volume floats above water. The Antarctic icebergs are usually flat-topped and table-like, but are far larger and of a deeper blue colour than those of the Arctic regions.

235. **Temperature of Ocean Depths.**—The hot surface water in the tropical zone is merely a film covering a vast depth of cold water. Even although the surface is at 70° or 80°, temperature of 40° or less is found at the depth of from 300 to 400 fathoms in almost all parts of the ocean. The

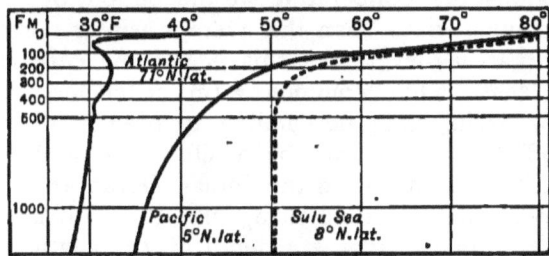

FIG. 33.—Curves of vertical distribution of Temperature in the ocean. Temperature is shown along the top; depth down the side. The middle curve, for example, shows 80° at the surface, 60° at 100 fathoms, 50° at 200 fathoms, 40° at 600 fathoms, and 35° at the bottom.

fall of temperature is consequently very rapid from the surface down to 400 fathoms in the tropics; but much less abrupt in the cooler regions to the north and south. Below 400 fathoms the fall of temperature to the bottom is everywhere very slight and gradual (Curve, "Pacific 5° N. lat.," Fig. 33), and no matter how great the depth may be the bottom temperature of the open ocean remains near the freezing point of fresh water. Five-sixths of the mass of the ocean has a temperature under 40° F., so that taken as a whole the hydrosphere is a body of cold water, its average temperature being probably about 38° or 39°. The prevailing low temperature of the hydrosphere is explained mainly by the great surface of water exposed in the Southern Ocean to the influence of the cold Antarctic ice-continent, and in less degree to the still colder winter weather within the Arctic Circle. The surface drift of warm salt water carried into the Southern Ocean from the north grows gradually cooler and therefore denser, and sinks about latitude 50° S. About this latitude also the comparatively fresh and cold water drifting northward from the Antarctic regions grows salter and sinks on account of the consequent increase of density. The sinking water appears to be drawn back by slow and massive movements to north and south, thus maintaining the circulation of the ocean to its greatest depths.

236. **Temperature in Enclosed Seas.**—Except in polar regions the temperature at the bottom of the deep ocean is much lower than the average winter temperature of the air at sea-level; but this is not the case for deep enclosed seas. The common form of enclosed seas is that of a basin, often descending to oceanic depths, but barred off from the ocean by a sill. The Red Sea, for example, is separated from the Indian Ocean at the Strait of Bab-el-Mandeb by a sill rising to within 200 fathoms of the surface, while it attains a depth of 1200 fathoms near the centre, and the Indian Ocean in the Gulf of Aden is still deeper. In the Red Sea the temperature at the surface varies from over 85° in summer to about 70° in winter. At the hottest season the rate of cooling is comparatively rapid to a depth of 200 fathoms, where the temperature is 70°, and from that level right down to the bottom the temperature remains uniform all the year through. The basin of the Red Sea is thus filled up to the lip with uniformly warm water, whereas, as shown in Fig. 34, the water of the Indian Ocean, nearer the equator, and with

the same surface temperature, sinks to 70° at about 200 fathoms, and falls as low as 37° at 1200, where it is prevented from entering the Red Sea basin by the ridge. The surface water in the Red Sea is densest when its temperature is lowest in winter, and the dense layers at 70° temperature sink to the bottom, so that the whole basin below the level of the barrier assumes and maintains the lowest average winter temperature of the air above. The hotter water in summer being less dense on account of its expansion, though it contains more salt, remains floating on the

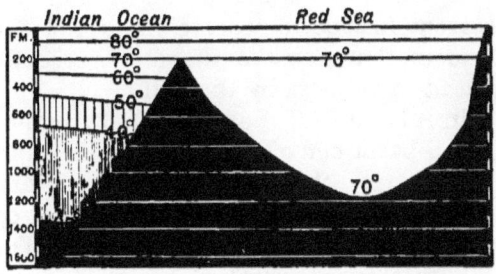

FIG. 34.—Temperature Section of the Red Sea and Indian Ocean; showing the action of a barrier in separating bodies of water at different temperatures. The shading is darker as the temperature is lower. Not drawn to scale.

surface, and its heat passes down only by the slow process of conduction. The Mediterranean furnishes another instance of the same distribution of temperature. The sill separating its basin from the Atlantic is 190 fathoms below the surface, and the water on it is at 55°, a temperature which prevails to the bottom of the Mediterranean, while in the Atlantic the temperature falls as low as 35° at the same depth.

237. **Circulation of Seas by Concentration.**—The great evaporation in the Red Sea raises the density of its surface water (at 60° F.) to 1·0300, and the salinity is 4 per cent. The level of the sea is lowered by evaporation to the extent of from 10 to 25 feet a year, and a surface current of the fresher but equally hot water of the Indian Ocean is consequently always pouring in. If there were no return current of dense salt water, it is calculated that the

Red Sea would become a mass of solid salt in less than 2000 years. Since there is no perceptible change in its salinity, it is certain that a deep undercurrent of salt water passes out through the Strait of Bab-el-Mandeb sufficient to carry back to the Indian Ocean all the salt received from it. The circulation of the Mediterranean is carried on in the same way, as the rainfall it receives is only equal to about one-quarter of the evaporation from its surface, and its water, although of higher salinity than the neighbouring Atlantic, is not growing salter. The outflowing current through the Strait of Gibraltar underneath the inflowing fresher current has been observed, and the deep water of the Atlantic in that neighbourhood is perceptibly warmed and increased in saltness by the outflow.

238. **Circulation of Seas by Dilution.**—The Black Sea is a deep basin cut off from the Mediterranean by the shallow Bosphorus, the Sea of Marmora, and the Dardanelles. This sea contains only about 2 per cent of salts, its water being very much freshened by the Don, Danube, and other great rivers which flow into it, supplying more water than is removed by evaporation, and raising its level about 2 feet higher than that of the Mediterranean. A steady surface outflow of brackish water from the Black Sea consequently sets through the Bosphorus; but a slower stream of very salt Mediterranean water forces its way along the bottom into the Black Sea, so that the sea is not permanently freshened. The cause of the undercurrent of salt water between seas of different salinity is that in order to produce equilibrium the pressure exerted by two adjacent columns of a fluid must be the same. A column of salt water exerts the same pressure as a column of fresh water higher in proportion to the difference of salinity. But (§ 38) water cannot stand at a higher level beside water at a lower level, and the fresher water pours over the surface of the salter column, upon which the pressure is thereby increased and the undercurrent is produced in order to equalise matters. As long as the supply of fresh water is kept up there can be no equality, and thus the circulation continues. The Baltic Sea has a somewhat similar circulation.

239. Wind-waves.—Difference of barometric pressure over a large sheet of water causes a slight change of level and sets up a to-and-fro surge, known as a *seiche* in the Swiss lakes, without the action of wind. The air, being more mobile, obeys the direct touch of solar energy much more readily and rapidly than water, to which motion is, however, imparted by wind. Part of the water surface yields to the stress of wind striking it obliquely, and is depressed, thereby ridging up the neighbouring portions and originating a wave, the form of which advances as a line of rollers before the wind. Only the form advances, for while the particles of water in the crest of a wave are moving rapidly forward, those in the trough move back to almost exactly the same extent. Thus rollers merely lift and lower the vessels that float upon them. Water being an elastic substance continues to swing up and down as a swell after the wind which produced the motion has died away, just as a pendulum continues to swing after the hand setting it in motion is withdrawn. Waves may be transmitted from a great distance, and as wind is always blowing somewhere the surface of the ocean is never quite at rest. When a wave enters gradually shallowing water the lower part is retarded by friction, and the upper part sweeps forward more rapidly. The wave becomes steeper and shorter, and finally the top curves over in a hollow sheet of clear water, which breaks with a roar into foam and spray, the roller becoming a breaker. Sailors are in the habit of speaking of waves as "mountains high," but this is only a metaphor. The highest wind-waves that have been measured have an amplitude of only 50 feet from trough to crest, and a length of about a quarter of a mile between successive crests. Earthquakes raise waves of much greater height and destructive power than either tide or wind. The wave form travels over the sea at a rate depending on the size of the wave and the depth of the water, the maximum speed being about 80 miles an hour. At the depth of 100 fathoms the greatest waves produce a movement too slight, as a rule, to affect anything but the finest mud, and probably wave-motion never penetrates to as great a depth as 500 fathoms.

240. Circulation of Water by Wind.—Apart from producing waves, the wind slips the top layer of water before it as one might slip a card from the top of a pack, and although it can act only on a very thin film a new surface is constantly exposed, and a steady breeze causes a great surface drift. Mr. J. Aitken appears to have been the first to point out the importance of this action in disturbing the deeper layers of water. Dr. Murray, by a series of temperature observations on Loch Ness, showed how rapidly wind acting in this manner on the surface of a deep lake could completely alter the distribution of the water.[5] The explanation of his observations seems to be as follows: On a calm summer day this lake contains a surface layer about 15 fathoms deep, the temperature of which is from 60° to 50°, floating upon 100 fathoms of water, the temperature of which is from 50° to 40°. When strong wind blows steadily along the length of the lake from A to B the surface water is driven toward B, where the wind heaps it up, but the greater pressure of the heaped-up water causes the lower layers at B to move off toward A, and thus the whole end of the lake-basin at B is filled with the warm water that had been resting on the surface, while the cold water formerly filling the depths rises against the shore at A, as represented by the arrows. If the wind lasts long enough the water will be thoroughly mixed and the temperature made uniform throughout (§ 228).

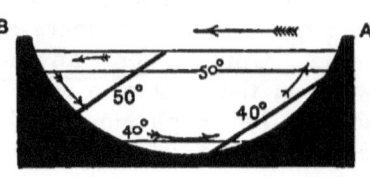

FIG. 35.—Circulation of water by wind. The light lines and figures show distribution of temperature before, and the dark lines and figures distribution of temperature after, the wind has been blowing in the direction of the long arrow.

241. Effect of On-shore and Off-shore Winds.—Bathers know that in summer the sea is colder when the wind is blowing from the land than when it is blowing from the sea. The reason is that the wind blowing from the sea (an "on-shore" wind) drives the surface water, which has been heated over a wide area, in toward the shore, on which warm water becomes banked up to a considerable depth, displacing the cold lower water, which slips seaward as an undercurrent (B, Fig. 35). During a prevailing sea-wind the water along the shore assumes what may thus be called an on-shore condition, just as by blowing steadily across a milk dish one might drive the cream to one side, and even blow it up on the shelving lip, completely displacing the milk on that shallow coast. A wind from the land in like manner drives the warm surface water seaward, and colder water from a great depth wells up to take its place (A, Fig. 35), this being characteristic of an off-shore condition. This enables us to understand how the permanent winds of the Earth which blow steadily off shore (like the trade winds from the west coasts of Africa and South America, § 179) cause cold water to well up from great depths. The upwelling off the coast of south-western Africa and off the coast of Morocco explains the exceptionally low sea surface and air temperatures observed in these neighbourhoods, and similar conditions are found on the west coasts of Australia and South America. Where the prevailing winds blow against the land, as on the north-east of South America into the Caribbean Sea, and toward Western Europe, the sea assumes a permanent on-shore condition, the warm surface water from the tropics being piled up against the land, while the colder deep water natural to the locality slips away seaward. The effect of the prevailing winds of the world is to set up a general skimming of the ocean from the equator poleward, sweeping the warm surface water away to one side and allowing cold water from the depths to rise up, completing the vertical circulation.

242. **Wind and Ocean Currents.**—In a strong gale the wind blows off the crests of the waves in spray or spindrift, and even a moderate breeze sweeps forward a thin layer of surface water over the ridged surface of the sea, giving rise to what is called a surface drift. The currents of the Indian Ocean and of the sea off the west coast of Central America change twice a year with the changing of the monsoons, and it is recognised that these currents are produced solely by the

wind. Ocean currents are very different from surface drift. They are usually narrow tracts of the sea surface, the water of which flows steadily and strongly in a definite direction, passing through the rest of the sea without appreciable mixing, as a river runs through a meadow. Some of these ocean rivers flow steadily in a constant direction at the rate of nearly 4 miles an hour; thus it is matter of importance to sailors to map out the ocean so that they may avoid or take advantage of the currents in making a passage. Solar energy in one form or another is undoubtedly the power that keeps the whole system of oceanic circulation in motion, and the rotation of the Earth (§ 89) together with the form of the coast-lines of the continents direct the flow of currents. Sun-power acts on the hydrosphere (a) by raising the temperature in the tropical regions far above that in the polar zones, thus causing expansion and altering the level; (b) by causing great evaporation in the tropical regions, great rainfall in equatorial regions and moderate rainfall in the temperate zone, thus altering both level and density; (c) by setting up the whole system of winds. Some difference of opinion exists as to the chief cause of oceanic movement, but it is usually allowed that the most powerful is the wind. All three, however, act together and reinforce each other. If the student compares the map of ocean currents (Plate XVIII.) with those of temperature, of salinity, and of prevailing winds (Plates XV. VIII. V. VI.), he will see that the currents circulate in the same way as the winds and around nearly the same centres, which lie close to the regions of maximum sea-temperature and salinity. All ocean currents are more or less irregular in form and speed; they usually flow as parallel streams separated by spaces of still water, and vary in position and strength, as the winds do, with the time of year. Plate XVIII. should be specially referred to in reading the following paragraphs.

243. **Equatorial Currents of the Atlantic.**—The trade winds blowing from the west coast of Africa drive the surface water before them in rapid currents. The *North Equatorial Current*, sweeping along the north-west coast of Africa past the Canary Islands, turns toward the west

about the latitude of Cape Verde, and while part of it is driven by the north-east trade winds into the Caribbean Sea, most of the current sweeps north-westward (as explained in Ferrel's Law), outside the West Indies, toward the coast of North America. The *South Equatorial Current*, originating in the *Benguela Current* of cool water, flows northward at first. In the latitude of the Congo it sweeps westward across the ocean and divides into two branches off the wedge-shaped front of South America. One branch (as explained by Ferrel's Law for the southern hemisphere) turns southward along the coast, and is known as the Brazil Current; and getting within reach of the brave west winds it is drifted east again to rejoin the Benguela Current. The other branch continues on its westerly direction and is driven northward by the south-east trades, most of it flowing into the Caribbean Sea. Along the north-east coast of South America there is a heaping up of water, produced by the convergence of the two great equatorial currents, and this does not appear to be fully compensated for by vertical circulation. Some of the banked-up water escapes eastward on the surface along the rainy zone of the equatorial calms, forming a narrow counter-current between the west-flowing North and South Equatorials. Near the coast of Africa this *Counter Equatorial Current*, consisting of extremely hot water of slight salinity, and known as the *Guinea Current*, sweeps along the north shore of the Gulf of Guinea, and is deflected southward by the coast to rejoin the South Equatorial.[6]

244. **The Gulf Stream.**—The level of the Caribbean Sea and Gulf of Mexico is raised considerably by the hot surface water continually pouring in from the south-east. Off the mouth of the Mississippi it is about 3 feet higher than off New York—an effect which may, however, be due in part to the attraction of the land (§ 252). The *Gulf Stream* forced out of this reservoir through the Strait of Florida is a river of salt and very warm water (surface temperature 81°), 50 miles wide, 350 fathoms deep, and flowing at the rate of 5 miles an hour. On emerging from the Strait it is swept to the north close along the American

coast by the branch of the North Equatorial Current, which had passed outside the West Indies and through the Bahamas. The Gulf Stream sweeps the bottom clear of mud not only in the Strait but for a considerable distance northward. As it flows on, it grows wider and shallower; off Cape Hatteras it curves away from the American coast and coming within the range of the prevailing south-westerly winds, it is carried eastward across the Atlantic, spreading out like a fan and growing cooler as it flows. The Gulf Stream passes to the south of the Grand Banks of Newfoundland with a velocity of about $1\frac{1}{2}$ miles per hour, and its rate gradually diminishes to about 4 miles a day in the general North Atlantic drift. This drift of comparatively warm water forks into three, diverging toward the coast of Spain, the British Islands, Norway, and the south-eastern coast of Iceland, stranding driftwood on that treeless island. The surface water of the tropics is thus being steadily poured into the temperate North Atlantic, where it drives the cold deep water toward the south, and gives rise to the highest temperature at great depths found in any part of the open ocean. The temperature of 40° occurs as deep as 900 fathoms off the west of Scotland, and seldom deeper than 300 fathoms in the tropics. This is the source which supplies the south-west wind with heat and moisture to modify the climate of Western Europe.

245. **Polar Currents of the Atlantic.**—Careful study of the drifting of ice-floes in the Arctic Sea gives some ground for believing that a current sets straight across from near the New Siberian Islands on the coast of Asia toward Arctic North America. Dr. Nansen has resolved to set out in 1892 on an expedition to the North Pole, believing that this current will drift his vessel to the point which so many explorers have hitherto attempted in vain to reach. A cold current, carrying icebergs in summer, when the frozen sea breaks up, flows south from the Arctic Sea between Spitzbergen and Greenland, strengthened by a cold drift from the north coast of Asia. It passes along the north coast of Iceland, where it strands driftwood

from the Siberian rivers, and as the East Greenland Current flows more rapidly, under the influence of prevailing north-easterly winds, along the east shore of Greenland, causing that side of the great ice-covered peninsula to be much colder and less accessible than the western. The Labrador Current is a more important cold stream, driven also by the northerly winds induced by the northern low-pressure region of the atmosphere (§§ 194, 197), and flowing southward along the west side of Baffin Bay, past the coasts of Labrador and Newfoundland. It meets the northern edge of the Gulf Stream off the Grand Banks of Newfoundland. Many geologists believe that this encounter led to the formation of the Banks, for the icebergs carried by the Labrador Current are melted on entering the Gulf Stream, and drop the stones and mud which were frozen up in them. The mingling of cold and warm currents undoubtedly produces the fogs for which this region is famous. Being comparatively fresh, the density of the cold Labrador Current is not greater than that of the Gulf Stream, by which it appears to be deflected along the coast of North America, where it is known as the Cold Wall. It disappears from the surface off Cape Hatteras, having partly mixed with the Gulf Stream and in part sunk under the less dense because warmer water. Recent observations point to the possibility that the cold current cuts horizontally through the Gulf Stream, like a paper-cutter through the leaves of a book, and mixes with the mass of Atlantic water. The limits reached by icebergs drifted from the north and south are shown on Plate XVIII., illustrating how the warm currents off Northern Europe keep the sea clear from this danger. The cool water of the Benguela Current is partly supplied by upwelling from beneath, but the steady flow of the current is maintained by cold streams sweeping north-eastward from the Antarctic regions.

246. **Circulation of the Atlantic.**—The Gulf Stream is often spoken of as if it were a phenomenon by itself; but it is really only part of a great system of surface circulation, the water whirling as if stirred in the direction of the hands

of a watch in the northern Atlantic, and as if stirred in the opposite direction in the southern part of the ocean. The centre of each whirl is nearly at rest, and immense quantities of floating sea-weed accumulate, especially in the North Atlantic, where the calm weed-hampered water is known as the Sargasso Sea. Mr. A. W. Clayden has devised an interesting model, in which a current of air sets up real currents on a water surface formed like the Atlantic. So far as can be gathered from the imperfect data more water is driven poleward by this circulation than returns in surface currents. Much of the surface water sinks off the British Islands (§ 244) south of the Wyville-Thomson Ridge (§ 258). Over this ridge the Atlantic water streams so strongly that the bottom is swept clear of mud to the depth of 500 fathoms. North of the Ridge the basin of the Norwegian Sea is filled up to its lip with ice-cold water from the Arctic region which finds no exit southward.

247. **Currents of the Pacific Ocean.**—The Pacific Ocean, on account of its vast extent and its remoteness from great trade routes until within recent years, has not been so carefully studied as the Atlantic. It is known, however, that the general system of its circulation is the same, and the map should be carefully studied in order to recognise the similarities. The Bight of Panama, extending along the west coasts of Central America and of the north of South America, serves, like the Gulf of Guinea, as the starting-place of the great equatorial current system. The south-east trade wind produces the *Peru Current* as a stream of cool water raised by the off-shore winds, precisely like the Benguela Current of the Atlantic. This stream, deflected westward by the Peruvian outcurve of the coast, sweeps as a *South Equatorial Current* past the Galapagos Islands on the equator, giving them a cooler climate than any other equatorial land. Setting westward before the steady trade winds, it sends off branches to the south, which wind amongst innumerable island groups, and ultimately reunite under the influence of the brave west winds, and drift eastward to rejoin the Peru Current. The main branch of the South Equatorial Current splits at New

Guinea, a small part passes through Torres Strait to the Indian Ocean, but the main body streams through the Malay Archipelago toward the Philippine Islands. Toward this goal the *North Equatorial Current* is also driven by the north-east trade wind, and as in the Atlantic, the piling up of warm surface water against the chain of islands gives rise to a strong *Counter Equatorial Current*, which sets straight eastward across the Pacific, along the line of equatorial calms, into the Gulf of Panama. The South Equatorial Current streams from the South China Sea into the Indian Ocean in winter, when the north-east monsoon is blowing, and mixes with a cold current flowing south from the Yellow Sea. But in summer, during the south-west monsoon, the pressure of water in the China Sea is increased by tributary currents from the Indian Ocean, and acts in many respects like the Gulf of Mexico. The extremely hot water (surface temperature 85°) escapes between Luzon and Formosa as a broad salt river. As it sweeps past the east coast of Japan, and begins to widen and thin out, the name *Kuro Siwo* or *Black Stream* is given it, from the deep colour of its clear water. The Kuro Siwo comes into range of the prevalent south-west winds, and, like the Gulf Stream, is carried at a diminishing rate eastward across the ocean, merging into a general surface drift, which washes the coast of Alaska and British Columbia. The North Pacific has its temperature increased throughout a great depth in this way. Cold currents resembling those of Greenland and Labrador, but much smaller in volume, set south from Bering Sea along the coast of Kamchatka and Sakhalin, passing between Japan and the Kuro Siwo like a cold wall. This cold wall is greatly increased by the north-eastern monsoon, and seems to prevent the oceanic part of the north equatorial current from entering the China Sea, by turning it aside to supply the Kuro Siwo, which would otherwise cease to flow at that season.

248. **Currents of the Indian Ocean.**—The south-east trade wind blows the surface water westward off the coast of Western Australia, causing an upwelling of colder water

similar to the Benguela and Peru Currents. The *South Equatorial Current* of the Indian Ocean is reinforced by affluents from that of the Pacific between Australia and Java, which give to the eastern shore of the ocean a partially on-shore character. Turning as it flows west, the South Equatorial Current washes the east coast of Madagascar, and turns south in several branches, which are drifted back to the West Australian Current by the brave west winds. A strong drift of warm water passing southward along the Mozambique Channel is known as the *Agulhas Current* off the south of Africa, from the fact that the Agulhas Bank turns the bulk of the stream from its south-westward direction back to the east. A narrow stream of the Agulhas Current rounds the Cape and joins the Benguela Current in the Atlantic. In winter, when the north-east monsoon is blowing, a *North Equatorial Current* appears, eddying westward round the Bay of Bengal and Arabian Sea, and setting southward along the coast of Africa to join the Agulhas Current. At this season there is also a well-marked *Counter Equatorial Current* across the ocean from Zanzibar to Sumatra, rather to the south of the equator. During the south-west monsoon the currents in the northern part of the Indian Ocean are reversed. The Somali coast assumes an off-shore condition (§ 241), with strong upwelling of cold water, and the currents flow in eddies eastward round the Arabian Sea and Bay of Bengal in the same direction as the Counter Equatorial Current, the force of which is increased.

249. **Currents in the Southern Ocean.**—The westerly winds of the Roaring Forties carry a continuous surface drift of water in an easterly direction round the world, thus serving to mix the surface waters of the three great oceans. In many parts of the Southern Ocean slow drift currents of small volume set northward, and this is particularly the case toward the west coasts of the southern continents; Drift ice is rarely found farther north than the latitude of 42° or 43°, but south of that line Antarctic icebergs are frequently met with.

250. **Functions of the Sea.**—The hydrosphere regulates

the distribution of energy, acting as a great fly-wheel to the world machine. Solar energy directly or indirectly is the cause of all its movements. The sea carries nearly half of the sun-heat falling in the tropical zone to higher latitudes, and from the high latitudes of the south it tempers the tropical climates of the western shores of the continents by cold updraughts. By the solution and restoration of carbonic acid, it helps to maintain the uniform composition of the atmosphere, and by its comparatively slow changes of temperature, it keeps up land and sea breezes and monsoons. It is an unfailing reservoir for supplying water-vapour to the atmosphere, and rain for the lakes and rivers. The smooth and level surface of the oceans allow the normal system of atmospheric circulation (§ 177) to be developed to a far larger extent than is possible on the land, and produces the steady winds which dominate the climate of the whole globe. In the sea also the material brought down by rivers from the land is redistributed and worked up into new forms.

REFERENCES

[1] J. Y. Buchanan, "On Ice and Brines," *Proc. Roy. Soc. Ed.* xiv. 129 (1887); or *Nature*, xxxv. 608, and xxxvi. 9.

[2] H. R. Mill, "On the Physical Conditions of the Clyde Sea Area," *Proc. Phil. Soc. Glasgow*, xviii. 332 (1887); or *Nature*, xxxvi. 37, 56 (1887). Also *Trans. Roy. Soc. Ed.* (1891).

[3] H. R. Mill and T. M. Ritchie, "On the Physical Conditions of Rivers entering a Tidal Sea," *Proc. Roy. Soc. Ed.* xiii. 460.

[4] H. R. Mill, "On the Salinity and Temperature of the Firth of Forth," *Proc. Roy. Soc. Ed.* xiii. 29 (1885); and xiii. 157; also *Nature*, xxxi. 541 (1885); *Scot. Geog. Mag.* ii. 20.

[5] J. Murray, "Effects of Wind on Distribution of Temperature," *Scot. Geog. Mag.* iv. 345 (1888).

[6] J. Y. Buchanan, "Physical Exploration of the Gulf of Guinea," *Scot. Geog. Mag.* iv. 177, 233 (1888). "On Similarities in the Physical Geography of the Great Oceans," *Proc. Roy. Geog. Soc.* viii. 753 (1886); also *Nature*, xxxv. 33, 76.

CHAPTER XI

THE BED OF THE OCEANS

251. The Lithosphere.—The wide smooth expanse of the hydrosphere is apt to give one a wrong idea of the surface of the Earth by veiling the true topography of the great hollows. Serious attempts to find out the whole form of the lithosphere only began when the vast hidden region acquired commercial value as a bed for telegraph cables. Since the commencement of submarine telegraphy accordingly the process of taking deep-sea soundings (§ 443) has been rapidly perfected, and hundreds of accurate measurements of depth have been made in all the oceans. During the magnificent expedition of the *Challenger* in 1872-76, many deep soundings were taken for a purely scientific purpose in parts of the oceans never likely to be visited by telegraph ships. In recent years numerous smaller expeditions fitted out by the British government and by the governments of the United States, Norway, Germany, France, and Austria-Hungary, have made detailed studies of parts of the sea-bed. The form of the floor of the ocean has thus been gradually felt out point by point, and though quite in the dark as to the scenery of the veiled part of the lithosphere, we are now able to compare its general features with the smaller portion which is open to the light of day. If the Earth, like the Moon, had lost its hydrosphere, and could be viewed from a distance, the surface would appear to be made up of two great and roughly uniform regions, both convex, following the curvature of the globe, but one about 3 miles higher than the other. The lower and larger is composed of broad gently undulating plains rising into gentle ridges, and broken by some abrupt peaks. It is divided into bay-like expanses by the higher region, the slopes up to which are almost everywhere steep and often precipitous. The higher region is smaller and more diversified, rising into numerous terraced plateaux and rugged peaks. The whole of the low-lying region and the lower slopes of the higher region are entirely covered by the hydrosphere, only the plateaux and peaks of the latter project above the water surface and form the land.

252. **Sea-Level.**—The surface which naturally presents itself for purposes of comparison in describing the configuration of the Earth is that of the Ocean. This surface is usually considered to be level, that is to say it is looked on as having the exact form of the geoid (§ 83) and being concentric with it. The level of the sea at any place is always varying on account of waves and tides. In constructing charts, all soundings of depth are corrected to their value for a calm sea at the average low water of spring-tides for the place in question. Heights on land are measured from a datum-level, which differs in different countries, but is usually the average height of the sea at some selected place. The heights marked on an Ordnance Survey map of Great Britain are quite accurate with regard to the datum-level (that of mean tide at Liverpool), but are 8 inches too high compared with the average sea-level round the island, and in certain places are as much as 2 feet too high or too low compared with actual mean sea-level. Many reasons exist for those small permanent differences of level, such, for example, as heavy local rainfall, or evaporation, the direction of prevailing winds or currents. The greatest distortion of the sea-surface is, however, due to the mobility of water and its readiness to yield to the attraction of gravity. If the surface of the lithosphere were smooth and its interior of uniform density, this property of water would ensure a truly similar surface in the ocean. The Elevated Regions projecting to unequal heights far

above the general level of the Earth, and composed of substances of different density, attract the water by gravity toward themselves, and thus prevent the uniform action of the central force, much as the sides of a tumbler attract the contained water by cohesion and heap it up slightly at the edges. The amount of distortion in the hydrosphere is as difficult to determine as the form of the Earth itself (§ 83), and must be found in the same way. It was shown by the survey of India that the sea-surface is 300 feet nearer the centre of the Earth at Ceylon than it is at the Indus delta, where the attraction of the Himalayas comes into play. According to Professor Hull's estimate, the attraction of the Andes is sufficient to raise the level of the sea more than 2000 feet higher on the west coast of South America than at the Sandwich Islands. The rocks beneath the bed of the ocean are, however, believed to be of greater density than those composing continents, and therefore their attraction on the sea should to a large extent counter-balance that of the land. In any case the sea-surface is undoubtedly not level in any strict sense, and all comparisons of height and depth of distant places are shadowed by uncertainty.

253. **Volume of Oceans and Continents.**—The most logical datum-level is the mean surface of the lithosphere, the surface which would be produced if the heights were all smoothed down and the hollows filled up uniformly to produce the geoid. The amount of distortion of the sea-surface must be ascertained, more soundings must be made in many parts of the ocean, and the yet unknown regions surrounding the north and south poles must be explored and surveyed before the position of this ideal surface can be found with certainty. A fair approximation to it has, however, been made in an exhaustive estimate by Dr. John Murray of the area of all the land and of all the oceans lying between certain limits of height and depth.[1] From these areas he calculated the total volume of the land which projects above, and of the oceanic hollows which extend beneath sea-level. The land is estimated to occupy 55,000,000 square miles, and its average height is

about 2200 feet above sea-level, while the sea covering the remaining 141,000,000 square miles of surface has an average depth of 12,600 feet, or 2100 fathoms (§ 355). The loftiest point of the land, Mount Everest in the Himalayas, reaches to 29,000 feet above sea-level, and the deepest parts of the Pacific Ocean descend to a depth of 28,200 feet below sea-level. The whole vertical range on the surface of the lithosphere is thus about 60,000 feet, nearly 12 miles, which is only $\frac{1}{700}$ of the Earth's diameter. The narrow crest of the Elevated Region forming the visible land has only $\frac{1}{14}$ of the volume of the ocean hollows, and thus the average level of the solid Earth evidently lies beneath the sea-surface, and the summits of the land rise higher above the mean level than the depressions of the ocean sink below it.

254. **Mean Sphere Level.**—From Murray's figures, the position of the mean surface of the lithosphere (mean sphere level) was calculated by the author to be about 10,000 feet (1700 fathoms) below the present sea-level, or more than half-way down the slope which separates the two great regions. If we imagine a transparent shell, similar in form to the Earth and concentric with it, to cut this slope at the level indicated, the volume of all the elevations projecting above the shell would be precisely equal to the volume of all the depressions extending below it. By a remarkable coincidence, one-half of the area of the Earth's surface is above mean sphere level and one-half below. The line of mean sphere level traced on a map (Pl. XIV.) thus serves to divide the surface of the lithosphere into a depressed and an elevated half.[2]

255. **Three Areas of the Lithosphere.**—The depressed half of the lithosphere is called by Dr. Murray the *Abysmal Area*, all parts of which are always covered by water more than 10,000 feet deep. The upper part of the elevated half of the lithosphere forms the *Continental Area*, which is always above water, and occupies rather more than one-quarter (28 per cent) of the surface. The remainder of the surface, measuring somewhat less than one-quarter (22 per cent), and always covered by water less than

10,000 feet deep, is called the *Transitional Area*. The Abysmal Area, or group of World Hollows, is capacious enough to contain exactly the whole volume of the group of World Ridges made up of the Transitional and Continental Areas. The position of the coast-line or boundary between the Transitional and Continental Areas obviously depends on the volume of the hydrosphere. It is convenient for most purposes to class the Abysmal and Transitional Areas together as the Bed of the Oceans. In originally proposing this division of the Earth's surface, Dr. Murray took the boundary line between the Transitional and Abysmal areas at the arbitrary depth of 1000 fathoms, or 6000 feet below sea-level.

256. **Elevated Half of the Lithosphere.**—The elevations and depressions of the Earth, although irregular in form and distribution, are arranged with a certain rough symmetry about the poles. A small detached elevation occupying about one-twelfth of the area of the elevated half has its centre within the Antarctic circle, and slopes down gradually on all sides to mean sphere level. The surface of the northern hemisphere is as a whole more elevated than that of the southern. A great Northern Plateau surrounding the pole to a distance of 2000 miles, and broken only by one depression (that of the Norwegian and Arctic Seas), is the centre of a continuous mass comprising fully nine-tenths of the whole elevated half, and extending toward the south in two vast World Ridges of unequal size. In reading the following paragraphs the student should refer constantly to the map (Plate XI.), and to Plate XIV. on which the line of mean sphere level is depicted. The *Western World Ridge* stretches from 60° N., where the Polar plateau splits, in a south-easterly direction to the equator, and thence southward, rapidly narrowing, to 60° S. The ridge, nowhere of great width, is narrowest between the Tropic of Cancer and the equator, where three small isolated depressions (the basins of the Caribbean Sea and Gulf of Mexico) nearly sever it. The crest of this ridge forms the connected continents of America. The *Eastern World Ridge* is of much greater size, and has

The Edinburgh Geographical Institute

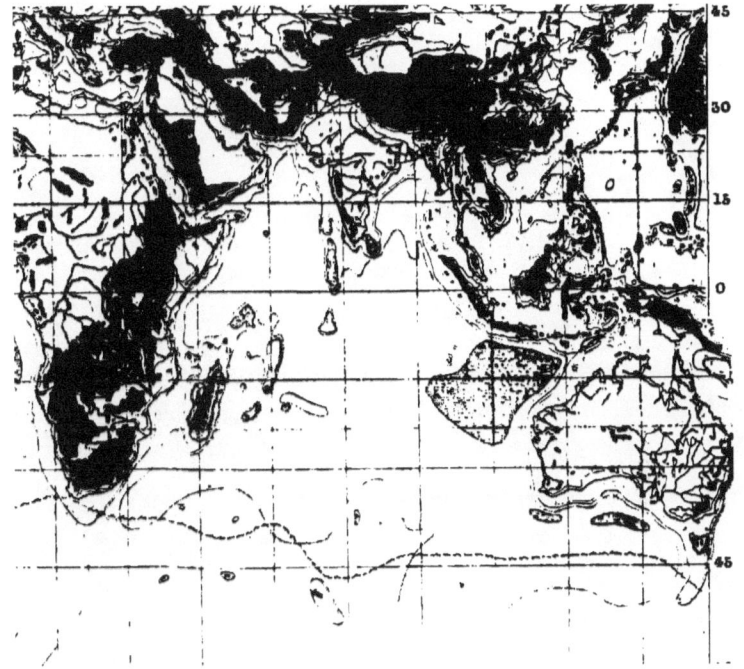

somewhat the form of a horse-shoe, the toe to the north. The western limb rises very steeply from the depressed area, and tapers southward to a point in 40° S.; it is crowned by the continent of Africa, and marked off from the European portion by two small depressions forming the deep basins of the Mediterranean. The eastern limb, marked off from the solid mass, which is the foundation of Asia, by a great series of deep depressions (the basins of the seas of the Malay Archipelago), runs south-eastward as a comparatively narrow ridge bearing Australia, and ends at 55° S. in two great spurs from which Tasmania and New Zealand rise. This limb lies exactly on the opposite side of the globe to the Western World Ridge.

257. **The Depressed half of the Lithosphere** or Abysmal Area forms a hollow ring round the south polar elevation, and runs northward in the form of nearly flat-bottomed troughs between the steep slopes of the World Ridges to the edge of the North Polar Plateau. It is ridged by long gentle rises and abrupt mountain-like peaks, and grooved by depressions infinitely various in size and form. Distinct hollows or basins of the Abysmal Area correspond to each ocean, and the slopes of the world ridges rising from them usually run parallel to the shore line which bounds the various oceans (§ 216). The basins of the Pacific, Atlantic, and Indian Oceans extend southward into the Southern Ocean, which has not a separate basin of its own. A typical section studied in conjunction with the map will impress the general form on the student's mind, although the scale of depth is necessarily exaggerated.

258. **The Atlantic Basin,** extending between the eastern edge of the Western and the western edge of the Eastern World Ridge, is long and comparatively narrow. It is deepest near the walls (Fig. 36) forming in fact two long sinuous troughs separated by the Dolphin Ridge along the centre, which reaches on the average to mean sphere level. The Azores, St. Paul Rocks near the equator, and Ascension all spring from this ridge, while the lonely islets of Tristan d'Acunha mark its southern extremity. Four great hollows or groups of hollows, the floors of which

descend to more than 3000 fathoms below sea-level, occur symmetrically, two in each of the lateral troughs, one north and one south of the equator. One of the north-western groups of hollows known as International Deep, contains in 20° N., just north of the Virgin Islands, the deepest sounding in the Atlantic, 4561 fathoms below sea-level, or nearly 18,000 feet below mean sphere level (see Fig. 36). The lateral troughs unite south of the Dolphin Ridge, and

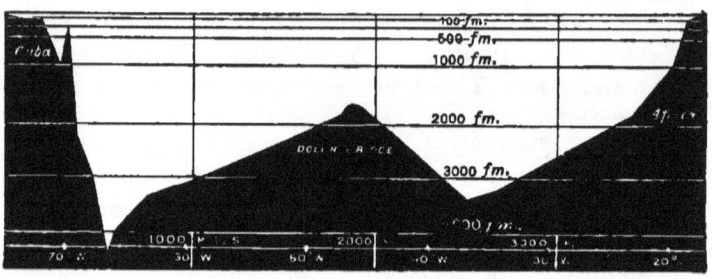

FIG. 36.—Section across Atlantic Ocean in 20° N. lat. The vertical scale is about 300 times greater than the horizontal; the slopes are thus shown 300 times as steep as they really are.

appear to form one vast abyss which deepens toward the south and extends far into the Southern Ocean. The deep basins of the Caribbean Sea, Gulf of Mexico, and Mediterranean communicate with the main Atlantic Basin over sills which rise nearly to sea-level. In the north the Wyville-Thomson Ridge, from an extension of which the Faroe Islands and Iceland rise, shuts off the deep basin of the Norwegian and Arctic Seas (§ 246).

259. **Pacific Basin.**—The Pacific Basin is far more vast than that of the Atlantic, and is still to a great extent unexplored; but the survey for a telegraph cable from Canada to New Zealand is at present (1891) revealing a chain of new and most important facts regarding it. The Pacific Basin appears to form one grand hollow extending from 60° N. to 60° S., between the western edge of the Western World Ridge, and the eastern edge of the Eastern. From 50° N. to 50° S. and right up to the steep walls to east and west, the depth is greater than 2000 fathoms,

and close under the edge of the Western World Ridge, off the west coast of South America, hollows more than 4000 fathoms below sea-level have recently been discovered. The map shows the nature of the slopes of the Pacific Basin to east and west, and brings out the fact that the Pacific and Indian Oceans are connected by shallow water across the top of a steep ridge pitted with small sea-basins of great depth. The floor of the basin slopes up very gradually in the south to form the gently swelling Antarctic Elevation. Numerous groups of long narrow ridges and isolated peaks, rising close to or above the surface of the water, with depressions of various forms between them, stretch roughly parallel to each other from south-east to north-west across the basin, becoming more numerous toward the west.

260. **The Tuscarora Deep.**—In the extreme north-west the steepest part of the bounding wall of the Pacific Basin rises abruptly, barring off the seas of Japan and Okhotsk, and bearing the chain of Japanese and Kurile Islands. In front of it lies the deepest abyss in the Earth's crust, the Tuscarora Deep. It extends from 20° N. to 50° N. in a crescent-shaped curve, deepening toward the steep slope of the World Ridge to the north-west, where a mighty gully 1000 miles long and 20 wide lies at a depth greater than 4000 fathoms (see Fig. 37). Here the United States surveying ship *Tuscarora*, obtained at least one sounding of almost 4700 fathoms below the surface, or 20,000 feet below

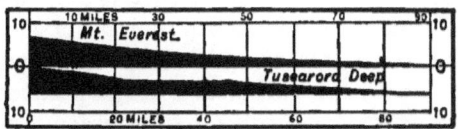

FIG. 37.—Steep slopes. The diagram is divided into squares representing 10 miles in the side. The upper black figure shows the true average slope from the summit of Mount Everest to sea-level; the lower shows the true average slope from sea-level to the bottom of the Tuscarora Deep.

mean sphere level. H.M.S. *Egeria* obtained an equally deep sounding in a very small depression south-east of the Friendly Islands; but there is no satisfactory proof of greater depths existing in any ocean.

261. The Indian Basin.—The Indian Basin, protected on three sides by the inner edges of the great Eastern World Ridge into which it penetrates, is only half the size of the Atlantic, and one-third of the Pacific, to which it bears some resemblance. The greatest depth, over 3000 fathoms, is found in the eastern angle between the coasts of north-west Australia and Java. The basin grows gradually shallower toward the south, most gradually toward the south-east. The western half is greatly diversified by narrow ridges running north-eastward from Madagascar to Ceylon, and rising in numerous groups of low islands above the surface of the water.

262. Islands and Shoals.—Those islands which are merely parts of the crests of the World Ridges separated by shallow water from the mainland, and composed of similar rocks, are termed *Continental Islands*. *Oceanic Islands* are those which rise from the depressed half of the Earth and have no geological relation to the neighbouring land. Many of them are composed of volcanic rocks, and must be viewed simply as the summits of ridges or submerged mountains. Others are built up of the remains of living creatures, and rise only a few feet above the surface of the water. These (§§ 280-282) require a foundation before they can be formed, and the foundation is usually a submarine peak or ridge. A submerged peak, rising within a few hundred feet of the surface, is called an oceanic shoal. It was supposed at one time that very few shoals of this kind existed, the bed of the ocean being looked upon as an almost unbroken plain, but the recent explorations of telegraph ships have revealed a large number of shoals in all the oceans, in some cases rising precipitously from vast depths.[3] Probably many more remain undiscovered, for unless the lines of soundings across an ocean are run at very close intervals, they might be passed over.

263. The Transitional Area.—From mean sphere level the upward slope of the World Ridges is at first gentle, but after a certain height in almost all places it becomes comparatively steep, in rare cases even forming a succession of rocky precipices. Fig. 37 shows that the average slope from the summit of Mount Everest to sea-level is very little steeper than the slope from sea-level to the bottom of the Tuscarora Deep; about 1 in 15 or nearly 4°. The steepness of sloping land almost always appears greater to the eye than it actually is. Only precipices of bare rock have an angle of slope greater than 45° or a gradient of 1 in 1, and the steep slope of the lower part of the world ridges probably rarely exceeds 35°, which on land would be felt a very steep hill to climb, a gradient of 1 in 1½. The steepest hill on a well-made road is 1 in 20 or an angle of 3°. Mr. J. Y. Buchanan found that in some cases where the slope was comparatively slight the original rocky wall had been covered by a mound of sediment brought down from the neighbouring land by great rivers (§§ 325, 326). In nearly all cases at the top of the acclivity, usually at the point where the depth of water is about 100 fathoms, the slope suddenly becomes much more gentle, and continues very gradual up to the coast line. This gentle slope has been termed the *Shore Flat*, or the *Continental Shelf*. The typical profile of the transitional area is given in Fig. 38, which represents the slope of part of the Gulf of Guinea. The outer curve shows the slope at a part of the coast where a pile of river-mud has been thrown down like an embankment in front of the ridge face, thereby reducing its gradient. These slopes are represented forty times steeper than they are in order to bring out the change of gradient, the vertical scale being forty times the horizontal.

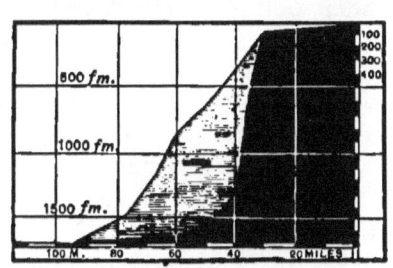

FIG. 38.—Slopes of the Gulf of Guinea. The vertical scale is 40 times the horizontal. Solid black shows average slope of the coast edge; the shaded part shows slope modified by river-borne deposits.

264. **The Continental Shelf.**—The world ridges forming the walls of the ocean-basins are flattened at the top like the rim of a pudding-dish, and beyond the flat edge

the continent itself rises. The breadth of the continental shelf varies greatly. In the map (Plate XI.) the area of the shelf is left white, and it will be seen to attain its maximum breadth off Western Europe where the British Islands stand upon it, off south-eastern America where it bears the Falkland Islands, around Florida, at intervals along the east coast of Asia, and off the north of Australia. Along the east and west coasts of Africa, and along the west coast of America, it is very narrow, and around some volcanic islands it is entirely absent. The total area of the continental shelf, covered with water less than 100 fathoms deep, is 10,000,000 square miles. This includes the whole of many shallow seas, such as the North Sea, the Baltic, the White Sea, Hudson Bay, and the Yellow Sea, and unites all the great continental islands, except Madagascar, Celebes, and New Zealand, to their nearest continent. The land bordering the coast-line is in most places a low undulating plain, which rises gradually inland until it attains an elevation of about 600 feet above the sea, and then rises more abruptly to much greater heights. The low plains (under 600 feet in elevation) measure altogether about 12,000,000 square miles. From the margin of the continental shelf to the end of the low plains there is therefore an expanse of 22,000,000 square miles, the level of which differs by only 1200 feet. Except possibly on the floor of the Abysmal Area there is no other part of the Earth's surface where so wide an expanse possesses such a slight range of elevation; and it is significant that the coast-line at present almost bisects it, occupying the only position in which a rise of 600 feet would submerge, and a fall of 600 feet would enable it to lay bare so large an area.

265. **Beach Formation.**—The upper margin of the Transitional Area is a region of great activity and rapid change. Tide and wind together urge the water against the land and withdraw it, dragging back the solid material it has seized. If the land is a low plain of very gentle slope the waves gradually encroach upon it, drawing the sand or soil seaward at every tide and building up the

continental shelf nearly to sea-level for a considerable distance, as, for example, along the east coast of India. Sandbanks or bars, sometimes locking in lagoons of salt water and forming a lace-like margin to the land, are produced where river deposits are brought down to the coast—for instance, on the south-east coast of North America, and in the vast mangrove-grown mudbanks of many parts of South America and Africa. Where the land is high and rocky the broken-off stones are rolled and rounded by the waves and used as battering-rams to break away the land; finally they are swept out to sea and spread in sheets over the bottom, the level of which is raised and the slope reduced. In this way a beach is formed, the upper part of it being quarried out of the solid land, and forming a notch or ledge (ABC, Fig. 39) on which the sea is always encroaching, while the lower part forms an embankment (CDE) built up of the excavated material which is laid down in flat beds one above another. The name *Beach* is restricted to the strip of land covered and laid bare by the tides. On a typical beach large stones are usually found heaped up near high-water mark;

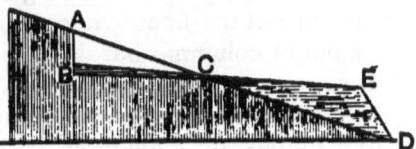

FIG. 39.—Formation of a Beach. AD, original slope of land; ABC, notch cut out by wave action; CDE, embankment of sand, etc. (worn-down rock); BC, gravel resting on beach.

smaller pebbles, rounded by the sea, form a steeply sloping bank at a lower elevation, and are rattled to and fro, ground against each other, reduced in size to fine shingle, and raked nearer the sea by every tide. Next there is a wide stretch of sand, which usually consists of quartz grains, resulting from the breaking down of the pebbles, the quartz being the densest and hardest ingredient of the rocks. Nearest the sea, and often only uncovered at the lowest spring ebbs, there are banks of mud formed of the softest ingredients of the rocks, which were ground to the finest powder and carried to the greatest distance. Sometimes perpendicular cliffs occur, at the base of which the rushing tides permit no fragments to accumulate.

266. Wave Action.—The measurements of Mr. Thomas Stevenson on the coast of Scotland show that during severe storms the waves may exert a force equal to 3 tons on every square foot of the cliffs they beat against. A force of 1 ton per square foot is commonly exerted by the waves of the Atlantic in winter, and 600 lbs. on the square foot in summer. This ponderous surge of the waves tears off loose pieces of rock, and the deluge of spray and pebbles which the breakers toss into the air has been known to break the windows of a lighthouse 300 feet above the sea. When a wave swells up against a cliff it powerfully compresses the air in all the cracks of the rock, thus striking a sudden blow throughout the whole mass. An explosive expansion of the air follows when the wave subsides, and the loosened fragments are sucked out along the lines of bedding or jointing (§ 290). This action and the bombardment by pebbles are the chief agents in forming sea-caves, of which one of the finest examples is Fingal's Cave in Staffa, carved out of columnar basalt. As the cave extends into the cliff it grows narrower, and finally a long diagonal tunnel may be drilled out, opening on to the upland far from the shore. Such openings or *blow-holes* are common along all cliff-girdled coasts, and throw up columns of spray during storms often with a noise resembling the outburst of a geyser. Blow-holes naturally widen as the sides are weathered (§ 310), and form deep isolated pools where the tidal water rises and falls at the bottom of a nearly vertical rocky shaft. When softer and harder rocks alternate along a coast, the former are in time cut back by the waves and form bays, while the latter project as headlands. Currents, or tidal eddies, attacking a narrow headland on both sides, and driving the pebbles against one part of the cliff, often break a cave right through, which when wide forms a tunnel, when high and narrow a natural bridge. Atmospheric erosion may cut as rapidly above as the waves do below, and the headland become separated from the mainland as an isolated rock or stack, round the base of which the water sweeps. Some of the finest examples of such cliff scenery occur on the north coast of Scotland and in

Orkney, where the Old Man of Hoy is a magnificent stack 450 feet in height.

267. Origin of the Continental Shelf.—The action of waves and tidal currents usually ceases to be perceptible at the depth of 100 fathoms. Beach deposits swept seaward by the waves assist in scooping out and deepening the shore, the final result being, possibly, to eat inward along the top of the wall of the world ridge until a depth of 100 fathoms is attained. The continental shelf is widest on the margins of the oldest continents exposed to the heaviest waves, and may be compared to the line which some chemical solutions etch on the glass bottles containing them. Harder masses resisting the attacks of the waves remain as islands or shoals on the continental shelf. Where currents sweeping mud and sand to and fro are checked by some inflection of the coast-line, sandbanks are formed. In many cases it is possible that the continental shelf is the end of a low plain submerged by subsidence; in others a low plain may be an upheaved continental shelf, and probably wave action is only one of the factors at work. Long furrows of great depth cross it in some places. These grooves and submarine cañons (§ 326) have a peculiar interest, because they seriously detract from the usefulness of the continental shelf as a guide to sailors groping their way to land by means of the sounding-line in foggy weather.

268. Marine Deposits.—Immense quantities of sediment are carried down by rivers into the sea (§ 331). M. de Lapparent calculates the amount as 33 times greater than all the sand, gravel, and pebbles worn off by tidal and solar energy acting through waves and currents on the coasts. Countless myriads of plants and animals living in the water affect the substance in solution (§ 222), forming shells or skeletons which at their death fall to the bottom, producing various kinds of deposits. Sea-water acts chemically on substances exposed to it, producing a further series of changes. In all parts of the ocean not precipitous nor swept by strong currents, the original rock is covered with a mantle of deposits of various thickness, to which the gently-rounded contour of the ocean-beds is largely due. MM. Murray and Renard in their report on the deposits collected during the *Challenger* expedition have adopted the following classification :—

MARINE DEPOSITS

1. **Deep-Sea Deposits** (*beyond* 100 *fathoms*).
   - Red Clay
   - Radiolarian Ooze
   - Diatom Ooze
   - Globigerina Ooze
   - Pteropod Ooze
   } I. PELAGIC DEPOSITS, *formed in deep water remote from land.*
   - Blue Mud
   - Red Mud
   - Green Mud
   - Volcanic Mud
   - Coral Mud

2. **Shallow-Water Deposits** (*in less than* 100 *fathoms*), sands, gravels, muds, etc.

3. **Littoral Deposits** (*between high and low water marks*), sands, gravels, muds, etc.

} II. TERRIGENOUS DEPOSITS, *formed in deep and shallow water close to land masses.*

269. **Terrigenous Deposits.**—Sediment, such as fine clayey mud, requires a very long time to settle to the bottom of fresh and still more of running water, but in sea-water, especially when the temperature is high, it settles out much more rapidly. The smaller a particle of mud and the deeper the sea, the farther from land will the particle be carried by currents before it falls to the bottom. As a rule, however, land-derived material all reaches the bed of the ocean within 100 or 200 miles of the shore; only in exceptional circumstances does it extend to a greater distance than 300 miles. The line of 250 miles from the coast shown on Plate XII. is practically the boundary of terrigenous deposits. Very large and swift muddy rivers like the Congo and Amazon (§ 230) form such exceptions. Congo mud has been found 600 miles from shore. The Arabian Sea and Bay of Bengal are carpeted for nearly 1000 miles from land by the mud of the Indus and Ganges river systems. Other exceptions result from icebergs,

which drop land-derived stones and mud all along the path of the ocean currents, which drift them into warm seas. Wind also blows sand or dust far out to sea. Volcanic eruptions throw up quantities of fine dust, which are carried far and wide by the winds and scattered over the whole sea surface. Pumice-stone, being porous, floats for months and probably years, and may be drifted to any part of the ocean before it becomes waterlogged and sinks. All terrigenous deposits, although soft and sticky when wet, fall into a loose powder on being dried, hence the term *Mud* is specially applied to them. Such deposits are characteristic of enclosed seas and of the upper margin of the Transitional Area, which they clothe much as snow clothes a tropical mountain, most thickly on the upper part of the slope. It is estimated that terrigenous deposits cover one-fifth of the area of the oceans, and it is distinctive of these deposits that they are made up of fragments of continental rocks, such as compact limestone, quartz, schist and gneiss.

270. **Blue Mud.**—The littoral deposits or shore formations sometimes extend in the form of sand or bars of fine gravel, enclosing hollows filled with mud, right across shallow seas. As a rule, however, deep enclosed seas, margins of islands and of continents for 200 or 300 miles from land, are carpeted with extremely fine mud, containing small grains of sand and the remains of shells and of marine plants. Where the material is derived mainly from rivers it assumes the form of a blue mud, which is the most characteristic of terrigenous deposits in every ocean, and is found at all depths. Blue mud owes its dark blue or slaty colour to chemical changes produced by decomposing vegetable and animal substances, in presence of the sulphates of sea-water, which appear to be reduced to sulphides, and decompose the ferric oxide abounding in all deposits into sulphide of iron and ferrous oxide. When there is much iron in the state of ferric oxide, as in the ochrey muds that redden the water of the Amazon, there may not be sufficient organic matter to reduce it all, and the mud retains its red colour. Blue mud contains variable quantities of carbonate of lime according to the abundance of shell-producing

creatures living in the water where it is deposited, but it accumulates so rapidly that shells as a rule form a very small proportion of the whole.

271. Green Mud.—Along cliff-bound coasts in which few rivers open, terrigenous deposits form very slowly, and to a distance only of 100 miles, or less, from land. The finely-ground particles of rocks are thus exposed for a long time to the action of sea-water and undergo extensive chemical changes. A greenish mineral called *Glauconite* is thus produced, which fills up the interior of dead calcareous shells, forming casts of the interior which remain when the shells themselves are dissolved away by weak acids.

272. Volcanic and Coral Muds and Sands.—Oceanic islands of volcanic origin are surrounded by *Volcanic* Muds or Sands, formed by the wearing down of volcanic rock and its subsequent partial decomposition by the chemical action of sea-water, the fragments of shells which are present being often coated with peroxide of manganese derived from the rocks. Islands of Coral origin are in a similar way surrounded by *Coral* Muds or Sands which consist almost entirely of carbonate of lime. The remains of calcareous marine plants (chiefly corallines) often make up a large part of this deposit.

273. Siliceous and Calcareous Organisms.—Certain minute moving organisms or living creatures, rarely visible except by means of the microscope, and possessed of the power of secreting silica from solution in sea-water, are found in the surface layers of all oceans, especially where the salinity is slight. One kind, known as *Diatoms*, abounds in cold seas and in estuaries, forming delicate cases or shells exquisitely marked. They probably obtain some of their silica by decomposing the clayey mud of rivers. *Radiolarians*, another class of silica-secreting organisms, frequent warmer water and are not found in estuaries; they form a minute framework or skeleton of glassy spicules often arranged in very complex and beautiful groups. The chief pelagic molluscs living on the surface far from land are a few kinds called *Heteropods* and *Pteropods* and they inhabit tropical seas. Their shells are thin papery cases of carbonate of lime, varying in length from half an inch downward. Innumerable forms of the simplest and smallest of living creatures abound in the surface water. They are most numerous in warm regions, and gradually disappear toward the poles. One class of these is called *Foraminifera*, as they construct dense microscopic shells of carbonate of lime pierced with innumerable little holes, through which the soft substance of the animal projects during life. The most common is a kind called *Globigerina*, on account of its globular form, the largest shells of which are about the size of a small pin's head. It has been proved that an animal requiring a shell of carbonate of lime can manufacture it out of any salt of lime, the carbonic acid coming from the creature itself, hence all the lime of sea-water (§ 222) is available to be drawn upon.[4] The death of countless millions of minute creatures produces a steady though invisible snowfall of dead bodies falling from the surface layers crowded with ever-renewed life, and gradually subsiding through the cold still depths of water. This takes place over every part of the hydrosphere, but within reach of terrigenous deposits the shells are covered over and buried in the rapidly increasing pile, of which they form a small proportion. Deposits of organic remains are more coherent and plastic than the muds, and have received the general name of *Ooze*. Living creatures, such as sponges which make skeletons of silica, calcareous sea-urchins, crabs, and corals, exist on the bed of the ocean to all depths, although they are incomparably more abundant in the shallow water near shore.

274. **Pteropod Ooze** is formed of the shells of all surface-living organisms in tropical seas, and contains a considerable proportion of pteropods, whence its name. It is never found below mean sphere level, but abounds on submarine ridges rising to within 1000 fathoms of the surface. The reason of this distribution appears to be that the delicate shells of pteropods expose a very large surface to the sea-water as they fall through it, and are dissolved away before they reach the bottom when the depth is great.

275. **Globigerina Ooze.**—The small dense shells of the

Globigerina can fall through a far greater depth than the thin pteropods before they are dissolved. Globigerina ooze accordingly covers a far greater part of the ocean bed. It does not occur in enclosed seas, nor under the cold currents of the north-east Atlantic, nor in the Southern Ocean south of 55° S.; but otherwise it is practically universal within certain limits of depth. Under the Gulf Stream its deposit is carried far to the north, as the surface water of that current swarms with globigerinæ. The ooze is a white or pinkish substance, which when dried is seen to have a fine granular structure, due to the little round shells of which it is composed. It varies in composition with the depth, that which has formed in the deepest water containing only the stronger and denser species, and the shells of these even being much corroded. The percentage of carbonate of lime varies from 30 to over 80, sometimes reaching 95; and if the carbonate is dissolved by a weak acid, the residue consists of a fine clayey substance mixed with the cases of diatoms and the spicules of radiolarians. At depths exceeding 2500 fathoms, with rare exceptions, none of this ooze occurs, the proportion of carbonate of lime in the deposit being reduced almost to the vanishing point. Globigerina ooze borders the upper zone of the Abysmal Area, and thins away toward the great depths (see Fig. 41).

276. **Radiolarian and Diatom Oozes.**—The siliceous skeletons of radiolarians and diatoms are present in small amount in almost every deposit. Silica is not nearly so soluble as carbonate of lime in sea-water; hence when the depth is greater than 2500 fathoms, and radiolarians abound on the surface, their spicules form a large proportion of the deposits reaching the bottom. The name of **Radiolarian Ooze** is given when they amount to more than 25 per cent. Radiolarian ooze is spread over a considerable part of the central Pacific, and the east of the Indian Ocean where the maximum depression occurs, but it is not found in the Atlantic or the Southern Ocean. **Diatom Ooze** contains about 50 per cent of diatom skeletons, mixed with from 10 to 20 per cent of carbonate of lime. It is the distinctive deposit of the Southern Ocean, where it occurs

at all depths; the small number of foraminifera living in the cold and comparatively fresh surface water accounts for the small quantity of carbonate of lime in the deposits of that region. The whole Southern Ocean is within the limits of icebergs drifting from the Antarctic region, and the Diatom ooze often contains a considerable proportion of terrigenous deposit, the nature of which proves the existence of continental rocks, and thus of an unexplored continent near the south pole.

277. **Red Clay.**—The deepest parts of every ocean are covered with a stiff clay of a deep brown or red colour, containing little or no carbonate of lime. Red clay is the distinctive deposit of the Abysmal Area, toward the upper margin of which it passes very gradually into Globigerina ooze; and where radiolarians abound on the surface the accumulation of their spicules gives to it the name of Radiolarian ooze. It covers more than half the area of the Pacific Ocean. Red clay is exactly like the residue of Globigerina ooze after the carbonate of lime has been removed. The snowfall of calcareous shells from the surface of the open ocean melts into solution before reaching the abysmal depths, but the horny remnants of those shells, siliceous relics of life, waterlogged pumice-stone, wind-borne dust from deserts and volcanoes, ultimately settle down and accumulate on the bottom. The rate of deposit is incomparably slower than that at which any of the oozes form. Microscopic examination has revealed as one of the constituents of Red clay cosmic dust from meteorites (§ 134), which falling uniformly over the Earth's surface is concealed by the rapid changes going on in every other region but the still Abysmal Area. The red colour of the clay is due to the formation of ferric oxide and peroxide of manganese from decomposing volcanic material. These oxides also become deposited upon any hard objects lying on the sea-floor, and form nodules composed of layer above layer and often attaining the size of a large potato, to which their usual shape is very similar. Manganese nodules were dredged up in great numbers by the *Challenger*, and in every case the nucleus on which they had formed was

found to be a piece of pumice, or the hard teeth or bones of the larger creatures inhabiting the sea. Sharks' teeth are very numerous, and also bits of the hardest bones of whales. Red clay also contains in certain localities small but perfectly formed crystals of the class of minerals known as zeolites (§ 286), which have evidently resulted from chemical changes in the material of the clay.

278. **Permanence of Elevated and Depressed Regions.** —From the scanty supply of materials out of which Red clay is elaborated, it is evident that if the deposit has attained any great thickness it must have been a very long time in course of formation. There is no evidence as to the thickness of the Red clay, but the teeth and bones found embedded in its nodules are known in many cases to belong to species of sharks which no longer live in the ocean, and must have been extinct for an immense period of time. Moreover, if the Abysmal and Continental Areas had ever changed places, some rocks would almost certainly be found on the land resembling a consolidated Red clay. None such have ever been discovered unless in volcanic oceanic islands that have been recently upheaved. Accordingly the existence of the Red clay is a strong argument that the elevated and depressed halves of the lithosphere have occupied their present positions during past geological ages.

279. **Corals.**—Many oceanic islands and reefs are composed of the stony framework of carbonate of lime which is secreted by animals known generally as coral polyps. These polyps belong to the same class as the sea-anemone, and are of many different species, each characterised by some peculiarity in the form of its calcareous support. Some secrete a wide disc, the surface of which is starred with their groups of waving tentacles; others form little cups on which they grow, these cups being either separate, as in the deep-sea corals, or united by a solid stony stem forming many branches. The branching corals of various species are of most importance in reef-building. The distribution of coral islands over the oceans depends on the suitability of the water for the life of the polyps and the existence of good foundations. The polyps flourish best in very salt,

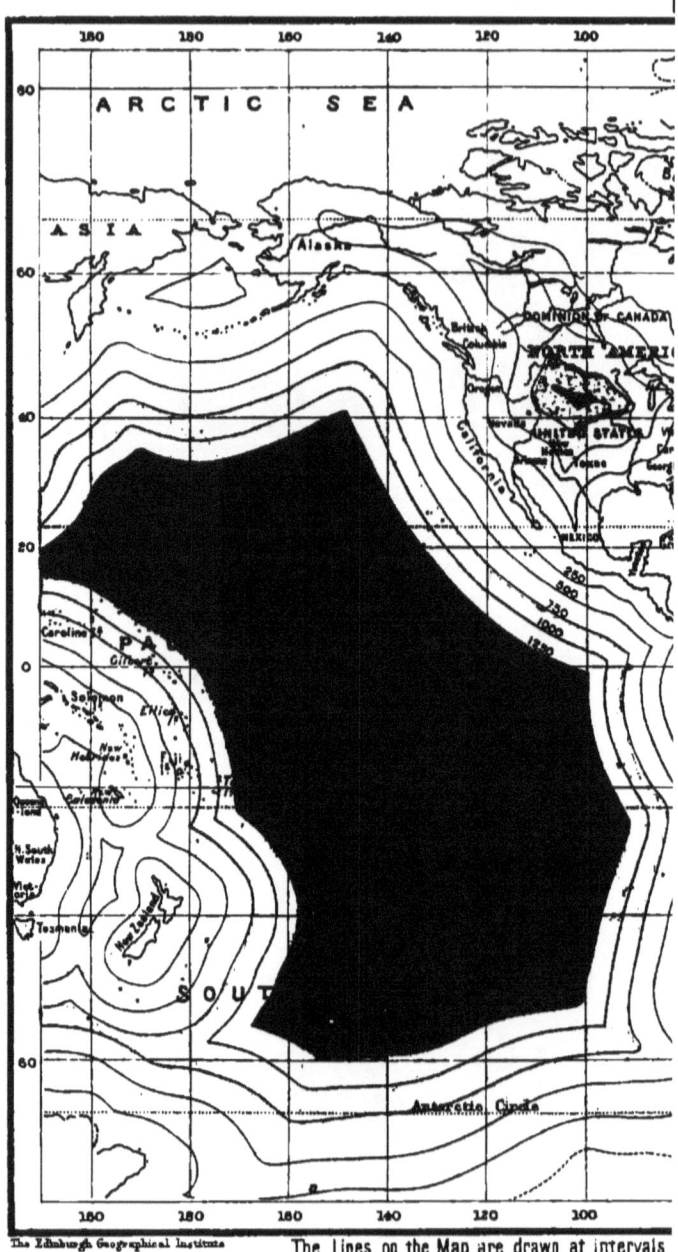

The Lines on the Map are drawn at intervals

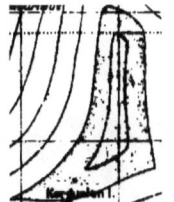

clear, and warm water; and, although they may live, they do not form reefs where the temperature is less than 70°, or has a yearly range greater than 12°, or a depth greater than about 20 fathoms. They are particularly active on the margin of the Red Sea, where the conditions of salinity, temperature, and depth are most favourable. The distribution of reef-building corals is given in the map of Plate XV. Corals are never found near the mouths of great rivers on account of the water being fresh and muddy. They do not build on the west coasts of the tropical continents because of the cold upwelling water (§ 241). The part of the Somali coast in the Indian Ocean against which the south-west monsoon raises cold water (§ 248) is free from corals on account of the great annual range of temperature which results. Corals are confined to the centre and western sides of tropical oceans, except in warm currents such as the Gulf Stream, which enables them to live luxuriantly far into the temperate zone, the Bermuda islands, in $32\frac{1}{2}°$ N., having the highest latitude where coral islands are now forming. There the polyps appear able to form reefs at a temperature as low as 68°, but these reefs are largely composed of calcareous sea-weeds and worm-tubes.

280. **Coral Reefs and Islands.**—The Gulf of Mexico and the west coast of Florida, the western Indian Ocean, and in particular the western Pacific, are the seats of very active and typical coral growth. There are three distinctive forms of coral structure. (*a*) The *fringing reef*, which closely surrounds the shore, forming on the seaward slope of the land in shallow water, and as it grows older gradually widening toward the sea. (*b*) The *barrier reef*, which usually lies at a distance from the land, running parallel to the coast, and on its seaward side often springing abruptly from great depths. On the landward side a shallow lagoon of still water is shut in by the reef, which is always broken by one or more narrow channels, allowing boats or even large vessels to enter. Innumerable volcanic islands in the Pacific, such as the Solomon Islands, the Fiji group, and Tahiti, are encircled with fringing and barrier

reefs. The great barrier reef of Australia, stretching for 1200 miles along the east coast of Queensland, is the finest example known. (c) The *atoll*, which is a reef in the form of a closed curve with no land in the centre. The lagoon encircled by an atoll is usually shallow, and the bed of it composed of coral which is either dead or not in vigorous life. Typical examples of the true coral islands or atolls are found in the Maldives, Laccadives, and Chagos groups in the west of the Indian Ocean. These reefs are usually very narrow compared with their length, and their surface never rises higher than from 10 to 20 feet above the sea. In most instances only a portion of the reef rises above the surface, giving the appearance of a chain of low islands separated by very shallow water. The coral polyp dies when it reaches sea-level, but blocks of coral are broken off by the waves and thrown on the reef, where they get broken down into sand, and this becoming compacted amongst the branches of living coral is raised by degrees until it forms dry land. Water percolating through the coral rock and sand gradually converts the whole into a solid mass of coral limestone, part of the carbonate of lime being dissolved and re-deposited in a crystalline form in the crevices. Drifting pumice strands on the beach and weathers into clay (§ 311) for the formation of soil. Ultimately the seeds of trees and other plants get drifted to the islands and take root, birds visit them, and the coral island becomes habitable.

281. **The Formation of Coral Islands.**—During the famous voyage of H.M.S. *Beagle* the naturalist Darwin made a detailed examination of several coral formations, and he came to the conclusion that the three typical forms were closely related to each other. He recognised that it was possible for atolls to form if they had a submarine mountain, the top of which was less than 20 fathoms below the surface, as a foundation, but he did not know that such peaks often occurred. He found also that the walls of coral rock on the seaward face of reefs sometimes rose from an enormous depth, and since coral polyps can only live and build in the warm surface layer, he concluded that the

corals had built in that layer, but that the foundations had been gradually sinking. Thus he supposed a fringing reef (I, Fig. 40) to form round a volcanic island, and as the island slowly subsided the corals built the reef higher and higher, keeping pace with the subsidence. In time, as the outer edge of the coral grows fastest on account of the greater abundance of oxygen in the breakers, the reef would widen and grow higher seaward, forming a barrier reef by the time subsidence has brought the sea-level to the position (II). Finally, subsidence submerges the whole mountain below the surface, and the barrier reef grows up to form an atoll (III) peeping above sea-level.

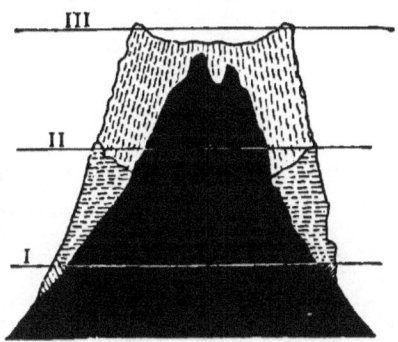

FIG. 40.—Darwin's theory of the origin of Coral Islands. I, II, and III show successive levels of the sea brought about by subsidence of a volcanic island (solid black). The corresponding coral formations, respectively fringing reef, barrier reef, and atoll, are shown by different shading.

In recent years a number of objections to this widely accepted theory have been made by many investigators. It has been pointed out that atolls are as common in areas which are being gradually elevated as in those that are subsiding. Dr. Guppy, in the course of a study of the Solomon Islands, where many reefs have been elevated far above sea-level, also found that the coral limestone is never of greater thickness than about 120 feet, and he thus casts doubt on the existence of vast submerged walls of coral. He found that the cake of coral rock rested either on volcanic rock or on rocks formed by the consolidation of pteropod or globigerina ooze.

282. **Murray's Coral Island Theory.**—During the cruise of the *Challenger* Dr. John Murray formed another theory, which has been strikingly confirmed by the observations of Dr. Guppy and others. He believes that the foundation for coral reefs is in every case supplied by submarine peaks. Some of these may have been formed

by volcanic upheaval and then reduced below sea-level by the eroding action of waves, and some may have existed originally at a suitable height. Others may have been raised to the coral zone by ages of submarine sedimentation, being covered first by globigerina ooze, then as the depth was gradually diminished by pteropod ooze, and finally brought comparatively rapidly within reach of reef-builders by the accumulation of the remains of sea-urchins, starfish, deep-sea corals, and the like. The reef-building polyps raise a flat table of solid rock, which, as it approaches the surface, grows more rapidly on the circumference on account of the abundance of food supplied by ocean currents. The rim finally reaches the surface and cuts off the supply of food from the polyps in the interior, which die, and the dead coral is partly dissolved by the water, partly scoured out by tide and waves, and so a lagoon is gradually hollowed. The outer slope of the reef is alive, and ever growing outward. As it becomes steep and wall-like, masses broken off by the waves roll down to the bottom and form a more gentle slope or talus on which the active corals continue to build seaward, always increasing the diameter of the atoll. Meanwhile the sea-water in the lagoon is at work dissolving and removing coral from the inner edge, and the island does not increase in width although its circle is continually widening. An atoll is thus supposed to grow like a "fairy-ring" in the grass. Fringing reefs growing seaward in the same way ultimately form barrier reefs, in which the same process of active growth seaward, and decay on the landward side has been observed.[5] In some cases barrier reefs have grown up directly far from the island on the edge of a wide and shallow continental

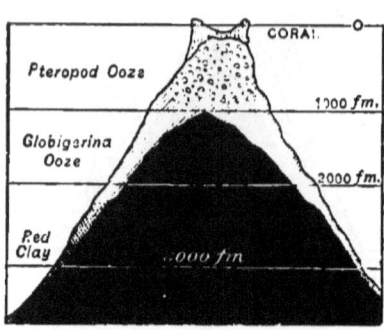

FIG. 41.—Murray's Theory of the origin of Coral Islands. The central volcanic rock (solid black) is shown covered by deep-sea deposits which build it up to the reef-building zone where an atoll is formed.

shelf, which is formed when a loose volcanic upheaval, such as Graham Island, which recently appeared in the Pacific, is rapidly worn away by the waves. Most geologists now recognise the greater probability and wider application of the solution theory of coral reefs over the subsidence theory.

REFERENCES

[1] J. Murray, "On the Height of the Land and the Depth of the Ocean," *Scot. Geog. Mag.* iv. 1 (1888).

[2] H. R. Mill, "On the Vertical Relief of the Globe," *Scot. Geog. Mag.* vi. 182 (1890); and "On the Mean Level of the Surface of the Solid Earth," *Proc. Roy. Soc. Ed.* xvii. 185 (1890).

[3] J. Y. Buchanan, "On Oceanic Shoals," *Proc. Roy. Soc. Ed.* xiii. 428 (1884); also *Nature*, xxxvii. 452 (1888).

[4] R. Irvine and G. S. Woodhead, "On the Secretion of Carbonate of Lime by Animals," *Proc. Roy. Soc. Ed.* xv. 308; xvi. 324. J. Murray and R. Irvine, "On Coral Reefs, etc." *Proc. Roy. Soc. Ed.* xvii. 79 (1889).

[5] J. Murray, "Structure, Origin, and Distribution of Coral Reefs and Islands," *Proc. Roy. Inst.* (1888); also *Nature*, xxxix. 424 (1889).

BOOKS OF REFERENCE

M. F. Maury, *Physical Geography of the Sea*. (The earliest and most interesting book on Oceanography, although the facts and theories are now out of date.) Edition 1883, T. Nelson and Sons.

Charles Darwin, *Coral Reefs and Islands*. (Several cheap editions recently published.)

J. D. Dana, *Coral Islands*. Sampson Low and Co.

H. B. Guppy, *The Solomon Islands: Geology*. Swan, Sonnenschein, 1837.

C. Wyville Thomson, *The Depths of the Sea*. Macmillan and Co.

J. J. Wild, *Thalassa*. Marcus Ward and Co.

"*Challenger*" *Reports, Narrative, Physics and Chemistry.* (Several volumes.)

Reports of the Norwegian North Atlantic Expedition (*Physics and Chemistry*).

A. Agassiz, *Three Cruises of the "Blake."* Cambridge, U.S., 1888.

Numerous papers on Oceanography will be found in recent volumes of *Nature*, the *Scottish Geographical Magazine*, and in publications of the Royal Societies of London and of Edinburgh, the Royal Geographical Society, and the Fishery Board for Scotland.

CHAPTER XII

THE CRUST OF THE EARTH

283. **Lithospheric Changes.**—In the Abysmal Area the hydrosphere protects the solid rock beneath, by the extremely slow formation of a covering of red clay or ooze. In the Transitional Area, where the hydrosphere is stirred more forcibly by solar energy, the formation of deposits is accompanied by the wearing away of rocks. All our knowledge of the substance and structure of the lithosphere is obtained by studying the processes of change going on in the Continental Area, which alone is open to our inspection. It is subject to much greater changes than the other areas on account of the strong action of solar energy, which through various agents is always crumbling down the heights and carrying the resulting detritus to the sea-margin. In the course of time this action, termed erosion, would, if not counterbalanced, reduce the whole Continental Area below sea-level.

284. **Elevation and Subsidence.**—The attention of tourists along the steep coasts of Norway and Scotland is often attracted by lines of horizontal terraces running parallel to each other at various heights above the shore. These when examined are found to be shelves or notches cut out of hard rock or soft ground, sometimes covered with pebbles and sand often containing sea-shells. Behind the terrace the cliffs are sometimes perforated by caves, which show every mark of having been excavated by wave action (§ 266). The terraces are in fact raised beaches, and their position

proves that the surface of the sea must have sunk, or the land must have risen since the waves eroded them. In the south of Scandinavia and the south of England there are many places where the sea now flows over what was dry land even during historical time. This encroachment cannot be due to erosion, as in some cases trunks of trees and walls of buildings may be seen still standing under the shallow water, and the necessary conclusion is that either the level of the sea has risen or the land has sunk. It is difficult to believe that for thousands of years the sea-level has been slowly sinking around Scotland and Norway, and at the same time slowly rising round England and Sweden, and the only satisfactory explanation of the facts is that the land must be undergoing gradual elevation in the north, and gradual subsidence in the south of Britain and Scandinavia. The regions of recent elevation and subsidence are marked on Plate XV. Since the average height of the land is much above sea-level, it is obvious that upheaval has been more rapid on the whole than erosion, and more general in its action over the Continental Area than subsidence. The interpretation of the appearances of the Earth's crust, and the utilisation of these to throw light on the past history of the planet, is the subject-matter of geology.

285. **Rocks.**—The *word* rock is usually restricted to the hard stony masses of cliffs and mountains, but the *term* rock has a wider meaning. Geologists class as rocks all substances which occur on or in the crust of the Earth and have not been recently formed by the decay of living creatures. Thus the term rock includes soil, sand, stones, etc., but not bones nor dead leaves. . Some rocks are uniform in structure like white marble or flint, but in most cases they appear to be built up of small separate portions which may be broken or rounded grains as in sandstone, large crystals of different compounds as in granite (§ 43), or minute crystals so tightly packed as to be indistinguishable by the unaided eye as in basalt. The grains of sandstone or clay are merely fragments of older rocks that have been broken and worn down before becoming cemented together again ; but the regularly formed crystals are portions of pure

substances, sometimes elements, although usually compounds, and they are known as *minerals*. While the term mineral is restricted to the pure constituents of rocks, the word mineral is often used to include everything useful found in the Earth's crust.

286. **Rock-forming Minerals.**—The crystalline minerals which make up many rocks must have formed slowly by the combination of their elements or the decomposition of other compounds. Some were evidently deposited from solution in water, as, for example, rocksalt and gypsum (calcium sulphate); in both these and in some other unimportant instances rocks may be composed of only one mineral. Other rocks have evidently crystallised from a state of fusion; in basalt, for example, the small crowded and imperfect crystals bear evidence to rapid cooling and solidification. Rocks, like obsidian, which show a vitreous or glassy texture, quite smooth, and it may be free from any appearance of crystals, have evidently been cooled still more quickly, so that crystallisation could not take place. After a rock has been formed its minerals may undergo chemical changes. The process of weathering, or slow alteration of rocks in air (§ 310), affects some minerals more than others. Many new kinds of mineral result from chemical change brought about by the absorption of oxygen (oxidation), by the absorption of water (hydration) producing zeolites, etc., or by the formation and removal of some product (decomposition). Mineralogists recognise about 800 different minerals, most of which, however, occur in very small quantities. Sixty or seventy only can be considered important as rock formers. Indeed the bulk of the rocky crust may be said to be composed of the following minerals, and those resulting from their alteration—felspar, quartz, mica, amphibole, pyroxene, and iron oxides.

287. **Igneous Rocks**, as a class, include all that have solidified from a state of fusion or have been formed by the accumulation of fragments thrown out by volcanoes. Most of them are dense and hard; they have a glassy or crystalline texture, and the minerals of which they are composed are almost invariably silicates or silica. Silica as flint,

agate, and chalcedony is also deposited from solution in water, but in that case its form is not crystalline. The way in which igneous rocks occur, whether poured out as lava on the surface or forced as intrusive sheets between beds of other rocks, greatly influences the part they take in determining the scenery of a country.

288. **Sedimentary Rocks** result from the consolidation of sediment deposited in lakes or on the margin of the sea. They are easily recognised by their structure, being built up of worn rock-fragments of all sizes. Fine muds consolidate into shales, sand into sandstone, gravel or pebbles into fine or coarse conglomerates, sometimes cemented together by the deposition of silica or carbonate of lime. In consequence of their formation in lakes or on the sea-shore sedimentary rocks show marks of bedding, the layers or strata having been laid down horizontally or nearly so. The beds of rock are not of uniform thickness throughout, but thin away as the original sediment formed a thinner layer of deposit far from the land. This class also includes rocks formed by the accumulation of remains of animal or plant life, such as decayed vegetation forming coal, and the shells of mollusca or of foraminifera, the skeletons of corals and other lime-secreting creatures giving rise to chalk and limestone.

289. **Metamorphic Rocks.**—Changes are produced by heat, pressure, and Earth movements so that it is difficult in many cases to decide the origin of rocks. It is convenient to class all such doubtful cases as metamorphic or changed. There is much difference of opinion amongst geologists as to the exact way in which metamorphism occurs, and we can only indicate here how some of the changes may take place. Limestone subjected to heat under pressure crystallises and forms marble; a bed of clay under similar influences is altered into slate. The temperature of rocks deeply buried under a mass of newer sediment is greatly raised (§ 291), and, as the pressure of the upper layers is extreme, changes of chemical composition and of structure are necessarily produced. When great Earth movements fold over and thrust forward masses of rock, the

friction produces heat enough to soften the substance which is rolled out, so that the original structure disappears, the minerals are altered chemically, and the rock acquires a flaky texture and is known as a *schist*. The change may produce a crystalline structure very similar to that of granite as in the rock called *gneiss*. Local or contact metamorphism is brought about by an intrusive sheet of liquid igneous rock forcing its way between other strata and altering their composition and physical state. The edges of sandstone may thus be fused into glassy quartzite, and soft clay beds baked into a hard porcelain-like mass.

290. **Dip, Cleavage, Joints, and Faults.**—Sedimentary rocks are sometimes raised by upheaval so steadily and uniformly that the strata remain horizontal, but far more commonly the strata are inclined in a particular direction. The inclination of a bed of rock to the horizon is called its dip, and is measured by the angle FAB (Fig. 42). Rocks are found dipping at all angles, sometimes as high

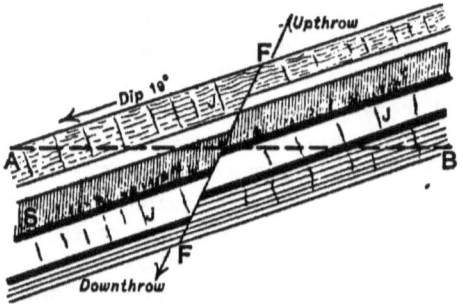

FIG. 42.—Illustration of rock structures. AB, horizontal line; FF, fault; S, slaty cleavage; J, joints. The long parallel lines mark planes of bedding making an angle of 19° with the horizontal.

as 90°; then the strata stand upright. The stresses which elevate rocks usually act horizontally as a thrust from two sides, and the particles of the rock sometimes yield and are flattened out. When this happens the rocks split up more readily along the flattened sides of the particles than along its original planes of bedding, and are said to have acquired cleavage-planes (oblique lines S in Fig. 42). The cleavage-planes of slate, by means of which thin slabs can be split off, are sometimes at right angles to the planes of bedding. In more rigid rocks the strain during upheaval is relieved by the strata cracking more or less nearly at right angles to the planes of bedding. When these cracks, which are originally extremely narrow, sometimes invisible, simply traverse the rock without any distortion (fine lines JJ) they are termed joint-planes, and it is on account of the existence of joint-planes in all rocks that the quarrying of stones is possible without continual blasting. Igneous rocks show joints, probably the result of contraction in cooling after solidifying. The fine hexagonal columns of basalt cliffs are outlined by joint-planes produced by the uniform cooling of a great mass of rock, the interior of which is brought into a state of tremendous tension by contraction until relieved by cracking into columns. A layer of wheat starch on drying is strained in exactly the same way by contraction throughout the mass, and similarly cracks into many-sided columns. The same phenomenon has been observed in partially solidified beds of moist sand and clay. When the rocks on one side slip along a crack so that the strata no longer correspond (F, Fig. 42) it is termed a fault; the lower side is called the downthrow, the upper the upthrow. Parallel lines of faults usually mark the borders of regions where upheaval has taken place and the strata preserve a low dip. When a fault shows at the surface no sudden rise of level marks the upthrow side, as the action of erosion is continually smoothing away such inequalities perhaps as rapidly as they form.

291. **Temperature of the Earth's Crust.**—Whatever be the nature of the surface rocks, the Sun's heat penetrates them slightly and slowly. By observations in Britain with thermometers fixed at various depths beneath the surface of the land, it has been proved that the difference of day and night temperatures vanishes at about 3 feet, and that the greater and more regular difference between summer heat and winter cold becomes less and less perceptible as the distance increases, and dies away within 40 feet. The average temperature shown by the rock thermometers on

the Calton Hill at Edinburgh during the eight years 1880-1887 were, at the depth of 2 feet, 39°·4 in February and 53°·0 in July, an annual range of 13°·6, with the minimum and maximum in winter and summer respectively; and at the depth of 20 feet, 46°·9 in January and 45°·4 in July, a range of only 1°·5, with the maximum temperature in winter and the minimum in summer, showing that it requires six months for the conduction of heat from the surface to the depth of 20 feet (compare § 229). A zone of invariable temperature lies beyond the reach of solar heat and is found at different depths in different places, being deeper in regions of great annual range of temperature. Beneath the invariable zone temperature increases with depth in all parts of the world. In deep mines the air is always oppressively hot, and the water from deep Artesian wells is warm in proportion to their depth. The Underground Temperature Committee of the British Association, after collecting all the observations of temperature at great depths which have been taken in mines and deep borings in all parts of the world, concluded that the rate at which temperature increases downward averages 1° in each 55 feet; Professor Prestwich, after a full discussion, puts it at 1° in 45 feet.[1] In some instances the increase is more rapid, in others less so, according to the conducting power of the rocks. The temperature 1 mile beneath the surface must be about 100° higher than that of the invariable layer, and at the depth of 30 miles the temperature must be high enough to melt all known substances. At greater depths than this the rate of increase of temperature must diminish, in accordance with calculations from the condition of small heated bodies. Professor Tait calculates, from the gradient of temperature and the conductivity of rocks, that through every square foot of surface the interior of the Earth is losing heat at the rate of 230 units (§ 65) per annum, or sufficient to warm 1¼ lbs. of water from the freezing to the boiling point.

292. **Interior of the Lithosphere.**—Surface rocks have an average density of 2·5, and the deep-seated igneous rocks a density of about 3·0, while the mean density of the

Earth, as a whole, is 5·5 (§ 85). Unless the enormous internal pressure of the weight of the Earth's mass were counteracted the rock substance would be compressed into less space, and the mean density of the Earth would be greatly raised. The high temperature of the interior causes the rock substance to expand against the pressure of gravity, and so maintains the comparative low mean density which is actually found. The great pressure in its turn counteracts the effects of high temperature by raising the melting-point of the rock substance (§ 72), and so preventing it from assuming the liquid state. Astronomical observations show that the Earth behaves as if it were a solid ball, and Sir William Thomson has calculated, from the imperceptible tidal effect produced in the lithosphere, that it must be as rigid as if it were composed throughout of solid flawless steel.

293. **Volcanic Action.**—Volcanoes are conical mountains in communication with openings in the Earth's crust, which continually or occasionally throw out steam, hot stones, or white-hot melted rock called lava. Professor Prestwich believes that there are hollows in the lower part of the Earth's crust full of molten rock, which is squeezed out by the pressure exerted by the Earth's crust contracting slightly as it cools. Mr. Mallett, on the other hand, thinks that the heated interior of the Earth in cooling contracts more rapidly than the crust, shrinking away from it and leaving hollows, into which the solid rocks subside with much straining and crushing. The motion of the rocks converted into heat melts some of them, and the cracked crust allows the hot fluid to escape. Other authorities point out that since the lithosphere is solid only on account of the pressure of the crust upon it (§ 292), any relief of pressure produced by the shrinking in of the central mass, or by the cracking of the strata above, must allow the rock substance to liquefy suddenly and with explosive violence. All volcanic activity is accompanied by the emission of great quantities of steam, to the expansion of which geologists believe the great power of volcanic explosions is due. It is probable that a good deal of under-

ground water (§ 313) creeps down by capillarity deep into the heated layers under the crust, there combining chemically with the rock under pressure, but always ready to resume the form of steam if the pressure is relaxed.

294. **Volcanic Materials.**—In addition to water-vapour volcanoes throw out other *gases* in great abundance. Hydrogen and oxygen, resulting from the dissociation of water at high temperature (§§ 71, 220), combine as they rush out, producing violent explosions and great flames. These flames, together with the reflection of glowing liquid rock on the overhanging vapour, gave to volcanoes the popular name of burning mountains. Sulphurous acid, sulphuretted hydrogen, nitrogen, carbonic acid, hydrochloric acid, and the vapour of boracic acid, also occur very frequently, being produced by the chemical action of heat and water-vapour on minerals in the volcano. *Lava,* or molten rock, is the most important of all volcanic products. Welling over the cup-like hollow at the summit it flows down the sides of the mountain in white-hot streams, which gradually solidify on the outside, and advance like a glacier of slow-moving viscous rock, ultimately hardening into crystalline igneous rocks, such as basalt and trachyte. *Pumice* is a sponge-like glassy rock which forms over the surface of certain lavas, being frothed up by the vapours which are continuously given off. *Scoriæ* are the rough cindery upper portions of very viscous lavas formed in the same way. During eruption immense quantities of these crusts of lava, together with stones torn from the throat of the volcano, are thrown out. The finer grained loose materials are known as *dust* or *volcanic sand.* A light gray powder, known from its appearance as *ash,* is the solidified spray of molten rock similarly thrown into the air by the explosion of escaping vapours.

295. **Volcanic Mountains.**—Wherever a crack or fissure of the Earth's crust allows volcanic activity to assert itself the material driven out from below accumulates and solidifies on the foundation of the surface rocks, which are usually sedimentary, and a cone or *mountain of accumulation* (contrast §§ 303, 329) is thus piled up. If the lava is very fluid and escapes from a long fissure it may flood

extensive tracts of land with nearly level sheets. Such lava floods now occur very rarely, although they were common in past ages. Volcanoes are usually connected with their subterranean lava-stores by a comparatively narrow pipe, in which the lava wells up and overflows on all sides. A very hot and fluid lava forms a hill of gentle slope; a cooler or viscous lava, which solidifies before it flows far, builds a steeper mound. In either case the centre is formed by a trumpet-shaped hollow called the crater, the rim of which is raised by each successive outflow. In some instances cones are built up round the orifice of a volcano before the flow of lava commences, and are composed of volcanic ashes, pumice, and broken stones, etc., the ejection of which is the prelude to an eruption. When compacted by the pressure of its own weight, and cemented together by the chemical action of rain, such a deposit forms the rock known as *volcanic tuff*. When fluid lava rises in the pipe of a tuff cone the pressure it exerts frequently bursts an opening in the side, through which a stream escapes. When the force of the eruption is small and the walls of the cone strong, the ascending lava may cool down in the funnel and seal the volcano by solidifying. The most common form of volcanic mountain is of composite structure, being built up of alternate layers of tuff and flows of lava. Such a cone grows slowly, and, as represented in Fig. 43, is the outcome of several periods of activity and quiescence. The explosions which herald a new eruption shake the mountain, and cracking the walls allow tongues of lava to penetrate in all directions from the central shaft. These sometimes force a way to the exterior and form small cones on its slopes, from which streams of lava flow. Sometimes they harden as *dykes* or walls in the fissures into which they were injected. The cone *cac* is represented as formed by a late outflow of lava, and occupies the middle of an old crater which had become plugged up and was then partially destroyed by an explosion.

296. **Volcanic Eruptions.**—Volcanoes are often classed as active, dormant, and extinct. Stromboli, in the Mediterranean, is the type of a continuously and moderately active

volcano. It serves as a natural lighthouse and also as an automatic storm warning, as its activity is always greatest when the atmospheric pressure is low and gales may be expected, while the violence of its eruptions is much reduced when the barometer rises. Volcanoes from which no erup-

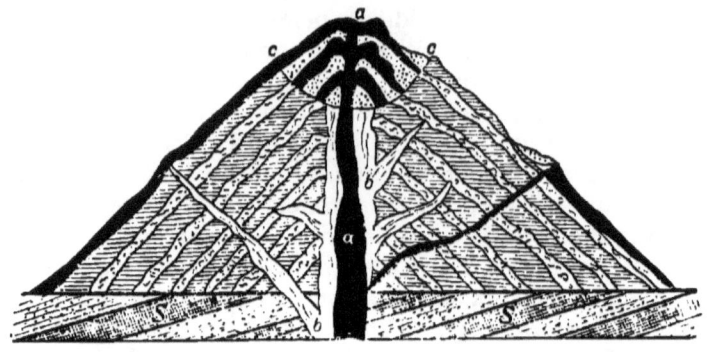

Fig. 43.—Ideal Section of a volcano. *SS*, stratified rocks of crust; *bb*, old lava solidified in throat of volcano and in dykes; *aa*, new outburst of lava; *cc*, old crater; *a*, new crater. (After J. Geikie.)

tion has ever been recorded are called extinct; those which break out at intervals are said to be dormant during their periods of tranquillity, but the distinction can hardly be drawn with confidence. Vesuvius is the type of volcanoes which are occasionally dormant and sometimes supposed to be extinct. The commencement of activity after a dormant period is usually preceded by earthquakes and subterranean noises, indicating that pressure is accumulating in the heart of the mountain. Hot springs break out on the slopes, and gases and hot vapour rise in increasing volume from the crevices in the crater. Then a terrific explosion occurs, shattering the solid lava plug and perhaps destroying the entire cone; volumes of water-vapour shoot up into the air, mixed with clouds of dust that darken the sky and fall like snow over the mountain slopes and surrounding country. Flashes of lightning dart from the overhanging cloud, the friction of dust and vapour on the air causing great electrical disturbance, and the noise of thunder is added to the roar of the escaping steam and volcanic explosions. The cloud reflects the fierce glare of the lava welling up in the crater, from which the explosions and bombardment of heated stones become more frequent, until finally the molten rock surges up to the lip and pours over as a river of fire. The vast quantities of water-vapour meanwhile condense into floods of rain, which convert the dust-strewn slopes into torrents of hot mud, more voluminous and often more important in obliterating the surface features of the scenery than the lava itself. Such a mud deluge destroyed the Roman town Herculaneum when the first recorded eruption of Vesuvius took place in the year 79. Snow-clad volcanoes like Etna and Cotopaxi send down still more serious floods on account of the sudden melting of their snow.

297. **Krakatoa.**—On 27th August 1883 the volcano of Krakatoa, a small island in the middle of the Strait of Sunda, terminated a set of comparatively quiet eruptions by the most terrific explosions which have ever been witnessed. A great crater had been previously formed, and sea-water gained access to the crater full of molten lava as the mountain walls were gradually broken down. The result was a temporary reduction of activity as the cold water chilled the surface, and then the grand explosion shot out a column of dust and vapour 20 miles high with a roar that was heard at Rodriguez 3000 miles distant, and attracted attention over one-thirteenth of the surface of the globe. The concussion caused by this explosion was severe enough to break windows and crack walls in Batavia 100 miles away, and the disturbance of the air was shown by the records of barographs to have expanded as an air-wave from Krakatoa until it spread round a great circle 180° in diameter, then contracted to the antipodes of Krakatoa, whence it was reflected back, and so continued pulsing round the world four times from the centre of disturbance to the antipodes, and three times back again. Two-thirds of the island were blown away, most of the material being deposited in the Strait of Sunda, where several new islands formed of piles of tuff and ashes appeared, and after a few months were washed away by the

waves. For weeks fields of floating pumice made navigation very difficult. The disturbance in the sea produced a wave more than 100 feet in height, which rushed upon the neighbouring coasts, overwhelming lighthouses and towns, and stranding ocean steamers in mountain valleys. More than 36,000 people were washed away and drowned. Part of Krakatoa was scattered as the finest dust through the air and carried to every part of the Earth, its presence being detected in rain, and by the magnificent red sunsets (§ 162) that were visible everywhere during the autumn and winter of 1883 and 1884.[2]

298. **Distribution of Volcanoes.**—Volcanoes are usually found in the line of great mountain chains and near the sea coast. They form a "ring of fire" round the Pacific Ocean, being very numerous in the Andes, and more widely spaced along the plateau of Central America, the coast ranges of North America, and the Aleutian Islands. Thence they increase in frequency along the island festoons of Asia, and come to a maximum in the Malay Archipelago and New Zealand. The West Indies, many of the small Atlantic islands, the Mediterranean coasts, Iceland, and Jan Meyen, also contain active volcanoes, but none are known with certainty in the heart of continents. The distribution of active volcanoes is shown in Plate II.

299. **Earthquakes.**—The crust of the Earth is elastic and readily transmits wave-motion. Any cause which produces a local disturbance of the crust sets up a series of waves, which may become apparent on the surface in the quick up-and-down or to-and-fro shaking of the land called an Earthquake. Earthquakes of considerable severity accompany volcanic action, and are accounted for by the jarring of the Earth's crust by successive explosions, but they are by no means confined to volcanic regions. The falling-in of underground caverns may give rise to earthquakes of slight intensity. Very severe shocks accompany the elevation of land when that process takes place in sudden steps of a few inches or a few feet at a time, in consequence probably of the strata, subjected to the powerful stresses set up by the contracting Earth, snapping under

the strain. Every large fault found in rocks must have given rise to earthquakes. Professor Milne points out that most shocks originate along the lower part of the slopes of the world ridges. This coincides with the lines along which the process of elevation is going on most rapidly, and where the strata are consequently subject to accumulating stresses. The regions in which earthquakes are common are coloured light blue on Plate II. and those where they are very severe and frequent are coloured in a darker shade. Many geologists believe that sea-water filtering through the bed of the ocean, or buried to a great depth in the lower layers of terrigenous deposits, causes explosions in the intensely heated region below, and that all great earthquakes originate from this cause and are essentially volcanic; the upheavals accompanying earthquakes would thus be reckoned as their consequences, not their causes.

300. **Propagation of Earthquakes.**—If the crust of the Earth were perfectly uniform in substance, and a shock were communicated to it at any point by a sudden yielding to stress, a wave would spread in concentric spherical shells from that centre like the sound-wave from a vibrating bell in air (§ 58). In the rock the wave travels more rapidly than in air, and the to-and-fro movement of each particle passing it on is very small. If the shock is given at A (Fig. 44) the circles I. II. III. show the position of the crest of the wave at intervals of 1, 2, 3 seconds. The wave is shown reaching the surface at B, directly over the centre of disturbance, in 3 seconds; there it strikes perpendicularly from beneath, although the force of the shock is greatest at a little distance from B. A second later the wave reaches the surface along a circular path (IV.-IV.) and strikes obliquely upward; at the posi-

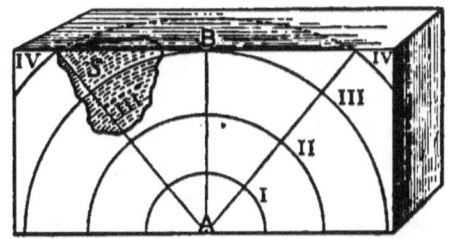

FIG. 44.—Earthquake wave, illustrating Mallet's method of finding the depth at which an earthquake originates.

tion reached in the next second, the stroke is still more oblique along a wider circle, and is more feeble on account of loss of energy due to friction among the rock particles. The distance of the centre of disturbance beneath the surface may be calculated by observing the angle from which the shock comes at different points and constructing a diagram somewhat like the above. It appears from many observations recorded by Mallet and others that the depth of origin rarely or never exceeds 35 miles. Although the crust of the Earth is probably homogeneous at a considerable depth, it is very far from being so in its upper part, and the earth-wave consequently travels at an unequal rate in different directions as it nears the surface. A thick bed of sand or loosely compacted and inelastic stones (S in Fig. 44) greatly retards and may entirely absorb the wave by friction between the particles, so that no shock would be felt on the surface, while houses built on the hard rock all round would be shaken severely. On the other hand, a small deposit of sand or alluvial soil occupying a shallow hollow would be jarred by confused earth-waves from every side and buildings on it damaged most severely.

301. **Earthquake Shocks.**—The area of the surface shaken depends on the intensity of the original shock and the nature of the Earth's crust at the place where it occurs. The memorable earthquake that destroyed Lisbon in 1755 shook a space four times as large as Europe, and probably made the whole Earth tremble; and that which damaged Charlestown in 1886 was felt over 3,000,000 square miles, from Cuba to Canada, and from Bermuda to the west of Missouri State. By the use of delicate seismometers the dying tremor of an earthquake-wave may be detected at a great distance, beyond the limit of unaided observation. Thus the tremor of an earthquake on the Italian Riviera in 1887 was distinctly recorded by instruments in Greenwich Observatory. The shaking of the Earth's crust throws down any slenderly supported rock masses like perched blocks, natural bridges, and earth pillars, and when such structures are conspicuous features of the scenery the district may be reckoned free from risk of serious shocks. Landslips, the

opening of great fissures, and other surface changes often result from earthquakes, which may thus alter the course of rivers and form or drain lakes. But the occasional destruction of cities and houses, and the peculiar sensation of terror and helplessness which earthquakes produce in most minds, are apt to give an erroneous and much exaggerated idea of the power of such shocks in forming the scenery of the globe. The researches of Professor Milne and other scientific men in Japan, and the extensive use of seismometers or earthquake measurers, have thrown much light on the nature of shocks and tremors. The to-and-fro or up-and-down motion of the Earth in a shock severe enough to throw down houses is probably not much more than an inch. A model constructed by Professor Sekiya (the professor of Earthquake Phenomena in Tokyo) of the path described by a particle during the passage of an earthquake shock resembles a tangled hank of twine.[3] It is the shaking produced by such a complex disturbance rather than the actual lifting of the surface that produces destructive effects. Some of the tremors detected by seismometers are not produced by the internal energy of the Earth. It has been proved in Italy that changes of atmospheric pressure jar the elastic and sensitive crust; and in Japan a gale blowing against a range of mountains has been found to set the greater part of the island quivering.

302. **Wrinkling of the Earth's Crust.**—The Earth necessarily contracts as it cools, and the crust composed of stratified rocks falls into wrinkles in order to adapt itself to the reduced area of the globe, just as the skin of an apple gradually becomes wrinkled in adapting itself to the drying and shrinking fruit. Reasons have already been given (§ 278) for believing that from a very early period the Abysmal and Continental Areas have occupied their present position, and probably they represent the troughs and crests of the earliest Earth wrinkles. The primitive furrows themselves must have disappeared as the crests were worn away by erosion, and the resulting sediment was deposited on the upper slopes of the hollows, to be consolidated in turn and form part of a new set of wrinkles, which shared

the same fate and passed on the process. Some geologists believe that as denudation lightens the ridges and loads the hollows, the Earth's crust is strained by the redistribution of the pressure on it; that consequently the strata snap with a succession of earthquake shocks, and the parts loaded with deposits sink, while those lightened by the effect of erosion are upraised. Other geologists take an opposite view of the result of sedimentation (§ 304). The typical form of an Earth wrinkle is a gentle ridge, A, accompanied by a gentle hollow, S (Fig. 45). The curved strata of the

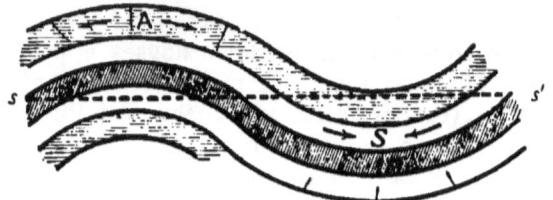

FIG. 45.—Strata bent into anticline A and syncline S.

ridge are said to form an anticline, because at the summit A the strata, as shown by the arrows, dip or incline away from each other. The curved strata of the trough are similarly said to form a syncline, as at S the strata dip together or toward each other. Even although the wrinkled crust should be worn smooth by erosion to form the surface ss', it is still easy to tell by observing the dip of the strata where the ridge and the hollow were situated. Thus rock structure is not concealed by surface change. Synclines and anticlines are ridged up in consequence of the lateral pressure or tangential thrust produced by the downsinking of part of the crust. The tremendous lateral pressure effected by a great subsidence throws the strata on both sides into sharp anticlines and synclines, while at a greater distance from the origin the wrinkles are low and uniform. The Geological Survey of Scotland has brought to light many remarkable proofs of the intensity of the thrust which ridged up the western margin of Europe in ancient times. Sometimes the compressing force was so violent that the strata, instead of puckering up into anticlines and synclines, cracked, and allowed one part to be lifted up and thrust bodily over the other, in certain cases for a distance of ten miles or more. The consequent crushing, faulting, and folding produced a very confused arrangement of the rocks, and extensive metamorphism. The structure of the region was extremely puzzling until Messrs. Peach and Horne traced out the thrust-planes along which the sliding movement took place. Figure 46 represents the production of thrust-planes, A, in a series of experiments on mountain structure recently carried out by Mr. H. M. Cadell.[4]

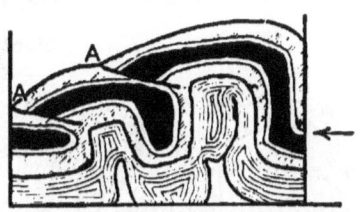

FIG. 46.—Production of thrust-planes. The strata represented are layers of clay and sand separated by cloth; they were laid down horizontally, and ridged into the position shown by a thrust acting in the direction of the arrow.

303. **Mountains of Elevation.**—When lateral compression of the Earth's crust takes place the strata pucker up along the line where they are weakest, and are thrown into a series of anticlines and synclines growing sharper and higher toward the central line. The rocks in the interior of the mass and those occupying the hollows of the synclines are necessarily compressed, heated, and altered, while those on the outer curve of the anticlines are stretched and split in the process. A mountain range is formed in this way, with anticlines as ridges and synclines as longitudinal valleys between them, the slopes of the surface corresponding to the dip of the strata. The true mountain ranges of the world are all of this character, the Alps, Himalayas, and Andes being typical examples, and it is significant that all such ranges are situated near the edge of great depressions, the subsidence of which probably accounts for their uplifting. Rocks of recent sedimentary origin always form the first gentle undulations on the slope of a mountain range, but toward the main ridge the strata are of greater age and more contorted, while in the centre there are masses of schistose or igneous rocks, probably produced either by the rolling and compression of the uplifted

strata or by volcanic action from below. Figure 47 represents a section across the chain of the Alps from north to south, the dotted lines indicating the anticlinal arch.

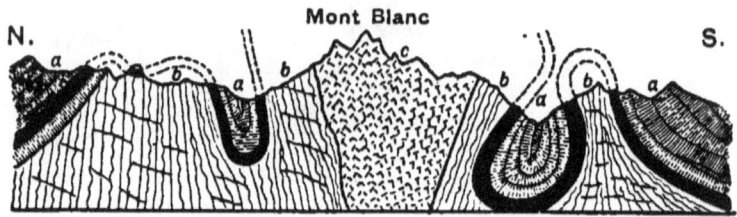

FIG. 47.—Section of the Alps. *a*, Tertiary rocks; *b*, secondary and primary rocks; *c*, central core of schistose and igneous rocks.

Erosion by solar energy probably accompanies the whole process of ridging up a mountain range, and after the elevation is complete the aspect of its scenery, the form of its slopes and valleys, are increasingly due to this cause. Streams flowing down opposite sides of the slope of the long mountain ridges hollow transverse valleys, and so cut the ridge into peaks. Two transverse valleys meeting in a col or pass allow of easy access between the longitudinal valleys which lie between the ridges. Anticlines are much more rapidly eroded than horizontal strata, even when the surface may have the same slope, for the direction of the joint planes and the dip of the rocks favour the formation of landslips. An anticlinal mountain may be viewed as geologically unstable, like a pile of inverted saucers. In many cases the low mountains of the Scottish Highlands, which in remote ages excelled the Alps in height, are now carved out by erosion (§ 329) from synclinal strata—a form of structure which gives great stability, like a pile of saucers set one within another right side up.

304. **Theories of Mountain Origin.**—The theory most generally held is that horizontal strata subjected to great thrusting stresses have wrinkled up along a line of weakness in the Earth's crust, by which the whole crumpling is confined to a narrow area, the actual lifting power being derived from the contraction of the heated interior of the Earth. Mr. Mellard Reade has brought forward another

theory of great ingenuity. Observing that all mountains of elevation are of comparatively recent formation and are ridged up out of thick sheets of sedimentary rock, he supposes that the accumulation of sediment produces the mountains. He points out that if a large and deep hollow in the Earth's crust is filled up with sediment to the line AB (Fig. 48) at the ordinary surface temperature, say 60°, the mass now forming part of the Earth's crust will grow warmer until, if the surface temperature remains at 60°,

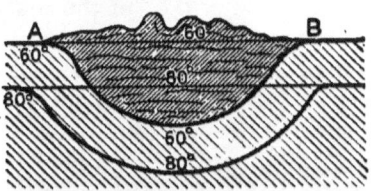

FIG. 48.—Mellard Reade's Theory of Mountain Building. Light shading shows original crust of the Earth, dark shading sediment; dark lines original isotherms, fine lines isotherms after deposition of sediment.

that at the depth of 1200 feet at 80°, and so on (dark lines in figure), the covering in of the cavity raises the temperature throughout by preventing the loss of heat through the crust, the new positions of the temperatures of 60° and 80° being shown by fine lines in figure. The warmed up strata necessarily expand, and as they cannot expand sideways or downward on account of the solid walls of the depression, they must expand upward, and the surface of the sheet of sediment is thrown into a series of ridges, true synclines and anticlines, like the surface of a cake as it rises in being baked. In this theory also the energy which does the work of elevating the mountain range is derived from the interior of the Earth.

REFERENCES

[1] J. Prestwich, "On Underground Temperature," *Proc. Roy. Soc.* xli. (1886).

[2] *The Eruption of Krakatoa*, edited by G. J. Symons. Trübner and Co., 1888.

[3] "Model of an Earthquake," *Nature*, xxxvii. 297 (1888).

[4] H. M. Cadell, "Experimental Researches on Mountain Building," *Trans. Roy. Soc. Ed.* xxxv. 337 (1888); or *Nature*, xxxvii. 488.

BOOKS OF REFERENCE

See end of Chapter XIV.

CHAPTER XIII

ACTION OF WATER ON THE LAND

305. **Land Sculpture.**—The crests of the world ridges upheaved by the internal energy of the cooling Earth in gently undulating strata, or in the sharp broken anticlines of mountain ranges, are subjected to erosion by solar energy acting through various agencies. Earth energy is continually at work raising the level of the elevated half of the globe, and depressing the Abysmal Area. Sun energy acts as a leveller, continually cutting down the high places and building up the hollows with the resulting detritus or crushed fragments. The process of uncovering old rocks by erosion of newer ones is termed *denudation*. The rate at which it proceeds depends to a very large extent on the chemical composition of the rocks, on their tenacity, their dip, and joints (§ 290), and it is to the variety of these conditions that the great variety and character of the existing scenery of every part of the world is due.

306. **Work of direct Sun-heat.**—One unit of heat (§ 65) when absorbed by one pound of an average rock raises its temperature about $4°$, compared with $1°$ in the case of water. In consequence of this low specific heat, although the heat does not penetrate far (§ 291), it greatly heats and expands the superficial layer. At night the temperature falls quickly by radiation and the chilled rock contracts. In dry tropical regions the alternate heating and chilling causes the surface layers to split off in angular pieces or thin sheets, which, when the face of the rock is

steep, slip down toward the base and form a talus or slope of detritus.

307. **Work of Wind.**—Air in motion (§ 175) is a powerful vehicle of energy for eroding rocks, sweeping away the fragments loosened by sun-heat in the tropics, and keeping the hard rock surface exposed to destructive radiation. The Sahara and some other deserts bear undoubted traces of having once formed the beds of shallow seas, so that their sand is partly of marine origin ; but the amount of sand is always increasing by wind action. Clouds of sand, driven by the wind like showers of hard angular hailstones against the face of the bare rock, cut into the surface as the artificial sandblast etches glass. In Kerguelen, situated in the Roaring Forties, all the exposed rocks are chiselled into grooves from west to east by wind-driven sand. Dunes, or wave-like ranges of sandhills, are piled up by the wind on deserts or broad sea-beaches, and attain the height of about 60 feet round the North Sea, and sometimes over 600 feet in the Sahara. The Bermuda Islands owe their configuration entirely to dunes of coral sand, some of which are 250 feet high, and have been hardened into a kind of limestone by the percolation of water.

308. **Wind-borne Deposits.**—Sand driven by the wind is an important ingredient in deep-sea deposits (§ 269), and rivers flowing across arid regions are kept charged with sand and dust in the same way. When the prevailing wind blows inland and the rainfall is scanty, sand and dust may be carried far before being deposited. The remains of many ancient cities in Egypt, Mesopotamia, and Central Asia have been covered by such dust, and their sites are now uninhabited deserts. The name *loess* is given to a deposit of very fine clay found first to the north of the Alps and amongst the Carpathians, where it often fills up valleys and covers large areas of ground at various levels. It is much more abundant in the north of China, where it covers thousands of square miles as a dense yellow earth to the depth of more than 1000 feet. The loess of Europe and of North America (Mississippi basin) is believed by most geologists to be the sediment of the greatly swollen rivers

of the glacial period (§ 352) subsequently modified by wind and other agencies. The great German geologist, Professor von Richthofen, who studied the deposit in China, came to the conclusion that there it resulted from the gradual accumulation of the fine dust carried by wind from Central Asia, and brought to the ground by the moister air near the coast.

309. **Water as a Sculpture Tool.**—Water is the agent by which the Sun's energy is usually brought to bear upon the land. The process consists in the Sun's heat evaporating the surface of the hydrosphere and depositing it as snow or rain on the land. The work done against gravity in raising water-vapour to the height at which it condenses to the liquid state, as rain, is converted into potential energy, all of which would be restored in heat to the hydrosphere if the rain fell without friction back to the sea again. Rain evaporated before it reaches the sea has a new store of potential energy imparted to it, like a clock wound up before it has run down. The height to which a quantity of water is raised by the Sun's heat is a measure of the dynamic power which the water can exert in its descent (§ 49). This power in the case of raindrops is expended in heating the air they fall through, and in friction against the channel down which the water flows, in breaking off portions of rock against the power of cohesion, and in dragging stones or gravel along. The expended energy finally takes the form of diffused heat in the water and rocks. The chemical properties of water and its effects as a solvent are also brought into action by sun-heat, which separates it from the salts in the sea, shakes it with the gases of the atmosphere, and pours this powerfully solvent and oxidising solution over the rocks. The hydrosphere might be compared to a bee-hive, whence the sunlight attracts swarms of workers in the form of raindrops, which after a longer or shorter journey return laden with spoil from the land.

310. **Weathering.**—Rain, assisted by the dissolved gases and surrounding air, acts chemically on rock surfaces, producing changes known as weathering. Next to beds of rock-salt and gypsum (calcium sulphate), limestone is the

rock which is dissolved most readily. The waste of the hard and massive surface is often shown only by the way in which it becomes studded with less soluble nodules or fossils originally hidden in its substance. Sir Archibald Geikie has calculated that by the acid-laden rain of towns one-third of an inch is removed from the surface of marble monuments in a century. Insoluble sulphides, such as that of iron, are rapidly oxidised by air in the presence of moisture to form soluble sulphates, and when this process goes on in the pores of a rock the expansion of the crystallised salt splits the block into thin layers. This action is the basis of the common way of making alum. In the case of granite (§ 43) and most other rocks the process of weathering is more complicated. Some of the minerals are decomposed. In felspar, for instance, the silicates of potash, soda, and lime are changed to carbonates which are washed away, while the silica and the more resisting silicate of alumina remain as a soft crust of kaolin or china clay, valuable for making porcelain. Granite has been found weathered in this way in South America to the depth of 600 feet. Rocks containing iron usually become brown or reddish in colour, although the freshly broken rock may be white or gray. The lines of stratification and joints (§ 290) of rocks are sometimes etched out by weathering, so that the face of a cliff assumes the appearance of a gigantic wall of masonry. The crumbling of rocks in rainy regions is assisted by the action of the Sun in drying and warming the surface, which may then be splintered into flakes by a shower of cold rain. Rain soaking by capillary attraction (§ 39) through the weathered crust and into the pores of the solid rock is frozen in cold weather, and the ice, expanding as it forms, acts like a multitude of minute wedges driven simultaneously in all directions. When the thaw comes, the bases of cliffs and banks are strewn with weathered crusts and stones, often of a great size, broken off in this way.

311. **Soil.**—Weathered rock is the basis of soil, which accumulates to the greatest depth on level or slightly-inclined land. When the rocks yield only angular grains of quartz

or silicates, the soil is pure sand, which allows water to drain away so rapidly that in a dry region no moisture is retained. When only the finely divided silicate of alumina results from weathering, the soil is a pure clay, forming when wet a sticky paste through which water does not easily pass. In rainy places clay land is consequently always wet and stiff. Sand and clay are both produced from the decay of most rocks, and the mixture of these constituents forms *loams*, which, according to the proportion of sand and clay, are either moderately porous or moderately retentive of moisture. Almost all rocks contain smaller or larger quantities of carbonate of lime, iron, and sulphates or phosphates of the alkalies potash and soda, all of which form part of the resulting soil. Rain contributes salts of ammonia (§ 152), partly derived from the air, partly from decomposing animal matter, and these are ultimately oxidised (§ 401) to nitric acid, which forms nitrates. Plants pulverise the rock fragments of the lower layers or sub-soil by their roots penetrating the crevices and acting as wedges. The decay of vegetation finally produces vegetable mould. Earth-worms have been shown by Darwin to assist in the formation of soil by dragging decaying vegetation into their burrows and by swallowing the earth, which is thrown out again on the surface as extremely finely-powdered worm-castings. Professor Henry Drummond points out that a similar service is rendered by the termites or white ants of tropical Africa.

312. **Work of Rain.**—Rain is the chief agent engaged in the slow but continuous moving on of particles of broken-up rock-crust and soil from high ground to low ground, and from low ground to the sea. When rain falls on beds of clay or soft rock mixed up with harder pebbles or boulders it washes away the softer material, except where it happens to be protected by a stone, which in course of time remains capping a pedestal. The largest examples of such earth pillars are those of the Sawatch region of North America, which attain a height of 400 feet. Mount Roraima, in north-eastern South America, a nearly perpendicular mountain of soft sandstone capped with hard conglomerate, and

rising 5000 feet above the plain, is believed by Mr. Im Thurn, who first succeeded in reaching its summit, to be simply a rain-wrought earth pillar on a gigantic scale; the soft sandstone, when freshly exposed, being rapidly washed away by the torrents of one of the rainiest regions of the world, while the harder conglomerate resists erosion and protects the rock beneath.

313. **Underground Water.**—Of the rain which falls upon the surface of the Earth in a region like Great Britain it is estimated that one-third is returned to the air by evaporation, one-third flows off over the surface, and one-third sinks into the ground. Where the rocks are impermeable by water, such as shales and stiff clays, more flows off over the surface, but where they are permeable, like sandstone, gravel, or many limestones, a greater proportion soaks through. The movement of water underground is slow or rapid, according to the facility with which the rocks allow it to work its way through them. In time some water undoubtedly filters downward, until, under the influence of great pressure and high temperature, it combines chemically with the rock substance (§ 293), but the greater part of it returns to the surface at a level lower than that it started from. Each variety of rock can absorb by capillarity (§ 39) a certain definite proportion of water, which remains in it as in a sponge, until enough accumulates to overcome friction, when it percolates through. The rate of percolation is often greatly increased by the presence of cracks or joints. Soft porous rocks becoming saturated may give rise to landslips, especially in cases where they rest on beds of stiff clay that become lubricated and slippery when wet. As the percolating water dissolves out narrow crevices between the grains of rock, the pressure of the strata above forces them together again, thus producing a slow general settling down of the land-surface.

314. **Wells and Springs.**—When a thick layer of permeable rock rests on an impermeable bed, water accumulates until the pressure of the liquid suffices to force a way between the rocks and so reach the surface on the slope of a hill or the side of a valley. This outflow of underground

water is termed a spring, and its origin is indicated at *s* (Fig. 49). If a pit is dug through the upper rock, as at W, deep enough to pass below the limit of saturation *l*,

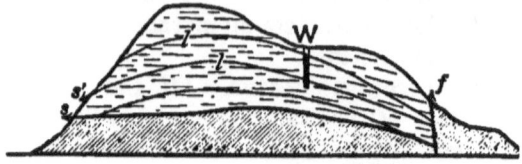

FIG. 49.—The origin of springs. (After Prestwich.) The darker shading represents rocks impervious to water, the light shading shows permeable rocks. W, a surface well; the curves on the shaded part show different positions of the limit of saturation; *s′s*, springs; *f*, fault.

water will ooze in from all sides, and a surface well will be formed from which water may be lifted by a bucket or pump. The limit of saturation rises in wet weather, but sinks in a dry season. When it rises from *l* to *l′* the water in the well deepens, when it sinks to the lowest curve shown, the well becomes dry, and if the height is not sufficient to overcome the resistance of capillarity the springs also cease to flow. When layers of permeable and impermeable rocks occur one above another, the water which soaks into the permeable rocks at the surface filters down along the junction with the impermeable layer, and if a fissure or fault occurs (*f* in the figure) so that the permeable layer is brought against an impermeable wall, the water will be forced up along the crack and will reach the surface as a fault-spring if the ground-level is below that of the limit of saturation. Artificial bores driven through an impermeable stratum of rock to reach the water-bearing strata below are termed Artesian wells, from the old name of part of the north of France where they were largely used. By this means a copious water-

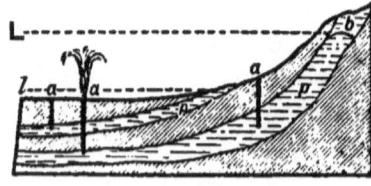

FIG. 50.—Artesian wells. *pp*, permeable rocks; L, *l*, limits of saturation, showing level beyond which water from the bores *aaa* cannot rise.

supply may often be obtained even in rainless deserts, as the deep layer of permeable rock may come to the surface at a great distance in a rainy region (Fig. 50).

315. **Thermal and Mineral Springs.**—When the dip of the permeable strata carries them far down into the Earth's crust the water is greatly heated (§ 291), and if it is brought back to the surface its high temperature entitles the outflow to the name of a thermal spring. Hot springs also abound in volcanic regions and along the slopes of recently upheaved mountains, in which cases they are not necessarily deep-seated (§ 289). Hot water dissolves much more of the rock substance than cold, and if it has traversed beds of very soluble salts, such as the sulphates, carbonates, or chlorides of the alkali metals or magnesium, it rises to the surface as a mineral spring, often possessed of valuable medicinal properties. When charged with carbonate of lime, dissolved in the presence of carbonic acid under pressure, the heated water on evaporating at the outlet deposits carbonate of lime in large quantities. Calcareous deposits from such springs often clothe whole hillsides with fantastic sheets of rock, which under the name of tufa or travertine furnish one of the most valuable building-stones in Italy.

316. **Geysers.**—Very hot water under high pressure decomposes the silicates in granite and similar rocks, dissolving large quantities of silica, which are deposited as a crust, termed siliceous sinter, when the heated water evaporates on the surface. Some of the most fairylike scenery in the world has been formed by such deposits of silica in New Zealand, where the dazzling pink and white terraces near Lake Tarawera were famous show-places until they were destroyed by an earthquake in 1886. Many hot springs depositing silica show the characteristic action of geysers—an Icelandic name expressive of the violent and explosive gushes of steam and boiling water which alternate with periods of quietness. At the bottom of the shaft of a geyser the temperature is far above 212°, but the water is kept from boiling by the pressure of the column above, and the uppermost layer is cooled by the air below the boiling-point. After a time the surface water gets sufficiently

R

heated from below to begin to boil (§ 72); this relieves the pressure on the layers beneath, which flash into vapour in a series of explosions, throwing up a column of water and steam with a terrific roar. The geyser remains quiescent until it fills up again, when the same process is repeated. In the Yellowstone region of North America (§ 364) the Giantess Geyser throws up a stately column of steam and water 250 feet high in each outburst, after which several weeks of tranquillity elapse; and "Old Faithful," throwing a column of 150 feet, explodes with wonderful regularity at intervals of about an hour.

317. **Caverns.**—Since the masses of tufa or sinter formed round hot springs have been taken from the rocks beneath, hollows or caverns must be left in the Earth's crust. These are usually enlargements of the natural crack or fault which allowed the spring to reach the surface. In limestone regions caverns are very numerous and often of great size, on account of the solvent action of rain-water charged with carbonic and other acids on the joints and faults of the strata. The roofs of caverns sometimes sink in, leaving a funnel-shaped hollow on the surface called a sink or swallow-hole, in which, if rubbish blocks up the outlet below, small isolated lakes may form. Part of a cavern roof may remain standing as a natural tunnel or bridge after the débris of the fallen portion has been carried away by rivers. Caverns are usually very picturesque on account of the formation by the dripping water of fantastic stalactites, white or tinted icicle-like appendages of carbonate of lime, hanging from the roof. Where the water-drop falls from the stalactite to the floor more carbonate of lime is deposited, and a stalagmite grows upward, and the two ultimately form a natural pillar. Small stalactites formed by the percolation of rain-water through the mortar may be seen hanging from the arches of bridges. The most extensive limestone caverns are those of Adelsberg in Austria, the Mammoth Cave in Kentucky (which comprises more than 150 miles of passages), and the Jenolan Caves in New South Wales. Some of these caverns contain lakes tenanted by blind fish, and underground rivers

flow through them. In all limestone regions rivers disappear beneath the surface, and although most of them, like the Guadiana in Spain and the Poik in the Adelsberg caves, reappear on land, several vanish altogether and ultimately well up through the salt water of the sea, sometimes from depths of 100 fathoms or more.

318. Surface Water.—During a shower, and for some time after it has ceased, little runnels of water flow down the steeper slopes of the land, uniting where opposed slopes meet to form streams, which ultimately converge in rivers and flow on to lakes or to the sea. If the land were composed of impermeable rock the whole of the rain-water not lost by evaporation would run off over the surface, and rivers would flow only during and immediately after the fall of rain; this is in fact the case in many mountainous regions where the smooth rock walls are too steep to allow soil to form upon them. On gentler slopes the rain first soaks into the soil, and the streamlets swell gradually and are kept flowing long after the rain stops by the subsequent oozing of moisture. About one-half of the water in large rivers enters them from springs which have pursued an underground course from higher levels, and being independent of local fluctuations of rainfall these give permanence to the flow. When the melting of snow takes place at one period of the year, or when heavy rains occur at definite seasons, the springs are replenished as a store to be drawn on gradually, and the increased supply of surface water produces a regular periodical rise in the level of the river. The Ganges always rises and overflows its banks in summer, when the melting snow of the Himalayas and the rains of the south-west monsoon fill its higher tributaries. Similarly the Nile (§ 375), after the monsoon rainfall of Abyssinia, overflows its channel in the rainless land of Lower Egypt every autumn, covering a narrow strip on each side with soft and fertile mud. The Amazon (§ 361), on the other hand, is almost always high, as the rainy seasons of its southern and northern tributaries occur at opposite times of the year with the shifting of the trade winds (§ 178), but its floods are greatest in June. Dr. John Murray calculates that of 29,350 cubic miles of rain falling on the land every year, only 6520 cubic miles reach the sea as the discharge from rivers, the remainder being re-evaporated or absorbed in the Earth's crust.

319. **River Systems.**—The connected streams which unite to form a river constitute a river system. The series of convergent slopes down which a river system flows—in other words the land which it drains—is called its *basin*, and is separated by a *watershed* or water-parting from the basins of neighbouring river systems. A watershed is always the meeting-place of the highest part of two diverging slopes. This is sometimes a mountain range, but often only the crest of a gently rising ground, on which the line of water-parting is difficult to trace (§§ 360, 362). It is usual to name a river system after the river into which the water is collected from the whole basin, the other streams being called tributaries or affluents. The basins of all river systems draining into one ocean are known collectively as the drainage area of that ocean. The beginning of a river is called its source, and must necessarily be the highest part of its course. When a large river flows from a lake it is often difficult to decide which of the short streams entering the lake is to be viewed as the ultimate source. The name of the main river in a great system, such as that of the Amazon or the Mississippi, is given by some geographers to the tributary which has the most direct course, by others to that of greatest length or to that with the highest source. This diversity of opinion accounts to some extent for the great difference in length assigned to rivers by different authorities. The area of the basins or the volume of discharge is a better measure of the size of a river. It is interesting to notice in the following table of the five greatest rivers that although the Nile basin receives one-third more rain than the Mississippi, its discharge is only one-fifth, on account of the great evaporation in crossing the desert. The Yang-tse-Kiang, Yenesei, Amur, and Mackenzie are intermediate in length between the Amazon and Congo, and the Yang-tse-Kiang and Orinoco have a discharge equal to the Mississippi.

Name.	Area of Basin. Square Miles.	Rainfall of Basin. Cubic Miles.	Average Annual Discharge. Cubic Miles.	Length of Chief Rivers.
Amazon	2,230,000	2834	528	3060
Congo	1,540,000	1213	419	2900
Nile	1,290,000	892	24	4000*
Mississippi	1,285,000	673	126	4200†
La Plata	995,000	905	189	2000

* Including Lake Victoria and its longest tributary.
† From Missouri source.

320. Torrential Track.—On account of the forms of the land-slopes (see sections of continents, Figs. 56-62) the course of a typical river falls into three natural divisions: the *Torrential Track*, with a slope usually exceeding 50 feet in a mile; the *Valley Track*, with a slope rarely greater than 10 feet, and often less than 2 feet; and the *Plain Track*, in which the change of level is only a few inches in a mile. Some rivers have only one or two of these characteristic divisions. Torrents dash down the mountain-sides with tremendous speed, often exceeding 20 miles an hour, leaping in cataracts from rock to rock and foaming through ravines. Little soil forms on the steep slopes, hence as a rule torrents swell quickly during rain and dwindle away to a mere thread of water at other times. The work of a river in its torrential track is purely destructive. When wholly immersed in water, rocks are practically reduced in weight from one-half to one-third, and are therefore moved with much less expenditure of energy than would be required in air. Huge boulders are thus hurled along by the flooded stream, and hammer out the hollows in which the water flows. The chips struck off at every concussion get broken into smaller pieces, forming pebbles, gravel, sand, and mud, or, to use a general term, detritus, which is swept away to lower levels. As the ravines are deepened, tributary torrents leaping down the rugged slopes carve out tributary ravines and increase the volume of water and of detritus in the river.

321. Valley Track.—The valley track of a river lies

over the more gentle slopes that separate mountains from plains, and the velocity of the stream rarely reaches 5 miles an hour, and is usually not more than 2 miles. The work of a river in this part of its course is at the same time destructive and constructive. A stream dashing along at 8 miles an hour can drag boulders 4 feet in diameter; at 2 miles an hour stones as large as a hen's egg are rattled along; at $1\frac{1}{3}$ mile an hour the current can just roll pebbles 1 inch in diameter; when gliding at half a mile an hour gravel as large as peas is swept forward; while at a quarter of a mile an hour a river cannot disturb fine sand. In the slackening current of the valley track heavy stones brought down by the torrent cannot be stirred, and the pebbles, gravel, and sand are successively deposited as the slope decreases; and, since a river is retarded by friction with the sides and bottom and flows slowest at the edges, the deposit of stones and sand takes place chiefly at the sides, where they form a shore or terrace. This is the constructive work of a valley river, and the terraces built up are termed alluvial deposits. The stones stranded in these terraces gradually get weathered and crumble to pieces; and during floods the river sweeps away the fragments which are readily broken by friction into sand or mud, and are deposited in new terraces farther down stream. The material swept along the bed of the river acts like coarse sand-paper, scouring the hard clay or rock which forms the river-bed; and as the stream sinks in its deepening channel it leaves its old terraces lining the valley at higher levels. The river also attacks the banks, pressing now against one side, now against the other, undermining cliffs and carrying away the fallen fragments, thus widening the flat bottom of the valley. Other conditions being the same, a valley cut through horizontal strata is equally steep on both sides; but if the strata dip across the stream, the bank toward which they dip becomes much less steep than the other on account of the greater erosive action of springs and percolating rain along the bedding planes.

322. **Plain Track.**—On the almost imperceptible slope of its plain track the work of a river becomes entirely con-

structive. Water in this case ceases to carve and commences to model the surface of the land. The alluvial deposits are composed of the finest sands, and finally of mud, which assist to raise the level of a wide area as the river wanders over the plain. The alluvial plains of the Mississippi cover 50,000 square miles, a space equal to all England. Remains of dead animals and plants swept away by the river in time of flood become embedded and buried in the alluvial deposits on the margin of rivers or in the mud and sand carried into lakes and seas, where they either decay away or are preserved by various processes. The work of a river has been compared to that of a mill which "grinds slowly, but grinds exceeding small," rough angular blocks being supplied in the torrential hopper, and the most finely powdered material poured into the great sack of the ocean.

323. **River Windings.**—When a swift-flowing river laden with sediment is checked by any obstacle the sediment is deposited, and a sandbank or mudbank is formed. When an obstruction of this kind is formed on the left bank of a river at A (Fig. 51) the current of the river is deflected from the straight line and strikes against the right bank, rapidly undermining it at the point, while the velocity of the stream is checked opposite on the left side, which becomes built up by the deposit of sediment. The current is reflected back to the left side at C, and so the process goes on, until the straight river forms a series of winding loops as shown by the dotted line.

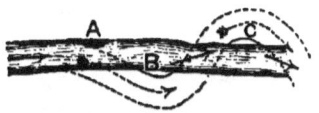

FIG. 51.—Origin of River Windings. A, obstruction on left bank; B, cutting in on right bank. The arrows show the direction of the stream.

The same effect is produced by the unequal hardness of parts of the bank, the softer being worn away and the harder left as obstacles deflecting the current. The windings once begun are perpetuated by the set they give to the current always against the concave side, which is made more concave, while the deposit of sediment adds to the convexity of the convex side. The narrow neck of land between two concave curves may ultimately be cut through by the river, which establishes a short direct passage, leaving an island; or the ends of the cut-off portion may be silted up, converting it into a crescent-shaped lake.

324. **Embanking of Rivers on Plains.**—During a flood the swift, muddy stream rises, and, overflowing the banks, immediately widens out on the level land; the current is checked at once, and most of the sediment is deposited close to the banks in the form of broad bars of alluvial soil. When the amount of mud in the water is very great, as in the Mississippi, the Po in Northern Italy, and still more the Yellow River (Hoang Ho) which traverses the loess deposits of China, the land on both sides of the stream is raised rapidly. The river-bed also gets silted up, and the great muddy river ultimately flows along the top of a gently sloping embankment, many feet above the level of the plain (Fig. 52). The natural mud walls, called *levees*, on the lower Mississippi are strengthened artificially in order to protect the dwellers on the fertile borders of the river. Floods frequently make a breach in the wall, and a stream, called a bayou in Louisiana, escapes, winding over the low plain, either to rejoin the main river at a lower level or to reach the sea independently. The Yellow River of China has repeatedly changed its course by the high banks bursting. One such disaster occurred in 1852, when the embankments burst about 500 miles from the sea, and the great stream, half a mile wide, formed a new channel, entering the Gulf of Pechili several hundred miles from its former mouth. In 1887 the banks burst again near the same place, leading to the most fatal catastrophe recorded in history, as the river, inundating hundreds of towns and villages, drowned several millions of people.

FIG. 52.—Embankment of a river. BB, original slope of valley. The light shading shows successive layers of deposit; AA, level of river.

325. **Bars, Banks, and Deltas.**—When rivers enter a tidal sea directly, the effect of the salt water is to cause a rapid precipitation of sediment, which may accumulate at the mouth of the river and form a bar. Bars are often purely

marine formations consisting of shingle or pebbles ridged by the waves, but most of them are due to a combination of river and sea action. When rivers enter a tidal sea by a comparatively wide shallow estuary, such as the Tay, Mersey, or Thames, sandbanks are formed, the size, position, and shape of which depend on the amount of sediment brought down and the form of the coast-lines which guide the tidal currents. Professor Osborne Reynolds, in a series of beautiful experiments, shows how, in a small flat-bottomed model of an estuary, the floor of which was strewn with fine sand, it was possible, by causing mimic tides to stream to and fro in rapid succession, to rearrange the sand in banks with channels between; precisely like those of the real estuary represented.[1] In lakes and seas not subject to strong tides, such as the Baltic, Black Sea, Mediterranean, and Gulf of Mexico, the sediment thrown down by rivers is not swept away, but accumulates like a railway embankment in course of formation until it rises to the level of the sea. The action of waves piles up the deposited mud into low islands on which vegetation takes root and assists to raise the level by forming vegetable mould. These islands split the river into numerous branches, which interlace with one another sometimes in a very complicated way. The typical delta of the Nile originated the name, for below Cairo the river splits into two main branches which enclose a triangular piece of land like the Greek letter Δ (*delta*) in form, the broad growing edge of the delta, 180 miles long on the Mediterranean, being the base of the triangle. The Mississippi delta grows much more rapidly than that of the Nile. It forms a long narrow peninsula spreading out into a series of branches, each traversed by an arm of the river and all constantly varying in size and position. When the amount of sediment is very great, deltas are formed even in tidal seas, as, for example, where the Ganges and Brahmaputra meet at the head of the Bay of Bengal. The Adriatic Sea is being filled up so rapidly by the sediment of rivers descending from the Alps and Apennines that the coast is lined by a broad belt of new land interposing a stretch of 14 miles between the present coast and the port of Adria, which originally gave its name to the sea.

326. **Submarine Cañons.**—Mr. Buchanan points out that along the margin of the Gulf of Guinea the soft mud brought down by the Niger and the Congo builds up the slope of the transitional area, diminishing its steepness; but that right under the broad, swift, and deep current of the Congo there is a deep submarine gully or cañon walled by the soft mud, but kept clear from deposit by a strong counter-current of sea-water setting along the bottom up the estuary. This counter-current is due to the same cause as that through the Bosphorus (§ 238). Professor Forel has pointed out a similar sub-lacustrine ravine under the impetuous Rhone as it enters the Lake of Geneva laden with glacier mud.

327. **River Work on Dry Plateaux.**—When a river flows across an elevated plateau it wears out a channel for itself, the form of which depends on the nature and arrangement of the rocks and on the rainfall over the surface of the region. The result of dip has already been referred to (§ 321). Lines of faults frequently mark out the sites of valleys and affect their formation. For the sake of simplicity and contrast, it will suffice to explain the extreme cases of river action on arid and on rainy plateaux composed of horizontally stratified rocks. In a dry plateau the river flowing from a snow-topped mountain range, over the steepest slope, receives few and small tributaries as it proceeds, and the action of the water loaded with wind-borne sediment is to wear its channel down through the rocks. Cutting now on one side, now on the other, it makes rapid progress through the softer strata, forming banks of comparatively gentle slope, and slower progress through the harder which are cut into steeper cliffs. The walls of the valley retain the original slope as the detritus, instead of accumulating in a talus, is swept away as it is formed, and weathering takes place very slowly in the dry atmosphere. The valley becomes eroded in a somewhat V-shaped curve, and forms a gorge narrow compared with its depth and sunk far below the level of the plain. Such gorges occur

on a magnificent scale in the plateaux west of the Rocky Mountains, where they have received the name of Cañons. The most wonderful example is the Grand Cañon of the Rio Colorado about 400 miles in length, in many parts from 4000 to 7000 feet beneath the level of the plateau, and with very steep terraced sides that strike the eye as vertical walls.

328. River Work on Rainy Plateaux.—A river flowing over a rainy plateau cannot form a cañon or V-shaped gorge because of the number of small tributaries it receives, each of which helps to reduce the slope of the valley walls. The action of rain on the cliffs leads to occasional landslips, forming a gently sloping talus which protects the lower rocks from erosion and gives the valley a U-shaped section. Only in places where the rocks are hard and vertically jointed and the river strong can the talus be swept away as it is being formed, and a steep-sided gorge result. The valleys excavated across a plateau in rainy regions become wider as they grow older; and according as the rate of denudation over the whole area is nearly equal to, quite equal to, or more rapid than the deepening of the river-bed, the apparent depth of the valley increases very slowly, remains unchanged, or actually diminishes.

329. Mountains of Circumdenudation.—To a traveller ascending the Colorado River the sides of the cañon appear like lofty and precipitous mountain ranges, and where a tributary cañon enters, the appearance of the two meeting slopes is exactly that of a mountain. On the summit instead of a peak there is a vast plateau stretching out as a boundless plain, broken by massive buttes, the remnants of more resisting rocks left as monuments of denudation. In a rainy region the valleys of adjacent rivers cut up the plateaux into rounded blocks of elevated land, the exact form of which depends on the composition and arrangement of their rocks. Most geologists believe that the mountains of Scotland and of Norway have been carved out in this way from a solid plateau of great height by the agency of rain, streams, springs, and ice, guided by the durability and structure of the rocks (contrast §§ 295, 303).

330. Rivers and the Land Surface.—When a river is fairly established in its valley it is the most permanent feature of a land surface. Upheaval, which acts very slowly, may even elevate a range of mountains across its course, while the river cutting its way downward remains at the same absolute level. The Uintah mountains were elevated in this way across the course of the Green River, one of the tributaries of the Colorada (§ 364). The range in such a case rises divided, like a bar of soap pressed upward against a horizontal wire. Where a river crosses soft and regularly placed rocks its valley is comparatively wide, the sides of gentle slope, and the gradient of the stream uniform; but where a strip of hard rocks is encountered the valley narrows into a steep-sided gorge, and the gradient of the river will be suddenly changed. In such circumstances the hard rock is cut through more slowly, and above it the gradient is reduced to what is termed the base-level of erosion, where no destructive action can take place but alluvial deposits are formed. The softer rock farther down stream being eroded more rapidly, a waterfall is formed over the hard ledge, which is worn through in time, and a line of rapids formed in the short portion of steep slope. Eventually the gradient of the bed becomes uniform and the rapids also disappear. The great waterfall of Niagara is caused by thick beds of hard limestone (black in Fig. 53) resting on soft shale. The river flowing over the cliff formed by the

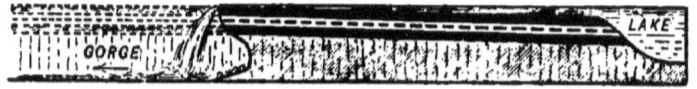

FIG. 53.—Ideal Section of Falls of Niagara.

edge of the limestone cuts away the soft shale from below and so produces occasional slips of the overhanging rock, causing the falls steadily to recede. The falls are now at the head of a gorge 7 miles from the escarpment of the limestone cliff, where the rock is being eroded much less rapidly by weathering. From recent surveys it is stated that the "American" falls have receded 30 feet, and the "Horse-Shoe" falls 104 feet in the last 48 years. If the structure of the rocks is the same all the way even at this rate the time, geologically speaking, is close at hand when the river-bed will be lowered along its whole length and Lake Erie will be drained. If the Niagara River had been muddy instead of being exceptionally clear, its erosive power would have been greater, and the falls would have been worked out long ago. The falls of St. Anthony on the Mississippi, for example, have been cut back about 900 feet since they were discovered in 1680.

331. **The Work of Rivers.**—The amount of sediment and of dissolved solids in the water of rivers gives a clue to their average effect in lowering the whole surface of their basins. From calculations of this kind it appears that in order to lower the average level of their basins by 1 foot the Danube must work for nearly 7000 years, the Mississippi for 6000 years, the Yellow River for 1500 years, the Upper Ganges for 800 years, and the Po only for 700 years. Dr. John Murray calculates that in 6,000,000 years river erosion at the present rate would reduce all the land of the globe to sea-level, and M. de Lapparent, observing that the deposit of sediment at the same time raises the level of the ocean, shows that at the present rate of surface erosion 4,500,000 years would suffice to equalise the level of land and sea.

332. **Lakes** are bodies of water occupying hollows of the land. As contrasted with rivers (§ 330) they are transitory features of a region, being subject to considerable fluctuations in extent and destined ultimately to disappear. Lakes often originate in the obstruction of a river valley. If blocked at a narrow gorge by drifting ice or an avalanche the river-bed below runs dry, and the water above rises, flooding the valley until it reaches the lip of the ice-wall. Ultimately the pressure of the accumulated water bursts the ice-barrier, and a terrific flood suddenly desolates the valley below. The famous parallel roads of Glen Roy in Scotland are believed to be beaches etched out at successive levels by the water of a glacier-obstructed lake, the

barrier of which gave way in successive steps separated by long intervals of time. A landslip, the melting of a glacier, or the flow of a lava stream, sometimes obstructs a valley by forming a barrier of earth, moraine stuff, or solid rock, through which the issuing stream cuts very slowly, and the lake so formed is permanent as far as the observations of a lifetime can discover. Hollows produced by the irregular deposit of boulder-clay left by the melting of an ice-sheet form lakes in regions where rainfall exceeds evaporation. Rock-basins (§ 339) contain the typical Alpine lakes of mountainous regions. Slow upheaval of the end of a valley, subsidence of a plain, or the collapse of caverns, are also processes of lake formation, and the craters of extinct volcanoes often collect a large quantity of rain, forming lakelets with neither inflow nor outflow.

333. **Great Lakes of the World.**—The Caspian Sea is the largest lake, and a typical example of a hollow isolated by upheaval of surrounding land. Lake Superior comes next in size and is the largest fresh lake. Lake Baikal in Asia, at an elevation of 1360 feet above the sea, is the deepest known lake, the maximum sounding obtained in it being almost 800 fathoms. The highest lake yet measured is Askal Chin in Tibet, 16,600 feet, and the lowest is the Dead Sea, 1290 feet below sea-level.

THE LARGEST LAKES

Name.	Situation.	Height above Sea.	Area. Square Miles.	Depth. Max. Fms.
Caspian	Eurasia	—90	170,000	500
Superior	North America	600	31,200	168
Victoria	Africa	3300	26,900	...
Aral	Asia	150	26,200	37
Huron	North America	580	23,800	117
Michigan	,, ,,	580	22,400	145

334. **Function of Lakes.**—When water begins to flow over a new land surface, either freshly upheaved from the

bed of the sea or remodelled by the deposit of boulder-clay, it necessarily forms a series of lakelets which overflow into one another by streams. As the river system cuts its channels more deeply the smaller hollows are either drained or filled up and remain as meadows along its course. The abundance of fresh lakes is a testimony to the comparative newness of the land surface and to the early stage of evolution of its rivers. A river issuing from a lake cuts down the lip it flows over very slowly, except when the barrier is soft clay, as all the sediment which gives to running water the properties of a file is dropped on entering the lake. Lakes thus act as filters for rivers. The exquisite deep blue colour for which the lakes of Northern Italy and Switzerland are famous is due to the scattering of light from the fine flakes of mica brought in by glacier rivers and suspended in the water. This deposit tends to gradually fill up the lake. The fans of alluvial deposit laid down by each inflowing stream grow into deltas; and flat meadows encroach on the water so rapidly that lawsuits are occasionally required to determine the ownership of the new land. Lakes regulate the flow of rivers by keeping up their supply in times of drought, and checking floods during rain. For example, if a river $\frac{1}{10}$ of a mile wide passes through a lake of 100 square miles in area, 10 miles from the sea, and a flood takes place in the upper stream which, if passed on directly, would raise the level of the lower 10 miles by 25 feet, and so produce a disastrous flood, the immediate effect is to raise the level of the lake 3 inches, causing a very slight increase of the lower stream.

335. **Salt Lakes.**—In arid regions, where evaporation is in excess of rainfall, rivers flowing into hollows of the Earth's crust may fail to fill them up to the brim, and lakes will thus be formed with no outlet. These are necessarily salt, on account of the evaporation of the river-water, and the salts contained differ (§ 221) from those of the sea. Analysis of the water of salt lakes shows this to be usually the case; but the salts of the Caspian are very similar to those of ocean water, indicating that it is part of the sea

cut off by a geologically recent elevation of the land. Yet its salinity is less than 2 per cent, while that of the sea averages 3.5. This is because the shelving shores, and particularly the wide shallow inlet of Kara-Baghas, act as natural salt-pans, evaporating the thin layer of water covering them and causing a deposit of crystalline salt, which is thus being gradually withdrawn from solution, while the evaporation is made good by a continual supply of fresh river-water. On account of the excess of evaporation the surface of the Caspian is now about 90 feet below sea-level, and its shores form a sunk plain. The Jordan Valley, in an equally rainless region, is a still more remarkable instance of a sunk plain. The Sea of Galilee is a small lake 600 feet below sea-level, and from it the Jordan flows for 100 miles along the line of a great fault in a valley averaging 7 miles in width, and enters the Dead Sea at a level 1290 feet below that of the Mediterranean.

336. **Ice Action.**—The snow-fields lying on the high parts of mountain ranges above the snow-line (§ 163) continually increase by the condensation of vapour from the atmosphere. The weight of the accumulating mass of snow compresses the lower layers, squeezing out the air, and forming compact ice, which, although one of the most brittle substances to a blow, is plastic when subjected to steady pressure. *Glaciers* or streams of ice flowing down the slopes prevent an excessive accumulation of snow on high mountains. The cause of the plasticity of ice under pressure is usually considered, following Professor J. Thomson's theory, to be that pressure lowers the melting-point (§ 72), allowing the lower layer of the mass to liquefy and adapt itself to the surface it rests on; the relief of pressure thus afforded allows the water to solidify again, and the process is repeated continually. But since ice at very low temperatures is plastic, though in a less degree, the theory of melting by pressure is not a sufficient explanation. Messrs. M'Connel and Kidd have recently made experiments which show that while crystals of ice are individually rigid and brittle, a mass of them frozen together is plastic even at low temperatures, the crystals apparently

sliding over each other.[2] Since glacier ice is known to consist of grains or lumps (from the size of a pea to that of a melon), each of which is a single crystal, the flowing of glaciers can be readily explained. Part of the accumulated snow on a mountain slope is got rid of by *avalanches* or snow-slips, which are powerful erosive agents, breaking through everything in their path.

337. **Glaciers,** although solid, flow like rivers, the centre and surface moving nearly twice as fast as the sides, which are retarded by friction with the valley. Compared with rivers their motion is very small. The Mer de Glace, the most famous glacier in Switzerland, creeps at the rate of about an inch an hour in the centre during summer, and only half as fast in winter. Some of the great glaciers of Greenland move much faster, advancing from 50 to 60 feet in a day, although 20 feet is a more common rate. The thickness of glaciers in the Alps often exceeds 1000 feet, and their length averages about 5 miles; the longest is the Aletsch Glacier, which measures 15 miles, including the parent snow-field. As a glacier descends along the valley, stones, clay and sand loosened by erosion fall from the slopes, and rest as huge heaps of rubbish, called *lateral moraines*, along each side of the ice. When two glaciers traverse convergent valleys the lateral moraines on one side of each coalesce to form a *medial moraine* (see Fig. 54) down the centre of the united ice-flow. In time a great glacier carrying the ice of many tributaries becomes roughened with numerous parallel ridges of rock rubbish along its length. The heat of the Sun in summer continuously melts the ice, except where it is protected by the overlying moraines, which thus stand out prominently on the surface. Isolated blocks of stone similarly protect and remain perched on ice pillars, while the general surface is being lowered. As a glacier forces its way along an irregular valley the ice is severely strained, and cracks or *crevasses* result, which are narrow and close at first, but gradually widen out in consequence of the centre moving more rapidly than the sides. Huge clefts are thus formed extending through the ice from surface to bottom, and swallowing up masses of moraine rubbish.

S

Some change in the channel alters the stresses, and as time goes on the old crevasses close up and new ones open.

FIG. 54.—Map of a Glacier showing the formation of medial moraines, by union of tributary glaciers. The arrows show the direction of flow, and the lines radiating from the edges represent crevasses.

The regions where glaciers occur are coloured dark blue on Plate VII.

338. **Glacial Work.**—Glaciers work both by transporting the moraine material that falls upon them and by powerfully eroding the ground they pass over. Moraine rubbish falling down crevasses gets wedged in the ice, which presses the angular stones firmly against the bed-rock as the glacier slides forward, the action exactly resembling the cutting of glass by a diamond. Immense quantities of sand and clay result from the grinding down of rock and stones, and are carried along the bed of the glacier, forming the *ground* or *bottom moraine* or boulder clay. When the climate of a glacial region grows warmer, as the Alpine district has been doing for the last twenty years, the glaciers melt away at the lower end, which shrinks up the valleys, while the boulders which may have been carried far by the ice are deposited on the slopes amongst rocks of an entirely different nature, and sometimes in very precarious positions. Such travelled and perched blocks are called *erratics*, or simply boulders. The rocks of the valley uncovered by the ice are seen to be deeply grooved or striated by the stones dragged over them, the run of the striæ showing the direction in which the glacier was moving. The surface scratched

by sharp stones is at the same time finely polished by the clay, and thus acquires a highly characteristic appearance. The general aspect of the smoothed and rounded rocks is supposed to resemble the backs of sheep, hence the peasants named them *roches moutonnées, i.e.* sheep rocks. The stones which took part in the polishing action, and remain embedded in the clay, are themselves scratched and smoothed in a similar way. The descent of a glacier in a steep valley is believed by some geologists to give it an impetus which causes the mass of ice to dig like a gouge when it enters suddenly on flatter ground. To this gouge action, strongest at the first shock, and then gradually diminishing, the peculiar form of the rock-basins of alpine lakes and fjords is usually ascribed. The deep weathered crust (§ 310) which forms on granite and other hard rocks is readily scooped out, and its presence doubtless helped in the formation of deep rock-basins. When the climate admits of glaciers reaching the sea they give rise to icebergs (§ 234), and distribute their deposits far over the bed of the ocean. At the end of a glacier on land the ground moraine forms a ridge of boulder clay, and the various moraine heaps carried along on the surface of the ice are thrown down above it, producing what is called the *terminal moraine.* A diminishing glacier in a climate that is growing warmer strews the whole valley, up which it has retreated, with consecutive terminal moraines made up of low hills of detritus. From the melting end of a glacier a rapid stream of ice-cold water flows away, milky with mud, which imparts to it great erosive power. The amount of sediment removed by the Isortek River in Greenland from the base of its parent glacier is calculated at 4,000,000 tons a year.

339. **Rock-basins** are usually long and narrow, and attain a maximum depth, often of several hundred fathoms, at a point about one-third of the distance from the head of the basin, as shown in Fig. 55. The lakes occupying rock-basins are characteristic of the valleys on the lower slopes of all mountains which once bore great glaciers. By subsidence of the coast-lands they form fjord-basins (§ 229) filled with

sea-water. On the west coast of Scotland Loch Morar, a fresh-water lake 178 fathoms deep, with its surface 30 feet above sea-level, is connected with the sea by a short river.

FIG. 55.—Section of Loch Goil, a typical rock basin, the slopes exaggerated 10 times. The upper line shows by its varying thickness the true slope of the bed of the basin.

Loch Etive, exactly similar in configuration but filled with sea-water, and only 80 fathoms deep, has its sill so near the surface that, although it is in free communication with the sea at high tide, the current rushing out at low tide forms a veritable waterfall. Loch Nevis, with a depth of 70 fathoms, has its sill 8 fathoms below the surface. The gigantic Sogne Fjord in Norway, more than 100 miles in length, is a rock-basin with a maximum depth of 700 fathoms.

340. **Ice-caps.**—In very cold climates, where the snow-line approaches sea-level, the whole surface of an extensive region may be covered by snow to such a depth that it is compacted into ice, filling up all the valleys and standing high over the mountains; such a covering is called an ice-cap. Greenland is covered with an ice-cap presenting a shield-shaped surface, which Dr. Nansen in his adventurous journey across the peninsula in 1888 found to be about 10,000 feet above sea-level, and nearly flat in the interior, sloping rapidly to the sea on each side. The weight of this shield of ice is always squeezing out its edges in the form of glaciers to the sea, and there is probably a constant though very slow outward movement of the ice from the centre over the hills and valleys of the deeply buried land. The Antarctic continent appears to be covered with a still larger and probably thicker ice-cap, regarding which little information has been obtained, except that the glaciers from it give rise to fleets of immense flat-topped icebergs (§§ 234, 276).

REFERENCES

[1] Osborne Reynolds, "On Model Estuaries," *British Association Reports*, 1889, p. 328, and 1890, p. 512.

² J. C. M'Connel and D. A. Kidd, "The Plasticity of Glacier and other Ice," *Proc. Roy. Soc.* (1888), xliv. 331, also (1890) xlviii. 259; *Nature*, xxxix. 203.

BOOKS OF REFERENCE

See end of Chapter XIV.

CHAPTER XIV

THE RECORD OF THE ROCKS

341. **Looking Backward.**—Two opposed agencies now at work on the Earth's surface—internal energy ridging up the crust, and solar energy cutting down the heights—are sufficient, if they have been long enough in action, to account for all the features of the land. The Uniformitarian School of geologists holds that the Earth has attained its present condition after passing through vast ages of change so slow as to be almost imperceptible. The other school, sometimes called that of the Catastrophists, affirms that the processes at work in past time were quite different from those of the present, being much more violent and uncertain in their action. They look on valleys as rent in the solid rock by Earth movements of titanic strength, and on mountain ranges as elevated to their full height in a single stupendous heave of the strata. Erosion is considered only to trim off the broken edges, as a plane smooths down the signs of the rough rending of a saw. Modern research shows that the truth lies between the two extremes. The Earth, like any other cooling body, must be cooling less and less rapidly as time goes on. When the crust was first formed its high temperature must have considerably increased the erosive power of water. So, too, tidal friction, now insignificant, must once have been a tremendously powerful agent in shaping the surface (§ 104). Thus, while the processes at work have been the same in kind as the Uniformitarians prove, the energy available for

DRAINAGE AREAS OF C

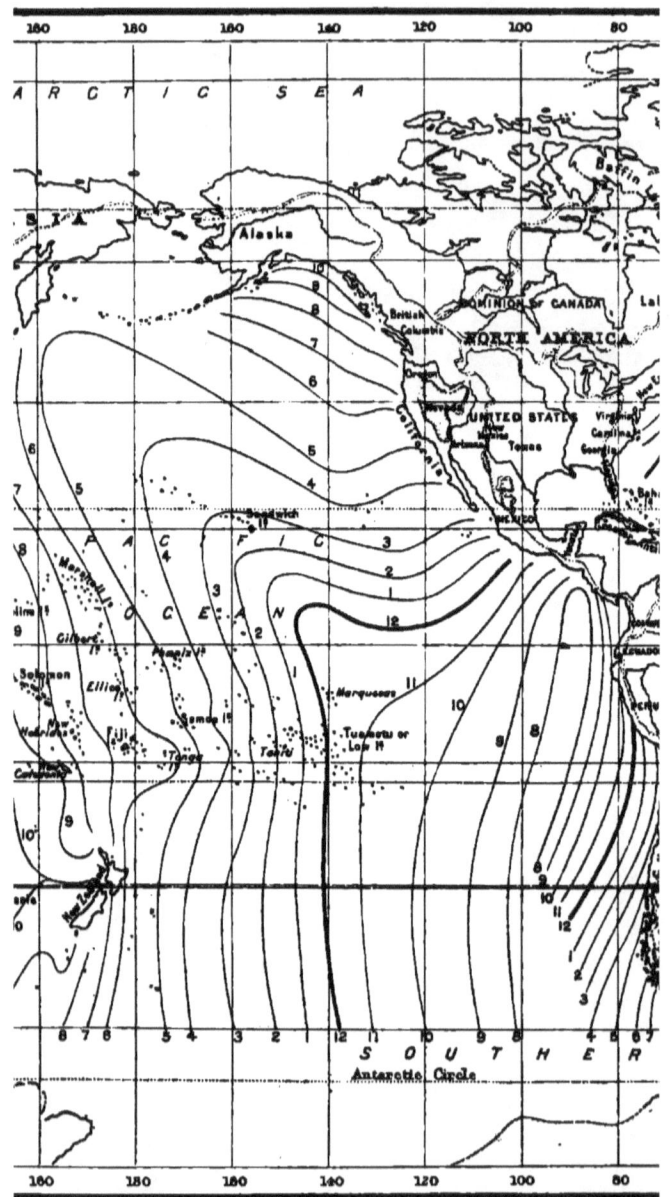

THE DRAINAGE AREAS are coloured according to the Oceans which they drain

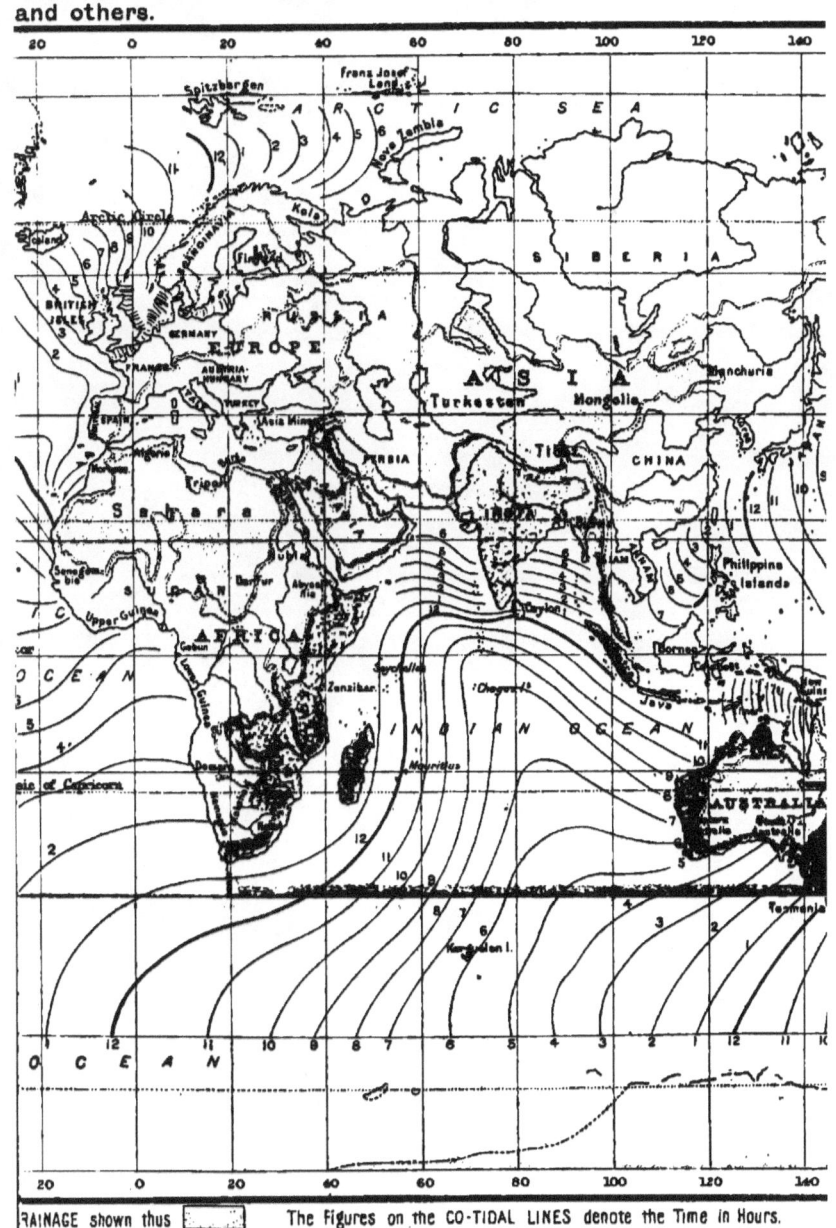

ND CO-TIDAL LINES OF OCEANS.
and others.

RAINAGE shown thus   The Figures on the CO-TIDAL LINES denote the Time in Hours.

the work in a given time was once much greater than now, as the Catastrophists maintain. Reasoning from the rate of cooling of lava, Sir William Thomson estimated that living creatures such as now exist could not have inhabited the Earth more than 100,000,000 years ago; and Professor Tait, calculating from the rate at which the Earth is losing heat (§ 291) and its present temperature, concludes that 20,000,000 years is more nearly the truth, while even 10,000,000 years may include the whole range of possible life on the globe.

342. **Reading the Rock Story.**—If exactly the same areas of the Earth's surface were always subject either to elevation or depression, we could not discover from the rocks laid bare on the surface any record of the process of their formation. The sedimentary rocks would remain in the subsiding hollows, the older layers being successively covered by newer ones. But it happens that the margins of the world ridges on which sediment is deposited are subject to frequent elevation and depression (§ 303), and the sedimentary rocks which are exposed bear traces of these changes which it is the special study of geologists to interpret. Where rocks are very much crumpled and folded, it often happens that the strata have been inverted, the bottom beds of a series having been folded back upon the upper beds. When a stratum occurs resting on a different sort of rock, which dips in a different direction or bears signs of ancient erosion, the two are said to be *unconformable*. This structure is clearly indicative of some time having elapsed since the formation of the older series, and before the accumulation of the overlying younger beds. The stratified rocks are like the sheets of an unbound book, some of which have been printed over a second time with a later part of the work; many have been crumpled, torn, and rubbed so that they are illegible; the numbering of all the pages except the last one has been destroyed, and there are evidently places where several pages together have been dropped out. By reading the legible portions of such a book one could find hints of the development of events if the mutilated work were a history, or of the unfold-

ing of the plot if it were a novel. A few consecutive pages found in their proper order would give a key to arranging the rest, and although uncertainty as to the precise sequence of some parts of the narrative would remain, the patient reader could in time obtain a fair idea of the nature and order of the contents. If it is possible to find a narrative showing a regular development of events written in characters with which we are familiar on the sheets of rock, the order and circumstances in which these rocks were formed can be got at, however confusedly they may now lie. Sedimentary rocks are full of picture-writings recording the history of successive races of living creatures, and the writings are very legible, being the actual mummies or casts of the creatures themselves.

343. **Fossils.**—All remains and traces of living creatures preserved in rocks are called fossils. Some of the traces are only footprints, or worm-tracks that have been impressed on an ancient surface of clay or wet sand, and after hardening have been filled in by finer sediment. Plants and animals are usually represented only by their hardest parts, such as bark, shells, teeth, or bones. But often the whole organism was surrounded by compact sediment, in which, as it decayed away, a hollow was left exactly corresponding with its outer surface. This mould became filled in turn with fine sediment, or impregnated with carbonate of lime or silica deposited from solution in the water which percolated through, and thus a perfect cast or model has been produced. The most complete fossils preserve not only the external form but the minutest internal structure, every part being individually turned into stone by the exchange of animal or vegetable substance, molecule by molecule, for some mineral such as pyrite (sulphide of iron), calcite (carbonate of lime), or one of the many forms of silica. Other fossils are simply shells or skeletons closely compacted together, such as chalk, made up of foraminifera like the deep-sea oozes (§ 275), coral limestone (§ 280), and siliceous earth composed of the cases of diatoms. Sometimes organic substance undergoes only partial decomposition while retaining much of its original form.

Coal, for example, is the residue of partially decomposed vegetation.

344. Interpretation of Fossils.—As a general rule it is assumed that the creatures whose remains occur in the rocks were similar in their habits to those now living, and were in an equal degree dependent on the climate. Rocks formed of the sediment of lakes and rivers may, by the greater abundance of land creatures amongst their fossils, be distinguished from those composed of marine deposits. These inferences are often confirmed by the nature of the rocks themselves, the fine mud of estuaries naturally yielding a shale, while the pebbles of an exposed seashore are compacted into a conglomerate. Rocks containing the remains of the same species of creatures have evidently been formed under similar physical conditions, and possibly at the same time; hence they are said to belong to the same geological horizon.

345. Divisions of Sedimentary Rocks.—There is so much scope for individual opinion in interpreting the record of the rocks that no minute classification of them meets the approval of all competent geologists, but a few comprehensive divisions are generally accepted. The most ancient sedimentary rocks are allowed to be those containing fossils exclusively of the simplest form of life. The variety and complexity of the organisms found, usually increase as the more recent strata are approached. The greatest thickness of a bed of sedimentary rock may in some cases give a rough measure of the shortest time it could have taken in formation, but all attempts at fixing a definite geological chronology have as yet been unsatisfactory. The great divisions of rocks and their more important subdivisions are given below in the order of antiquity, and some typical forms of life are mentioned.

QUATERNARY
RECENT—Now forming.
PLEISTOCENE—All modern plants and animals. *Man*.

TERTIARY
PLIOCENE—Most modern plants. *Elephant, Ox.*

MIOCENE—Tropical plants. *Ape, Antelope.*
OLIGOCENE—Tropical plants.
EOCENE—Tropical plants. *Palæotherium, Lemur.*

SECONDARY

CRETACEOUS—Flowering plants. *Foraminifera, Marsupials, Toothed Birds.*
JURASSIC—Ferns. *Saurians, Marsupials, Archæopteryx, Corals, Ammonites, Cuttlefish.*
TRIASSIC—Cycads. *Ammonites, Reptiles.*

PRIMARY

PERMIAN—*Amphibians.*
CARBONIFEROUS—Lycopods. Tree-ferns, Conifers, *Crinoids, Fishes, Amphibians.*
DEVONIAN AND OLD RED SANDSTONE—*Fishes, Brachiopods,* Lycopods.
SILURIAN—Sea-weeds. *Graptolites, Trilobites, Fishes.*
CAMBRIAN—*Trilobites, Sponges.*
ARCHÆAN—No forms of life known with certainty.

346. Older Primary Rocks.—The primary division is called the *Palæozoic*, as in it the fossils of the earliest living creatures are preserved. The Archæan, which forms the foundation rocks, consists mainly of crystalline schists. Wherever these appear on the surface we know that the land is of extreme antiquity, for it must either have remained above the sea while all the other formations were being deposited elsewhere, or if it was upheaved after being covered with younger rocks, the period must yet be sufficiently remote to have allowed all the more recent strata to be eroded away. No fossils are known with certainty in Archæan rocks. The Cambrian, Silurian, and Devonian systems, named after the districts in south-western Britain where they were first studied, were formed in successive periods. Fossils of sea creatures are abundant in these rocks; a peculiar crustacean called the trilobite swarmed in the Silurian seas, and seems to have become altogether extinct before the end of the Primary period. The earliest land-plants, which were cryptogams, leave a record in the Upper Silurian rocks. In the Old Red Sandstone rocks which were laid down as sediment in fresh-water lakes in the Devonian period, fossils of fishes clad in enamelled bone and of scorpion-like creatures appear.

347. **The Carboniferous System** is composed of thick beds of limestone, which must have been deposited at the bottom of a clear shallow sea, of sandstones laid down on ancient beaches, and of shales which represent the solidified mud of estuaries. The name Carboniferous comes from the beds of coal which result from the decay of bark, fronds and spores of club-mosses, and tree-ferns of giant size, on the swampy margin of the ancient sea. Clay-beds usually underlie coal-seams, and represent the soil in which the carboniferous plants grew, being often full of the fossil roots. The formation of coal is an interesting example of chemical decomposition. The action of heat and pressure on vegetable matter in the absence of air is to drive out more and more of the oxygen, nitrogen, and hydrogen it contains, combined with very little carbon. The following table gives the average composition (omitting the ash) of dry wood; peat, which results from vegetation decaying in recent formations; lignite, a woody form of coal found in tertiary rocks; true coal; and anthracite, which is apparently derived by heating coal. It is conjectured that the final product of this process is the diamond, which is pure crystallised carbon.

CARBONIFEROUS MINERALS

	Wood.	Peat.	Lignite.	Coal.	Anthracite.
Carbon	50	60	67	85	94
Hydrogen	6	6	5	5	3
Oxygen and Nitrogen	44	34	28	10	3
	100	100	100	100	100

The great limestone beds of the Carboniferous period are composed of the remains of crinoids, mollusca, and many other marine creatures. Amphibians mostly small, but some of great size crawled through the marshes, but the only true land animals preserved are of the nature of scorpions, insects, and snails.

348. Newer Primary Rocks.—In the Permian period, named after the Russian government of Perm, where the rocks of this age are greatly developed, plant life appears to have been less abundant and varied than in Carboniferous times, but remains of great amphibians abound, and those of true reptiles appear for the first time. Palæozoic rocks sometimes exert a considerable local influence on a freely suspended magnet (§ 98). In the course of a magnetic survey of the British Islands, Professors Thorpe and Rücker recently found a line of magnetic disturbance running across the comparatively recent strata of southern England, coincident with a deeply buried mass of Palæozoic formation running from the old mountains of Wales toward the Carboniferous region of the continent of Europe, the existence of which had previously been inferred from geological evidence.[1] In 1890 this conclusion was strikingly confirmed by the discovery of coal in a very deep boring through the tertiary rocks of eastern Kent.

349. Secondary Rocks.—The secondary rocks are termed *Mesozoic*, because they contain evidence of the existence of living creatures intermediate between those of the Primary period and of the present time. In the Trias there are signs of gigantic amphibians, reptiles of the crocodile kind, and of the simplest forms of mammals, the marsupials. The Jurassic system takes its name from the Jura Mountains, and is sometimes known as Oolitic (egg-stone), from the granular limestones resembling the structure of a fish-roe, by which it is characterised. Many beds of limestone of this period are fossil coral-reefs. The most abundant mollusca were the ammonites, with wonderful rolled shells, and cuttle-fishes. Saurians—reptile-like animals—grew in those days to an enormous size, and inhabited air, sea, and land. The Pterodactyls were small reptiles with wings not unlike those of a bat in appearance. Ichthyosaurus and Plesiosaurus were swimming reptiles, sometimes 40 feet in length, and the land reptiles were probably the hugest animals that ever inhabited the globe—the remains of the Atlantosaurus, discovered in North America, indicating a length of 100 feet and a height of 30 feet.

Archæopteryx, the first bird-like creature, appears in the Jurassic period. The Cretaceous or chalky rocks are largely composed of solidified globigerina oozes, and innumerable shell-bearing sea creatures occur amongst them. Fishes like the herring and salmon appear for the first time, and huge reptiles and birds with teeth were common. Traces of the flowering plants also appear amongst the prevailing ferns.

350. **Tertiary Rocks.**—A great gap generally separates the period of the Mesozoic rocks from that of the *Cainozoic* or Tertiary. During the interregnum the great reptiles and ammonites became extinct, and forms of life appeared more closely resembling those of the present day. The divisions of tertiary rocks—Eocene, Oligocene, Miocene, and Pliocene—were originally arranged in the order of the abundance of the fossils of mollusca, resembling those now existing. As the period progressed plants and animals which approached more and more closely to those we now know appeared on the Earth. Foraminifera attained a great size and were extremely numerous, one being the large coin-shaped nummulite which makes up many of the limestones. Mollusca like the oyster and snail began to predominate over those of the cuttle-fish kind. Amongst the mammals the marsupials became less numerous, and many transition forms of the Eocene approach the carnivorous type. Later, gigantic ant-eaters, the elephant-like Mastodon, pig-like animals, antelopes, and apes appeared. A succession of animals of increasing size, approaching nearer and nearer the nature of the horse, runs through the series, culminating in the true horses of the Pliocene age. The fossils of these large animals are never so complete as those of mollusca or ferns, some teeth, or a few shattered bones, being all that is usually found. The Tertiary period was characterised by great volcanic activity in all parts of the world, and the existing scenery of many lands is due to the effects of denudation on the basalt sheets and lava dykes of the old volcanoes.

351. **Quaternary Rocks.**—The post-Tertiary or Quaternary rocks are the least ancient of all. They are rarely

even consolidated, consisting chiefly of clays and sands. The Pleistocene formation in Northern Eurasia and America consists almost entirely of boulder clay, the result of ice-action, and the period has been termed the Great Ice Age. Many exposed rock surfaces on the mountain-tops as well as in valleys, in places where glaçiers have never been seen, closely resemble the *roches moutonnées* of Switzerland (§ 338). Perched blocks are scattered thickly over all parts of Northern Europe and America, and from their nature many of them are known to be far travelled. The conclusion is irresistible that after the formation of the last tertiary rocks these lands were subject to ice-action. Great and wide-spread subsidence, and subsequent elevation of the land took place during this period. Some writers, among whom is the Duke of Argyll, maintain that the boulder clay, perched blocks, and ice-scratchings were brought about by this subsidence permitting fleets of icebergs sailing southward to strand or rub against surfaces which were afterwards elevated. To most geologists, however, the evidence of true glacier action having occurred over the whole area is overpowering, although the period is so remote that atmospheric erosion has in many cases obliterated the work of ice.

352. **The Great Ice Age.**—Glaciation probably occurred on the grandest scale, the ice marching over mountain and valley with little regard to the form of the surface. In the Glacial period it appears that all Northern Europe and Northern America (see light blue tint on Plate VII.), were covered by vast ice-caps, thicker than that now overspreading Greenland, which polished and smoothed off the mountains, and covered the valleys and plains with layers of boulder clay. The ice seems to have spread beyond the margin of the land, to have hollowed out deep furrows across the Continental shelf, and sometimes even to have ploughed up the shallow sea-bed and scattered the sand and shells on the coast-lands. Professor James Geikie points out that the Great Ice Age was divided into periods during which the climate was very severe, while between them a genial climate prevailed, and interglacial beds of peat were formed containing a varied vegetation and the remains of insects

and mollusca. The cause of changes of climate, sufficient to produce such effects, has been the subject of much speculation. The late Dr. Croll, whose theory is now most widely received, pointed out that the changes in the eccentricity of the Earth's orbit (§ 109) combined with the precession of the equinoxes (§ 115), must have produced a severe climate in the northern hemisphere at the period when aphelion occurred in the northern winter, and the eccentricity was at a maximum. If Croll's theory is true, cold periods must have occurred in all geological epochs. Erratic blocks and glaciated stones found in many different formations seem to confirm it, but no sign has been found of such extensive ice-action as characterised the Pleistocene. This may be accounted for by the probable absence in those remote periods of continental areas sufficiently extensive to support a great ice-sheet. Some geologists account for the changes of level during this period by supposing that the great ice-sheet depressed the elastic strata by its weight, producing extensive subsidence, followed by upheaval when the ice-cap melted. Others explain raised beaches (§ 284) on the assumption that the land remained rigid and the mass of ice raised the level of the ocean by attraction (§ 252). In the river and cave accumulations of the Pleistocene age the first undoubted signs of the human race appear in the form of coarse chipped stone implements and rough etchings on bone of extinct or no longer indigenous animals.

353. **Evolution of Continents.**—Rocks of Archæan and Palæozoic age cover a greater area on the Earth than those of Mesozoic age, which are in turn more extensive in their distribution than those of the Tertiary system. This shows that more of the elevated half of the globe was covered by the sea, in which sediment accumulated, in Palæozoic than in Mesozoic, and in Mesozoic than in Tertiary times. It is pointed out by Professor J. Geikie that the elevated and depressed halves of the World have been growing more and more distinct throughout geological ages, and as the Abysmal Area has grown deeper and the World Ridges higher the superficial extent of the hydrosphere has been steadily diminishing, although its volume remains the same.[2] This change must

be looked on as a general result of innumerable minor elevations and depressions. The following hypothesis of the growth of continents is not to be looked on as an established theory, but as a probable conjecture of the relative order in which the various land-masses were formed. Plate XIV., adapted from Professor J. Geikie's maps, shows in the deepest tint the areas of the World Ridges that are believed (although the evidence is far from complete) to have projected above the hydrosphere during the greater part of the period when Palæozoic rocks were being formed. They composed groups of great islands clustered on the northern and scattered over the southern parts of the World Ridges, between which warm ocean currents would flow from the equatorial seas, and an equable climate would reign over the whole land. In the Mesozoic period the lands (shown in the second tint) were far more extensive, but insular conditions still prevailed. The deepened Abysmal Area drained the oceans from the summits of the World Ridges, and the up-ridging of the Continental Area raised wide tracts far above the sea. The western and eastern edges of the great Eastern World Ridge were clearly outlined, but the sea spread across its central portion from east to west, and from north to south. The Western World Ridge was developed similarly, land extending along its western and eastern edges in North America, separated by a wide sea-channel from south to north, while in the South American portion the central part of the existing continent had appeared running almost from north to south. In the Tertiary period there was an enormous increase of upheaval over the World Ridges, and the crests of them (lightest brown on map) everywhere emerged. The sea still swept over the central part of the Eastern World Ridge from north to south and south-west, so that the Indian Ocean was united with the Arctic Sea, and through the wide Mediterranean with the Atlantic. Africa and Australia were almost as extensive as at present. Britain was separated from Scandinavia, and the south of Europe formed a mountainous archipelago, amongst the islands of which the Alps and Balkans were conspicuous. The

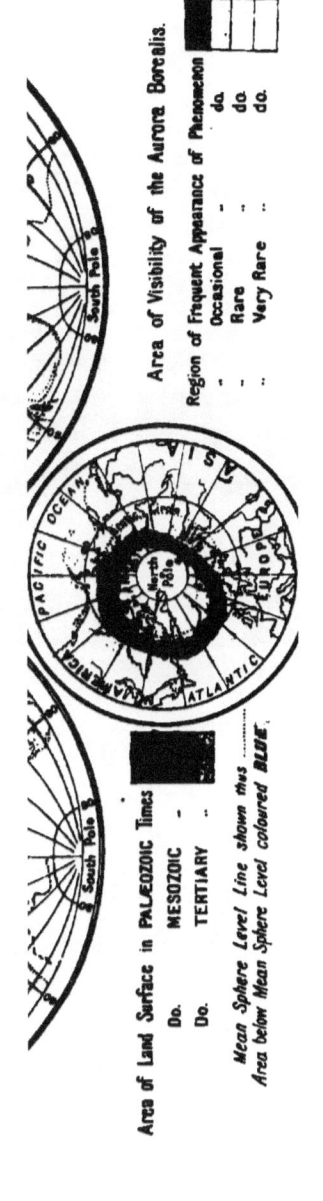

Indian peninsula was still an island, and the Himalayas were beginning to appear. The Western World Ridge was nearer completion, North America was almost all above water, and the line of the Andes was commencing to give outline to South America. By the close of the Tertiary period the elevation of the continents had been practically completed.

REFERENCES

[1] Thorpe and Rücker, "Magnetic Survey of British Islands," *Philosophical Transactions*, vol. clxxxi. (1890, A), 53. See also *Good Words*, 1890.

[2] J. Geikie, "The Evolution of Climate," *Scottish Geographical Magazine*, vi. 57 (1890).

BOOKS OF REFERENCE

A. Geikie, *Text-Book of Geology*. Macmillan and Co. (A complete discussion of Geology from the modified Uniformitarian standpoint, with references to important original papers.)

J. Prestwich, *Geology*. Clarendon Press. Two volumes. (An admirable treatise from the modified Catastrophist standpoint.)

J. Geikie, *Outlines of Geology*. Stanford.

A. Geikie, *Class-Book of Geology*. Macmillan and Co.

A. H. Green, *Physical Geology* (second edition).

A. Geikie, *Scenery of Scotland*. Macmillan and Co. (A fascinating account of the origin of surface features.)

J. Geikie, *Great Ice Age*, and *Prehistoric Europe*. Stanford.

N. S. Shaler, *Aspects of the Earth*. (Suggestive essays.) Smith, Elder and Co.

A. J. Jukes-Browne, *Building of the British Islands*. Bell and Son.

J. W. Judd, *Volcanoes*. International Science Series.

J. Milne, *Earthquakes*. International Science Series.

T. Mellard Reade, *Origin of Mountains*. Taylor and Francis.

J. Croll, *Climate and Time*. A. and C. Black.

CHAPTER XV

THE CONTINENTAL AREA

354. Crest of the World Ridges.—(Read §§ 214, 251, 255, 256.) The five largest islands or peninsulas in which the crests of the World Ridges break through the uniform covering of the hydrosphere are termed continents, and designated by the names Eurasia, Africa, North America, South America, and Australia. They are distinguished from other islands and peninsulas by size alone, Australia being ten times larger than New Guinea, and Africa ten times larger than Arabia, these being the greatest island and peninsula not called continents. The elevated region round the South Pole is crowned by the unexplored and scarcely discovered continent of Antarctica. The land mass of Eurasia is conveniently supposed to consist of the two "continents" of Europe and Asia, and if this be allowed, we find that the six known continents group themselves into three pairs. North and South America share the Western World Ridge; Asia and Australia, on the eastern limb of the Eastern World Ridge, lie diametrically opposite; while Europe and Africa occupy the western limb of the Eastern World Ridge, diametrically opposite the great Pacific basin. Until the Tertiary period, when the heights of Central Asia were upheaved, the Indian Ocean stretched to the Arctic Sea; and even in Quaternary times Europe and Asia were separated by a broad channel of water between the Mediterranean and the Arctic Sea. The prevailing continental form is a south-pointing triangle. In each pair of continents the northern has a wide extension from east to west, a deeply indented coast, and a great group of islands on the south-east stretching toward the unindented coast of the southern member, which, as a rule, extends from north to south, and has an island or island group lying to the south-east.

355. **Comparison of the Continents.**—By studying the maps (Plates XI. XII. and XIII.) and the following tables the student will be able to compare the characteristics of the separate continents. The average heights in Table A are those calculated by Dr. John Murray, from whose figures also the relative areas at various elevations (Table C) are derived.[1] The distance from the sea of the continental centre or position farthest from the coast is that calculated by the Russian, General von Tillo; the figure for Europe is not strictly comparable with the others, since Europe is widest at its junction with Asia. Professor Krümmel, a leading German oceanographer, has calculated the percentage of surplus coast given in Table A. Since a circle has the smallest boundary of any figure of the same area, if we imagine the coast-line stripped off a continent like braid off a coat, and the continent moulded into a circular outline without change of area, a smaller length of coast would serve to surround it. The length of coast left over, is expressed as percentage of the original length, and serves as a measure of the surplus available for bordering peninsulas and bays. In the three northern continents, it will be noticed, more than two-thirds of the coast-line are thus available; in the three southern continents less than one-third. Table B, calculated by Dr. Rohrbach,[2] gives the percentage of each continent lying within certain zones of distance from the coast, and is thus a measure of their accessibility from the sea (compare Plate XII.) The chief mountain ranges of each continent are marked by red lines on Plate XVIII.; this should be compared with the orographical map (Plate XI.), on which plains and plateaux are more clearly shown.

Comparison of the Continents

Continent.	Asia.	Africa.	N. America.	S. America.	Europe.	Australia.	All Land.
TABLE A.—Area, Elevation and Coast-line							
Area (million sq. miles)	16.4	11.1	7.6	6.8	3.7	3.0	55.0
Average height (feet)	3000	2000	1900	2000	940	800	2100
Highest point (feet)	29,000	18,800	18,200	22,400	18,500	7200	29,000
Surplus coast (per cent)	61.7	28.3	64.6	32.6	87.6	30.6	...
Distance of continental centre (miles)	1616	1119	1057	1057	810	591	...
TABLE B.—Percentage of Continental Areas within Equidistant Zones from Coast							
0-125 miles	22.9	18.4	31.6	24.1	46.4	35.7	25.8
125-250 ,,	14.8	16.2	21.1	20.1	21.0	25.5	17.9
250-375 ,,	11.7	14.5	15.7	15.5	14.2	21.7	14.4
375-500 ,,	9.3	13.1	11.5	12.4	7.7	13.5	11.1
Mean distance from coast (miles)	485	420	295	345	210	215	380
Percentage of area under mean distance	60	53	58	56	62	55	...
Do. over mean distance	40	47	42	44	38	45	...
TABLE C.—Percentage of Continental Areas within Zones of equal Altitude above Sea							
Below sea-level	1.4	0.1	0.05	0.0	1.8	0.0	0.6
0-600 feet	23.3	12.5	32.25	40.0	53.8	29.8	26.7
600-1500 ,,	16.0	34.8	32.1	26.8	27.0	64.3	27.8
1500-3000 ,,	21.7	27.6	13.3	16.8	10.0	4.1	19.3
3000-6000 ,,	21.8	21.8	13.2	7.0	5.5	1.5	17.0
6000-12,000 ,,	10.0	2.8	8.4	5.0	1.7	0.3	6.0
Above 12,000 feet	5.8	0.4	0.7	4.4	0.2	0.0	2.6

356. **Continental Slopes.**—The simplest conceivable continent would consist of two land-slopes meeting, like the roof of a house, along a central line or axis, so that a section across it would resemble A, Fig. 56. The axis of a continent is usually formed by a mountain range of eleva-

FIG. 56.—Typical Section of a Continent. In BCD the short slope is shown to the left, the long slope to the right.

tion (§ 303), which most frequently occurs near the edge of the slope of the world ridge, and consequently near one side of the continent, so as to produce a short slope on one side and a long slope on the other, giving a section like B. A mountain chain is rarely single, and is about equally steep on both sides. It occupies a narrow strip of a continent; so while the short slope of the continent is nearly uniform to the sea, the long slope is broken into a steep and a gentle portion, giving the section C. But since both sides of a continent have been ridged up, a lower and broken mountain range usually intervenes between the long slope and the sea, converting the central part of the continent into a wide valley, and forming a second short slope to the seaward side, as shown in section D. The various slopes form parts of river-basins (§ 319), and the course of rivers in an ordinary map serves to mark out the direction of the slopes. Where there are no rivers, or when rivers flow into a salt lake, a region of internal drainage results. Such regions occur in every continent wherever the arrangement of the heights cuts off rainfall and allows full scope to the action of evaporation. One-quarter of the Earth's land surface is thus situated. The long slopes of all the continents are directed toward the Atlantic Ocean and its seas, which thus receive the drainage of more than half the land (Plate XIII.) All the continents turn their backs, so to speak, on the Indian and Pacific Oceans. The following table is calculated by Dr. John Murray.[3] The small area draining into the Southern Ocean is added, in the table, to those of the Atlantic and Pacific Oceans.

PERCENTAGE AREA OF CONTINENTS SLOPING TO EACH OCEAN

	Eurasia.	Africa.	N. America.	S. America.	Australia.	World.
Atlantic, including Mediterranean	13.9 ⎫	49.0	36.0 ⎫	86.4	...	34.3 ⎫
	⎬ 37.9		⎬ 76.5			⎬ 50.8
Arctic Sea	24.0 ⎭	...	40.5 ⎭	...	...	16.5 ⎭
Pacific	19.6	...	20.3	6.3	9.3	14.4
Indian	15.3	20.0	...	...	40.0	12.8
Inland	27.2	31.0	3.2	7.3	50.7	22.0
	100.0	100.0	100.0	100.0	100.0	100.0

357. **South America** being the most typical continent may be first described. The triangular outline is modified by a large outcurve of the northern half of the west coast north of 20° S., and on the middle of the east coast by a more prominent outcurve culminating in Cape San Roque. Its greatest length, nearly along the meridian of 70° W., is 4800 miles, from Point Gallinas on the Caribbean Sea in 13° N. to Cape Horn on the Southern Ocean in 56° S. The greatest breadth from west to east is 3300 miles along the parallel of 5° S., between Point Parina (82° W.) and Cape San Roque (35° W.) A group of rocky islands, the Chonos Archipelago, runs for 1200 miles close to the fjord-grooved west coast at its southern extremity, and a tortuous channel separates the south-eastern tip, Tierra del Fuego, from the mainland. The average elevation of the continent is almost exactly that of the whole continental area.

358. **The Andes.**—The main axis of South America lies close to the west coast along the crest of the Andes, which form the longest mountain system, unbroken by passes of low elevation, in the world. The short slope to the Pacific varies from 30 to 150 miles in breadth; the long slope to the Atlantic is in parts 3000 miles wide. A mountain system is not a ridge, but a region showing diversities of

structure and scenery from point to point. The highest peak of the Andes is Aconcagua, 22,400 feet, in 33° S.; but at least thirteen other summits rise more than 19,000 feet above the sea. Many of the passes, which mark the meeting of the heads of transverse valleys of opposite slopes, are elevated more than 14,000 feet, and the lowest in a stretch of 4000 miles is 11,400 feet above sea-level. Tertiary sedimentary rocks form the slopes of the Andes, and are overspread in many places by sheets of volcanic rock, while the loftiest volcanic cones in the world shoot up in solitary grandeur above the ridges. The Andes are young mountains, geologically speaking, and are still growing. Every little step of upheaval is accompanied by earthquakes (§ 299), which occur more frequently along the western margin of South America than anywhere else. South of Aconcagua the system consists of a single rugged ridge, which gradually diminishes in height and in steepness toward the south, where the sea has invaded its valleys forming the Chonos Archipelago. From Aconcagua northward to the equator the system forms two mountain ranges, one keeping close by the Pacific coast, the other sweeping inland. Where they diverge most widely the two mountain walls encircle a high plateau of internal drainage, which is as large as Ireland, and its lowest part, 12,000 feet above the sea, is occupied by the great Lake of Titicaca. Converging at the northern extremity of the Titicaca Plateau the two ranges wall in a longitudinal valley of great length, sloping northward and traversed by rivers which escape by wild gorges through the eastern ridge. From the equator northward the ridges of the Andes diminish in height, unite in the "Knot of Pasto," and then branch into three spurs, separated by the long valleys of the Magdalena and Cauca sloping to the north. The eastern spur sweeps round the north coast of South America, completing the framework of the continent. Along its whole length the eastern ridge of the Andes slopes down to the central low plain by a succession of great terraces, and sends out many short diverging mountain buttresses. Ores of silver, mercury, and copper abound in these mountains, and coal-beds occur in the south. On the rainless short slope in the centre nitrate of soda forms extensive deposits.

359. **Eastern Mountains and Low Plains.**—The long slope of South America from the base of the Andes forms one vast low plain stretching from north to south, the portion of which, at a less elevation than 600 feet, is equal to two-fifths of the continent. It is broken into three divisions by two very gentle ridges stretching eastward from the Andes. The northern and smaller swells up into the *High Plain of Guiana*, which is cut into lines of heights, known as the *Sierra Parima*, the *Sierra Pacarai*, culminating in Roraima (§ 312), and the *Sierra Acaray*. The larger or *High Plain of Brazil* fills the whole eastern outcurve. It is an upheaval of very ancient rock, which has been cut by the valleys of numerous great rivers into a medley of mountain masses, few of which exceed 3000 feet in height. The *Sea Range*, under many names, runs along the coast from 10° S. to 30° S., forming the steep seaward slope of the High Plain. The eastern mountains contain

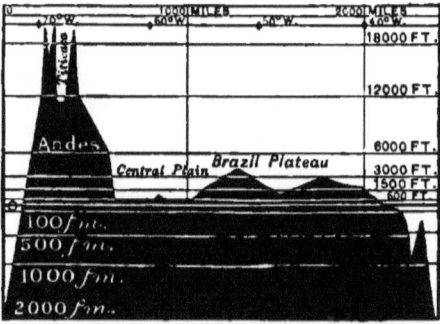

FIG. 57.—Section across South America on parallel of 18° S. Vertical scale 300 times the horizontal. Sea-level marked O.

deposits of gold and of diamonds, and are covered in many parts by fertile soil. Fig. 57 gives an idea of the form of the slopes of South America on the parallel of 18° S.

360. **Orinoco Basin.**—The northern division of the Low Plain is known as the *Llano*, and forms the basin of the Orinoco River, which is kept supplied with water by tribu-

taries descending from the mountain borders. In the rainy season, June to August, the plains are flooded, driving the inhabitants to take refuge in houses built in the trees. The Orinoco, from its source on the south-west of the Guiana High Plain, flows along the watershed which parts its basin from that of the Amazon. One branch, retaining the name Orinoco, eventually flows down the northern slope and sweeps east to the sea, while another, known as the Casiquiare, breaks away down the southern slope and flows rapidly into the Rio Negro, a tributary of the Amazon. The two great river systems are thus connected by a natural canal.

361. **Amazon Basin.**—At a distance of 1900 miles from the sea the vast central plain only reaches an elevation of 600 feet, and the basin of the Amazon presents the gentlest land-slope in the world. Nearly the whole plain is covered with dense tropical forests, and it is therefore called the *Selvas* or Woods. On each side the Amazon and its tributaries overflow in the rainy season (§ 318), covering the land for 20 or 30 miles from the banks, so that the forests appear to be growing in the water; and depositing fine alluvial soil which, over the whole region, does not contain a stone as large as a pea. Numerous great tributaries, many exceeding 1000 miles in length, converge to the main river from the slopes and high valleys of the Andes. Of these the Maranon is generally considered the head stream, although the Ucayali is longer. Other rivers flow in like veins joining a leaf-stem, from the Guiana High Plain in the north and the Brazil High Plain in the south. Two of the largest rivers of the latter region flow north in wide valleys but do not reach the Amazon: one, the Tocantins, enters the sea close to its mouth, and the other, the Rio San Francisco, curving sharply to the east, reaches the Atlantic about 10° N.

362. **La Plata River System.**—From the temporary lake which forms west of the flat low plateau of Matto Grosso in the rainy season, and gives origin to some of the southern tributaries of the Amazon, the river Paraguay flows south

along the low plain, receiving numerous tributaries from the Andes slopes on the west, and the great River Parana from a southern valley of the Brazilian High Plain on the east. The united river swerves eastward and enters the wide shallow estuary termed the Rio de la Plata at 34° S. The undulating grassy plain of its lower track is called the *Pampas*, and is one of the flattest low plains in the world. South of the La Plata several rivers flow to the Atlantic from the Andes; all are subject to floods on account of the abrupt change of slope at the base of the mountains, the inclination of the low plain toward the east being too slight to let the water drain away when the torrential track is flooded. Patagonia, the southern extremity of the continent, is for the most part a desert of shingle, and much of it is an area of internal drainage on account of the drying of the brave west winds by the Andes (§ 201).

363. **North America** presents the typical form and configuration of a continent, but it resembles South America passed through a mangle, being larger, wider, lower, with less contrast between its heights and plains, and a much more broken coast-line. Fig. 58, a section across the continent on the parallel of 36° N., and Fig. 59, on the

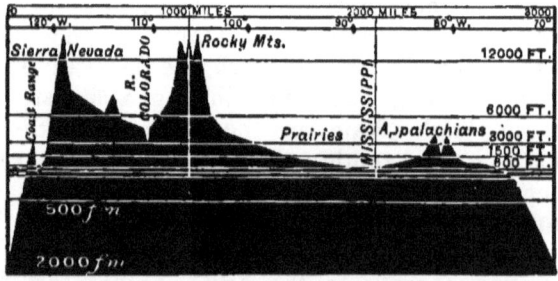

FIG. 58.—Section across North America in 36° N. Vertical scale 300 times the horizontal. Sea-level marked O.

meridian of 90° W. along the central low plain, are on the same scale as that of South America. The total length of the continent, nearly on the meridian of 100° W., is

4000 miles from the ice-bound Parry Islands in 75° N. to the tropical isthmus of Tehuantepec in 17° N. The greatest breadth, on the parallel of 52° N., is 3000 miles. In the extreme north-west Cape Prince of Wales on Bering Sea comes within 40 miles of the north-eastern extremity of Asia; and on the north-east Greenland is bound to America by continuous ice in winter. The west coast and northern part of the east coast of North America are high and rocky, but the south-east presents the longest stretch of gently shelving shore in the world.

364. **Western Heights of North America.**—From Tehuantepec to Alaska the axis of the continent runs along the *Rocky Mountains*. This range is often considered to be a continuation of the Andes, but it is less lofty, the passes across it are lower, and the two slopes into which it divides the continent are more nearly equal than those of South America. The average distance of the range from the west coast is about 400 miles, except where the great Pacific outcurve increases the distance to almost 1000 miles. Mount Brown, near 52° N., is the highest peak, 16,000 feet; and Pike's Peak (14,200 feet), in 39° N., is one of the next in elevation. Midway between these summits one of the grandest portions of the range has been set apart as a permanent museum of physical geography on a grand scale, under the name of the Yellowstone National Park (§ 316). On the east the Rocky Mountains slope down in wide terraces comparatively gently to the central low plain. On the west their slope is abrupt but short, terminating in a wide plateau, averaging 5000 feet in height, which runs along the entire length of the continent, and is buttressed on the west by a less continuous series of ranges. The *Sierra Madre* is the western buttress of the plateau in the south, where it forms the watershed, and near the point where it diverges from the Rocky Mountains the volcanic peaks of Orizaba (18,200 feet) and Popocatepetl (17,500 feet) rise as majestic summits, which with Mount Wrangel (17,500 feet) in Alaska are the loftiest in North America. Farther north the plateau is supported by the rugged snow-clad *Sierra Nevada*, which presents a very

steep front to the west, cut into by rugged transverse valleys, with scenery of the wildest grandeur. Its highest peak is Mount Whitney (14,900 feet), and at Mount Shasta it passes into the *Cascade Range*, which runs northward, diminishing in height, to Alaska, its chief summit being Mount St. Elias (13,900 feet). Between latitudes 35° and 40° N. a lower mountain ridge, the *Coast Range*, joined to the Sierra Nevada on the north and the south, encloses a remarkable low plain, the Californian Valley, the rivers of which find access to the sea through an abrupt gap near the middle of the range. The eastern part of the centre of the plateau between the Rocky Mountains and the parallel Wahsatch Range, in longitude 112° W., forms the most elevated region, and is crossed by the Uintah Mountains, running from west to east. Cutting right through the Uintah range, and southward across the plateau to the Gulf of California, the great Colorado River and its tributaries lay bare the structure of the rocks, showing the horizontal sedimentary strata, interspersed with outflows of basalt, based on a bed of Archæan gneiss. The other great river of the Pacific slope is the Columbia, the tributaries of which converge from all parts of the Rocky Mountains, from near Mount Brown in the north to the Wahsatch Range. In the north-west, where the low bordering ranges spread out, the great Yukon flows down the northern slope of the diminished plateau into Bering Sea. Gold, and the ores of silver, lead, mercury, and copper, occur very abundantly in the valleys and mountains of the plateau.

365. **The Great Basin.**—Between the Wahsatch Mountains and the Sierra Nevada the plateau sinks slightly into a vast triangular area of internal drainage, known as the Great Basin. It is most depressed near the sides, and rises in the middle in a series of mountain ridges. In the Quaternary period a wide sheet of water—called Lake Bonneville—occupied the eastern depression, and its shrunken remnant now forms the Great Salt Lake, at the base of the Wahsatch Mountains. A smaller expanse—Lake Lahontan—filled the western depression, which is

now dotted by a series of small salt lakes under the eastern slope of the Sierra Nevada. The soil of the Great Basin is encrusted with borax and other alkaline salts deposited by the shrinking lakes. In recording their researches on this region, the officers of the United States Geological Survey have produced a series of the most fascinating memoirs on physical geography. The volumes on the exploration of the Colorado River by Major Powell, and on Lake Lahontan by Mr. Russell, are especially interesting.

366. **The Appalachian Mountains**, running parallel to the east coast, form a broad chain of moderate height, Mitchell's Peak, 6700 feet above sea-level, being the loftiest. They are true mountains of elevation, the alternate anticlines and synclines forming parallel ridges and longitudinal valleys, and their rocks are much more ancient than those of the western heights. In the south, Carboniferous strata and coal seams are laid bare in the transverse valleys, and the extension north of the St. Lawrence, in the broad low ridge of the Laurentides, is composed mainly of Archæan rock. The Appalachians, which are sometimes called the Alleghanies, form a complete minor axis, giving the east of North America a short slope to the Atlantic and a long slope westward. The watershed follows the eastern ridge of the chain in the south, and the western ridge in the north ; the Hudson River, however, cuts right across the entire chain.

367. **Mississippi Basin.**—One great valley, formed by the meeting of the long slopes of the two mountain axes, occupies the whole centre of North America. The southern and northern halves of this valley dip in opposite directions

FIG. 59.—Section of North America on the meridian of 90° W. Vertical scale 300 times the horizontal. Sea-level marked O.

from a broad flat transverse ridge of very slight elevation in 48° N. The southern south-sloping half of the valley forms the basin of the Mississippi River. The Mississippi rises on the crest of the gentle transverse slope, and after a winding course of more than 1000 miles receives on its right bank the Missouri, a river of much greater length, formed by the union of tributaries from 900 miles along the Rocky Mountain Range. Farther south the Arkansas, another long river, flows in from the Rocky Mountains. The steep eastern slope of this range, unlike that of the Andes, stops at an elevation of nearly 6000 feet above the sea, and thence the rivers flowing to the Mississippi cross a slope so gentle that the land is spoken of as the Great Plains. As the elevation diminishes the slope decreases also, and the lowlands of the basin become known as the Prairies. The Ohio River, flowing down the slope of the Appalachians, is the largest tributary reaching the Mississippi on its left bank.

368. **Arctic Basins.**—In the northern half of North America several nearly level terraces, of from 200 to 300 miles in breadth, separated by narrow zones of steeply sloping land, descend from the Rocky Mountains toward Hudson Bay. The lower terraces are covered with boulder clay, and the terminal moraine of the great Pleistocene ice-sheet has been traced in the form of a huge ridge called the Grand Coteau des Prairies. This ridge turns the Missouri River to the south, and the Saskatchewan, flowing from near Mount Brown in the Rocky Mountains, to the north, thus separating the northern and southern slopes. Upon the lowest terrace, where the glacial remains are thickest, a line of wide shallow lakes stretches from 49° N. to the Arctic Sea. Lake Winnipeg in the south receives the Saskatchewan, and has an outlet by the Nelson River to Hudson Bay. This lake is the centre of a great but ill-defined drainage area, some of the hundreds of small lakes surrounding it being connected with several river systems, on account of the confused ridges left by the melting ice-sheet. Traces remain of a much larger ancient body of water, called Lake Agassiz, which included Lake Winnipeg, and many smaller lakes and river-valleys. The Athabasca, rising near Mount Brown, flows north-eastward to Lake Athabasca, which has an outlet northward to Great Slave Lake, whence the wide

Mackenzie River flows parallel to the Rocky Mountains to the Arctic Sea, receiving the outflow from Great Bear Lake on the Arctic circle.

369. **St. Lawrence System.**—The gentle transverse ridge separating the northern and southern slopes of North America is nowhere higher than 2000 feet, and it only attains this elevation in the east. Its surface is slightly concave, the northern edge, called the *Height of Land*, being a continuation westward of the Archæan plateau of the Laurentides; while the southern edge, known as the *Great Divide*, is a prolongation toward the east of the moraine heaps of the Coteau des Prairies. The central hollow contains a remarkable group of lake basins, which are claimed, with some probability, to contain half of the fresh water in the world. Before the Ice Age they were probably in connection with the Mississippi river system, and from ancient raised beaches surrounding them they were evidently at one time much more extensive than now. The western group—Lakes Superior, Michigan, and Huron —are closely connected, and their surface stands about 600 feet above sea-level. From the south of Lake Huron they discharge into Lake Erie, whence the Niagara River (§ 330) leads northward into Lake Ontario, from which the broad St. Lawrence sweeps on to the Atlantic.

370. **Australia**, the only known continent entirely in the southern hemisphere, is 2300 miles in extreme length along the parallel of 26° S. (see section Fig. 60). The greatest

FIG. 60.—Section across Australia in 26° S. Vertical scale 300 times the horizontal. Sea-level marked O.

breadth is 2000 miles along the meridian of 143° E. from Cape York in 11° S., which is the most northerly point, to Cape Otway in 39° S. Incurves of the north and south coasts reduce the width to 1100 miles in the narrowest part of the continent, while both the east and west coasts form bold outcurves. Tasmania rests on the Continental Shelf to the south, and New Guinea to the north. The average height of the land, as far as can be judged from the imperfect exploration of the interior, is about 800 feet. In spite of this low elevation the proportion of land less than 600 feet above the sea is small, while the proportion between 600 feet and 1500 feet in elevation is greater than for any other continent.

371. **Configuration of Australia.**—The continent is apparently one low plateau, rising into a line of hills along the west coast, and ridged irregularly here and there by mountains in the nearly unknown interior. It sinks in the south-east to an extensive low plain (the Australian Basin) less than 600 feet above the sea. Half of Australia is made up of areas of internal drainage. The *Great Dividing Range*, forming the axis of the continent, rises along the eastern edge. It sweeps round the south-east coast under the name of the Australian Alps, and culminates in Mount Townsend or Kosciusko (7300 feet). Thence it runs northward under different names as a chain of short ranges, scored by deep transverse valleys, sending short full rivers to the Pacific. Diminishing in height toward the north, it merges into the general elevation of the plateau. The ranges were, as a rule, ridged up out of primary rocks, the Silurian system being now most prominent in the south, and the Carboniferous, with thick seams of coal cropping out, farther north. Gold, and the ores of silver, tin, and lead occur in great abundance.

372. **River Basins.**—The southern part of the Dividing Range slopes down very steeply westward to the low plain of the Australian Basin. The Murray River flows westward across the Basin from its source near Mount Townsend, and after receiving the Lachlan and Darling it swerves to the south and enters the sea. Many long rivers are marked on maps converging from the east and north to the Australian Basin, but most of these are stony channels only occupied by water after rain, and many of the streams dry up as they

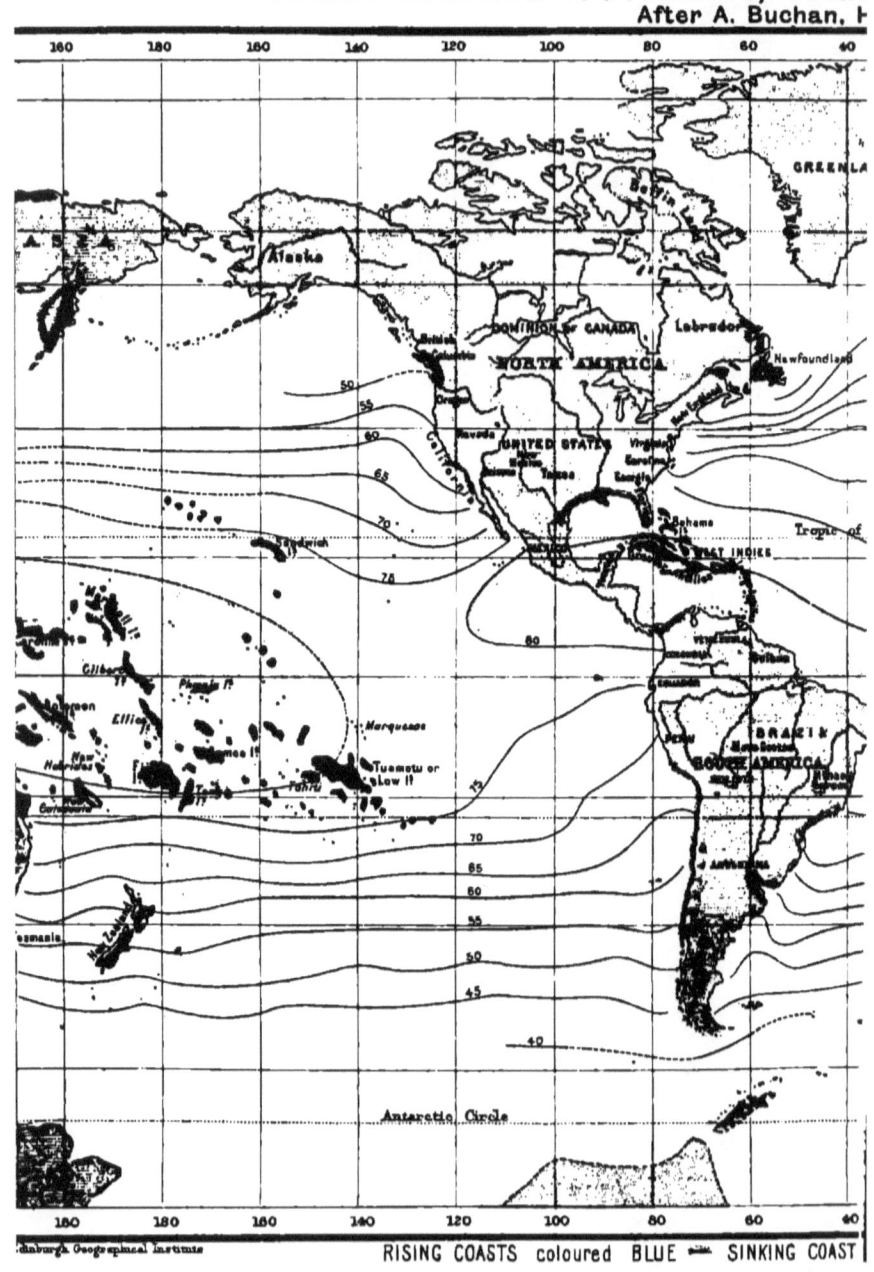

REEFS, RISING AND SINKING COASTS.
Guppy, and others.

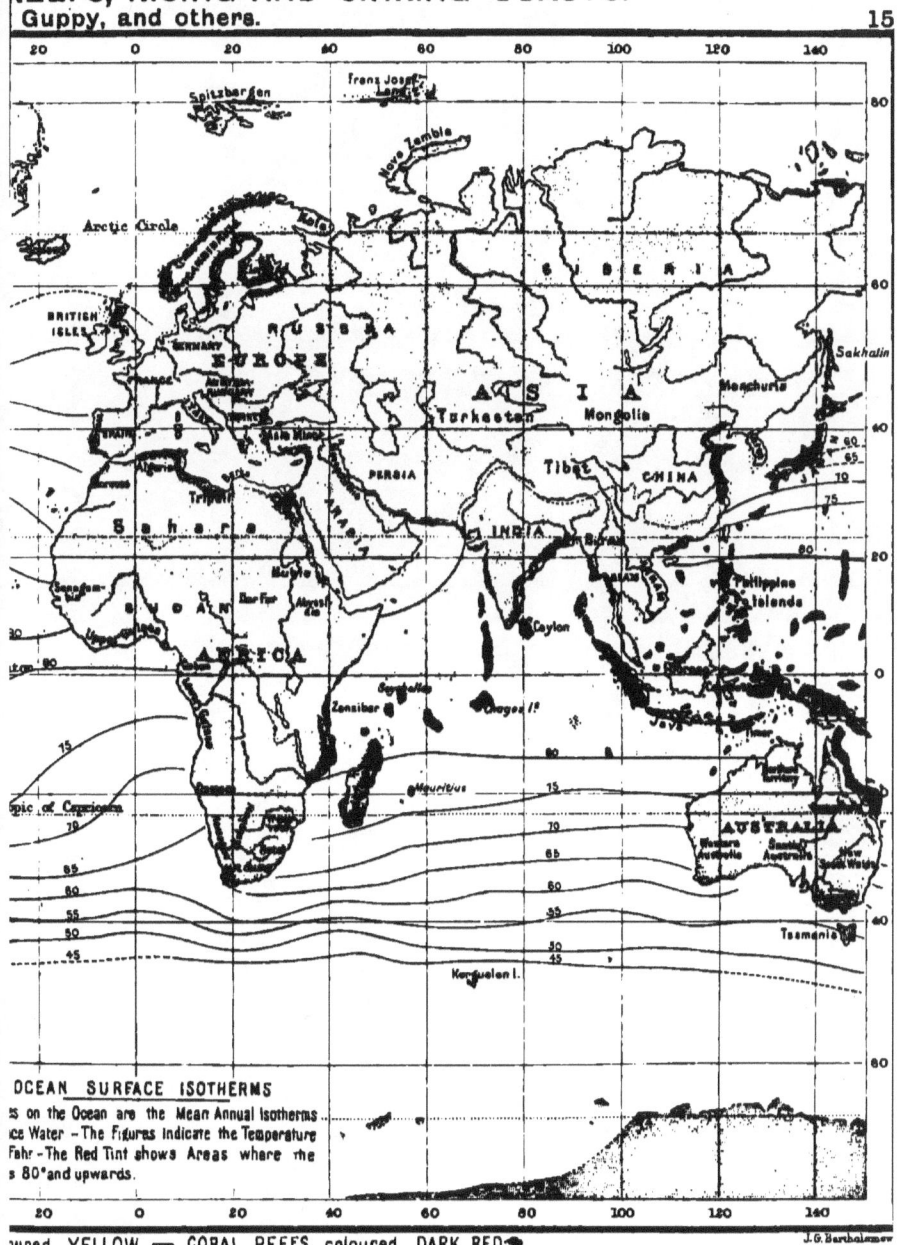

OCEAN SURFACE ISOTHERMS
...s on the Ocean are the Mean Annual Isotherms
...ce Water – The Figures indicate the Temperature
Fahr – The Red Tint shows Areas where the
... 80° and upwards.

...ured YELLOW — CORAL REEFS coloured DARK RED

flow. The Basin is divided by the Flinders Range west of the Murray, and its western part forms a depression scarcely raised above sea-level, in which lie Lake Torrens and Lake Eyre—salt lakes with no outlet. The whole depression is rimmed round with coral limestone of Tertiary age, and appears to have formed a wide shallow bay long after the rest of the continent was upheaved. The plateau to the west is a great desert not fully explored, and composed of the rock known as desert sandstone, fringed to the south by grand cliffs of tertiary limestone which line the Great Australian Bight as a wall about 400 feet high, unbroken by a single river for 1000 miles.

373. **Africa** presents a typical triangular outline resembling that of South America, but the north-western outcurve is much more pronounced, while the north-eastern outcurve is broken by the depression of the Red Sea. Round Africa the Continental Shelf is extremely narrow, and the islands it bears are few and small, while the coast-line is less indented than that of any other continent. The greatest length, nearly 5000 miles, lies along the central meridian of 20° E., and the greatest breadth, 4500 miles, is on the parallel of 10° N. Africa is the only continent crossed by both tropics, the equator passing nearly through the centre. The average elevation of Africa is nearly that of all the land; but no other continent has such a small proportion of land below 600 feet in height (one-eighth of its area), and none has so great an extent (nearly two-thirds) between the heights of 600 and 3000 feet.

374. **Slopes of Africa.**—A section drawn across the continent, along the equator (Fig. 61) hardly shows how completely the typical continental structure is departed from, as Mount Kenia is only an isolated peak, not part of a range. All the rivers pursue singularly curved courses, unlike those of any other continent, and where they drop over the edges of the plateaux form great cataracts. The watersheds are not dominated by mountain ranges, but by the broad backs of plateaux, out of which the main features of the land-slopes have been carved by erosion. The Atlas mountains run along the coast in the north-west and rise

into a succession of snow-crowned peaks, the loftiest of which was estimated by Mr. Joseph Thomson to be 15,500 feet above the sea. All round the coast, except in the north and north-west, the edges of the plateau present a

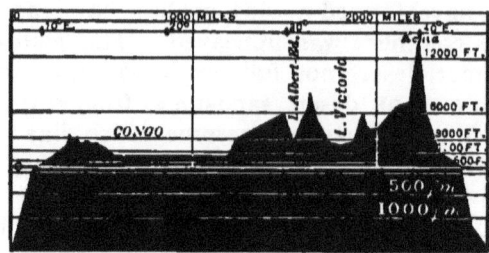

FIG. 61.—Section across Africa on the equator. Vertical scale 300 times the horizontal. Sea-level marked O.

mountainous aspect, and several great volcanic summits rise from their highest levels. Kenia, Kilima-njaro, and Ruwenzori reach heights approaching 19,000 feet above the sea. The loftiest elevated belt, which may be termed the *Great Plateau*, runs from the Red Sea southward and westward across the continent, and may be looked on as forming the main axis. Its greatest elevation is in the rugged valley-riven plateau of Abyssinia, and it continues highest on its eastern side. A strip of eastward-sloping land, down which the *Zambesi* pursues a cataract-broken course to the Indian Ocean, separates the Great Plateau from a smaller plateau which fills the southern extremity of the continent. This Southern Plateau sinks to the sea in steep terraces bordered on the south and east by curved mountain ranges, the most important of which is the Drakenberg. It dips to the west and is drained by the Orange River, a rapid stream flowing through a deep cañon far below the general level. The Great Plateau sends off three long branches of high land toward the north-west, which cannot be clearly traced on a map unless the contour-line of 1500 feet is shown. The first or Coast Ridge runs round the west coast and descends to sea-level in terraced mountain slopes. It bears the high Cameroon

Mountains near the angle of the Gulf of Guinea, and slopes down very gradually inland. Its western extension is pierced by the great River Niger, flowing into the Gulf of Guinea. The second or Central Ridge runs from the equator toward the Atlas Range across the northern high plain. Uniting with the Coast Ridge in latitude 5° N. and again in 20° N. it forms two great basins, of which the southern or equatorial is, on the average, higher, and the northern lower, than 1000 feet. The third or Red Sea Ridge runs along the Red Sea coast from the northern extremity of the Great Plateau. A very remarkable hollow furrows the whole length of the Great Plateau for nearly 2000 miles from north to south, and contains a succession of four great lakes connected with three distinct river systems. These are Lakes Albert and Albert Edward draining to the Nile in the north, Lake Tanganyika attached to the Congo in the centre, and Lake Nyasa united to the Zambesi in the south.

375. **Nile River System.**—Lake Albert collects the head-waters of the Nile, receiving the Semliki River from Lake Albert Edward lying at the base of Ruwenzori, and fed by the ceaseless torrents from that mountain. It also receives at the northern extremity the outflow of the largest lake in Africa, the Victoria Nyanza, which is situated on a higher part of the plateau east of the Great Hollow at an elevation of 3300 feet. This branch, the Victoria Nile, is broken by a succession of falls as it descends the steep edge of the plateau. From Lake Albert the White Nile flows northward to the Mediterranean across the desert which stretches between the slopes of the Red Sea Ridge and the Central Ridge, receiving many tributaries from both. The rainy heights of Abyssinia send down the Blue Nile and the Atbara, on which the periodical flooding of the Nile depends, but after the junction of the latter stream the Nile flows in three great bends across the parched low plain to its delta (§ 325) without receiving another drop of water, and subject to continual evaporation (§ 318). The six famous cataracts which occur in its lower course are produced by its bed crossing bars of hard rock, and they thus

differ in their nature from the cataracts of the plateau rivers of the south.

376. Congo Basin.—Shut in between the Central and West Coast Ridges, the equatorial basin was probably at one time a great inland sea several times larger than the Caspian. Its waters found an outlet across a comparatively low part of the West Coast Ridge, which they eroded into a deep gorge and so drained the lake into the Atlantic, leaving a basin of fertile soil now covered in great part with dense forests. Rivers flow into the circular basin from the high ground on every side and become tributaries to the giant Congo. This river descends from the Great Plateau at the equator foaming over the cataracts of Stanley Falls, sweeps through the basin in a magnificent curve as a navigable stream for 1000 miles, and bursts in a far grander chain of cataracts over the plateau edge through the gorge of Yellala. The source of the Congo lies somewhere in the Great Plateau about 13° S., Lake Bangweolo, 4000 feet above the sea, serving as a reservoir to collect the head-waters. In its northward course the river is joined by the Lukuga from Lake Tanganyika in the centre of the Great Hollow, 2600 feet above the sea; but it is only when the level of that lake is raised considerably above its average height that it overflows. Tanganyika, like most continental lakes, was once much larger, and appears to be shrinking into a basin of internal drainage, destined ultimately to become a small salt lake.

377. Tsad Basin and Sahara.—The northern basin enclosed by the West Coast and Central Ridges is even larger than that of the Congo, and so far as this very inaccessible region has been explored it appears to have no outlet. Lake Tsad, a sheet of shallow water varying in size from 4000 to 10,000 square miles according to the rainfall, and 800 feet above the sea, receives a number of great rivers from the south, and overflows in the rainy season to a much lower enclosed basin in the north-west, where excessive evaporation leaves only a crust of salt upon the ground. To the north nearly the whole breadth of Africa forms the internal drainage area of the Sahara, a sandy high plain broken by the rugged mountains of the Tibesti

Range which cap the Central Ridge, and dipping in the west and north to a low plain with some small depressions, called *shotts*, below sea-level.

➤ 378. **Eurasia**, containing one-third of the land of the globe and occupying the central part of the Eastern World Ridge, when looked at largely, shows the typical features of a triangular outline and a mountainous axis giving a long and a short slope to the land, and supporting a plateau of internal drainage. It is the least tropical of the continents, only the three south-eastern peninsulas crossing the tropic of Cancer. The greatest length of Eurasia is about 7000 miles, from Cape Roca in 9° W. to East Cape on Bering Sea in 170° W., the continent extending more than halfway round the Earth. The greatest breadth is about 5000 miles, along the meridian of 105° E., from Cape Chelyuskin in $77\frac{1}{2}$° N. to Cape Buru in $1\frac{1}{4}$° N. at the extremity of the Malay Peninsula. More than one-quarter of this vast area slopes together, forming basins of internal drainage, and almost a quarter slopes north toward the Arctic Sea, giving a peculiarly inaccessible character to half the continent and tending to increase the severity of its continental climate. The low plain of Eurasia forms a great triangle with its base along the Arctic Sea. This is divided into a smaller western and a larger eastern portion by the low belt of the Ural Mountains in 60° E., and may be taken as forming the boundary between Europe and Asia. A section of the continent, along the meridian of 90° E. (Fig. 62), gives a general idea of the structure. The main features of the west coast of Europe correspond on a smaller scale with the east coast of Asia—the Scandinavian peninsula answering to Kamchatka, the Baltic to the Sea of Okhotsk, the British Islands and North Sea to the islands and Sea of Japan. Similar resemblances connect the south coasts. Spain and Arabia are both square and massive plateaux; Italy and India are both separated from the continent by a low plain under a lofty mountain wall, and taper southward, ending in a large island; and the Balkan peninsula, like Indo-China, is mountainous, deeply indented, and terminates in an archipelago.

379. **Asia,** the highest as well as the largest of the continents, has an average elevation of more than 3000 feet. The zone of heights between 600 and 1500 feet is narrower than in any other continent, and more than one-sixth of the surface stands more than 6000 feet above the sea. The orographical centre of Eurasia is formed by the lofty plateau of *Pamir* (in 38° N. and 73° E.), as large as Ireland, and rising to 25,800 feet above the sea in its highest summit, while its lowest point is 9000 feet; it is called by the dwellers in the region "The Roof of the World." From this centre, mountain chains spread out like the ribs of a fan to the east and to the west. The lofty range of the *Hindu Kush*—cleft by a few snow-blocked passes and rising into summits 24,000 feet high—runs south-west from the Pamir, separating the low plain of India from the low plain of Northern Asia. It branches in lower ridges to the south and west, enclosing the internal drainage area of Iran (Persia), which lies at an average height of 3000 feet. The northern mountain ridge, sweeping round the south shore of the Caspian as the Elburz Range, merges into the broken Plateau of Asia Minor. Here the southern ranges also converge, walling the Plateau of Iran from the low plain down which the Tigris and Euphrates pour into the Persian Gulf. Mount Ararat, 17,000 feet above the sea, is the grandest summit in Asia Minor. The plateau spreading southward occupies Arabia, most of which is an internal drainage area. One of the most perfect types of a mountain chain of elevation is presented by the *Caucasus*, which runs from the Black Sea to the Caspian as a magnificent barrier between the high plateau of Asia Minor and the low, level plain of Europe, and culminates in Mount Elbruz, 18,500 feet high. In the calculation of elevation in the tables of § 355 this chain is assigned to Europe.

380. **Eastern Asiatic Mountain System.**—The mountain chains which radiate eastward from the Pamir converge at two centres, one near the north of the Indo-China peninsula, the other near the Sea of Okhotsk. Between these three knotting points the long mountain ranges seem on the map to droop in graceful folds. They define an

area which is as large as South America, and is occupied by the highest and most extensive plateaux in the world. The southern front of the whole system is the triple chain of the *Himalaya*, sweeping in a noble curve south-eastward from the Pamir, rising from the plain in stately slopes and ridges, and crowned by innumerable snowy summits, amongst them

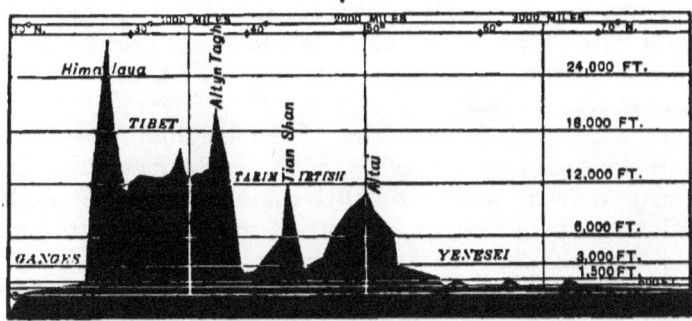

FIG. 62.—Section across Asia on the meridian of 90° E. Vertical scale 300 times the horizontal. Sea-level marked O.

Mount Everest (29,000 feet), the culminating point of the Earth's surface. It is cleft by no passes less than 15,000 feet above the sea. The *Karakorum*, a short but very lofty range (its chief summit, Dapsang, is 28,700 feet above the sea), runs parallel to the Himalaya from the Pamir. Thence also the long and lofty range known as the *Kuen Lun* stretches east, and sends off the Altyn Tagh and Nan-shan range in a north-easterly curve. Between the Himalaya and the Kuen Lun extends the high plateau of Tibet, 13,000 feet above the sea, and measuring 2000 miles from east to west, and 1700 miles from south to north. The plateau slopes downward to the east, and the mountains and valleys which ridge its surface converge into a series of close parallel ranges at the Indo-China knotting point. Thence some ranges diverge southward into the peninsula, some descend eastward toward the plain, and some sweep north-eastward to the Okhotsk knotting point as the Khingan Chain. Most of the Tibet high plateau is free from snow in summer owing to the extreme dryness of the air, and is a region of internal drainage. The great rivers Indus and Brahmaputra rise in the most northerly longitudinal valley of the Himalaya, and break a way round the northern and southern extremities of the range to the southern plain. Other rivers, amongst them the Irawadi and the Mekong, flow south in the longitudinal valleys of the Indo-China Peninsula. Rising on the eastern margin of the plateau, the Yellow River (*Hoang Ho*, § 324) sweeps north-eastward until it breaks a passage through the Khingan Range and turns south again over the eastern plain. The Yang-tse-Kiang rises close to the Yellow River; at first it rushes southward through one of the longitudinal valleys, but making a gap through the bordering mountain, and piercing in turn several parallel chains, it swerves northward and emerges from its gorges on the plain, to once more approach the Yellow River near its mouth.

381. **Tarim and Gobi Basins.**—Standing with one foot, as it were, on the northern edge of the Tibet High Plateau, the Kuen Lun and Altyn Tagh reach down with the other to the much lower High Plain of the Tarim River and the Gobi Desert, which averages a little more than 3000 feet above the sea. The vast range of the *Tian Shan* ("The Mountains of Heaven"), with some summits 24,000 feet above the sea, stretches north-eastward from the Pamir, and walls in the northern side of the Tarim basin. Many rivers from the slopes of the amphitheatre, formed by the converging mountains, unite in the Tarim, which flows east for 1300 miles to dry up in the swampy salt lake of Lob Nor. The Tian Shan is continued north-eastward by a number of ranges, including the Altai, the Sajan, and the plateaux of Vitim and Aldan, all of which rise much higher than the Gobi, and are separated from each other by mighty valleys sloping into the northern low plain. They are united by the Yablonoi and Stanovoi Ranges to the great Khingan Chain at the Okhotsk knotting point, and continue in the diminishing Stanovoi Range to East Cape. There is abundant evidence that the Gobi High Plain, now covered in most parts with drifting sand, was once a vast inland sea, discharging its surplus waters by the great valleys between the northern heights into the Arctic Sea. Now

the few rivers which flow into it from the surrounding mountains end in the sand or in small salt lakes, and the process of desiccation seems still going on. Rising on the western slope of the Khingan with tributaries from the eastern slope of the Yablonoi, the great Amur River cuts through the Khingan Chain and several parallel ranges, and finds its way into the Pacific.

382. **Northern Low Plain.**—The Tian Shan and other northern mountains descend in terraces to a narrow belt of undulating land about 1000 feet in elevation, which sinks into the wide low plain less than 600 feet above the sea. Lake Balkash, without outlet and intensely salt, occupies a depression north of the Tian Shan, from which it receives several rivers, but its area is steadily diminishing by evaporation. Lake Baikal (§ 333), to the north-east at a higher level, receives much water from the Altai and surrounding heights, but its outflow is comparatively trifling. The northern plain bears evidence in its gravel beds of having emerged from the sea in the Quaternary period, and the gradual desiccation of Asia probably dates from the time when its upheaval cut off from the interior the tempering influence of the sea. Three vast rivers, only matched for length of course and area of basin by the giant streams of Africa and America, flow from the mountains across the plain to the Arctic Sea during the few months when they remain unfrozen. The Lena, farthest east, rising near Lake Baikal, terminates in a wide delta. The Yenisei flows due north from the Sajan Mountains, and receives no considerable tributaries on its left bank, but the Angara, from Lake Baikal, and the two Tunguskas flowing from the east, join on its right bank. The Ob and Irtish flowing north from the Altai unite, and after receiving tributaries from the eastern slope of the Urals, enter the head of a long narrow gulf of the Arctic Sea. From the Pamir the Amu Daria and Syr Daria (Oxus and Jaxartes) flow across the desert low plain, rapidly dwindling by evaporation, to Lake Aral, the area of which is shrinking. In the time of the early Greek geographers the Oxus swerved to the west and entered the Caspian, and its old bed, from which it seems

'to have been diverted by sand-dunes (§ 307), may still be traced. At a more remote period the Aral Lake was part of a large sea which covered the Caspian basin and communicated with the Mediterranean, and in Quaternary times spread over the low watershed of the Ob to the Arctic Sea.

383. **Indian Peninsula.**—A great low plain extends along the base of the Himalaya, separated by a gentle ridge into a south-western slope traversed by the Indus on its way to the Arabian Sea, and a gentler eastern slope, along which the Ganges (§ 318) flows to the vast delta which it shares with the Brahmaputra. An ancient and much denuded plateau largely built up of volcanic rocks fills the southern part of the peninsula. This plateau is loftiest on its western edge, where it sinks in abrupt terraces to the sea, presenting a mountain-like wall known as the Western Ghats. The more gentle slope to the east has been cut by numerous rivers into wide valleys, and the broken plateau edge forms a lower and less regular line of heights more remote from the sea, called the Eastern Ghats. The coast-line on both sides is remarkable for its unbroken character and the gentle shelving of the beach.

384. **Europe**, a bunch of peninsulas thrust out into the Atlantic Ocean, is the only great land-mass not crossed by one of the tropics, and from its well-marked sea-climate it may be appropriately termed the Temperate Continent. An axis of true mountains of elevation runs through Southern Europe, and another forms the low belt of the *Urals* on the boundary with Asia. A rim of ancient plateaux worn into mountains of denudation marks the north-western border in Scotland and Scandinavia. Within this elevated frame the land is a wonderfully uniform low plain, fully half of the continent being less than 600 feet above the sea. Only one-sixth of the surface has an elevation greater than 1500 feet. The lines of elevation have a comparatively slight share in determining the slopes, which exhibit none of the typical continental simplicity.

385. **Southern Mountain System.**—The *Alps*, the most thoroughly studied mountain system in the world, form the orographical centre of Europe. The main chain (§ 303) runs east and west in a series of ridges separated by longitudinal valleys and cleft by transverse valleys into distinct mountain blocks. Mont Blanc (15,800 feet) is the loftiest summit. On the south the main range slopes down steeply to the low plain of Lombardy, which is enclosed to the south by the *Apennines*, an extension of the western Alps. The northern range of the Alps descends to a plateau sloping gently to the north and east, and buttressed by the limestone ridges of the *Jura*. To the east the system runs southward through the Balkan Peninsula as the Dinaric Alps, also a limestone chain, full of the characteristic scenery wrought by erosion and subterranean solution. The *Balkan Range* stretches east and west across the peninsula, sloping down to the low plain of the Danube in the north. The granite heights of the *Black Forest* Mountains run north of the Jura, and are continued by a broad ridge of Palæozoic rock, which dips down into the northern plain in an outcrop of the coal-measures. A broken hill country extends north-eastward from the Alpine plateau, sinking in elevation toward the north, and terminating in the Harz Mountains in 52° N. The hilly region rises in the east into the steep heights of the *Bohemian Forest*, which runs north-west from the eastern extremity of the Alps. The Bohemian Forest Range turns sharply north-eastward as the *Erzgebirge* or Ore Mountains, the rocks of which are traversed by veins of many metallic ores, and these in turn run eastward as the *Sudetic Range*. Supported between the three ranges the irregular plateau of Bohemia rises toward the south, and is terminated by the higher land of Moravia. Eastward the Sudetic Range adjoins the fine curve of the *Carpathian Mountains*, which sweep steeply round the low Hungarian plain, and sink down gradually to north and east into the great Northern Plain. The Carpathian Range terminates in the *Transylvanian Alps*, which first run parallel to the Balkans, and then converge in the west until they almost

meet that range. West of the Alps the Vosges Mountains run northward, separated by a wide flat valley from the parallel range of the Black Forest, and terminating in the same belt of ancient rock. Separated by the narrow valley of the Rhone on the west, the Auvergne plateau, studded with extinct volcanic cones, rises in a steep terraced slope known as the Cevennes Mountains, and sinks more gently to the low plain on the north and west. The rugged high plain of the Iberian Peninsula is shut off from Northern Europe by the straight line of the *Pyrenees*, one of the steepest mountain ranges, and presenting some of the finest examples of erosion in the form of cirques or round valleys.

386. **Rivers of Western Europe.**—In Western Europe the main watershed (see Plate XIII.) lies, as a rule, nearer the south coast than the north, following roughly the Pyrenees, the Cevennes, the Vosges, the Alps, the Black Forest, the Franconian Jura, the Moravian Plateau, and the Northern Carpathians. Thus the northern slope is long, and the southern slope short. In Eastern Europe the watershed is nearer the north coast, crossing the low plain on a ridge of very slight elevation, which stretches from the Carpathians north-eastward to the Urals, and swells up into the Vàldai Hills about the centre. This gives a comparatively short slope to the north and a long slope to the south. The rivers of Western Europe, the Guadiana, Tagus, and Douro in the Iberian Peninsula, and the Garonne, Loire, and Seine from the Auvergne high plain, flow to the Atlantic Ocean directly. The rivers of Central Europe all originate in the Alps and its connected ranges. The Rhone and Rhine flow in opposite directions along the great longitudinal valley which bisects the Alps. The Rhone, descending from its source near the great central mass of the St. Gothard, enters the Lake of Geneva, escapes westward between the Alps and Jura, and sweeps south to the Mediterranean, beneath the steep front of the Cevennes. Flowing east, the Rhine turns northward into Lake Constance, passes out westward between the Alps and the Black Forest, turns north through the wide valley between the Black Forest and the Vosges, crosses the

ancient rock plateau by a series of grand gorges, and, flowing over the low plain, oozes into the North Sea along several branches embanked above the sunk plain of Holland. The Elbe drains the Bohemian plateau, and breaking through the mountain barrier in "the Saxon Switzerland," between the Erzgebirge and the Sudetic Range, winds across the low plain north-westward to the North Sea. The Oder and Vistula, from the northern slopes of the Sudetic Range and Carpathians, flow northward to the Baltic. The Danube is remarkable for its disregard of mountain barriers. It rises on the eastern slope of the Black Forest, flows eastward across the plateau north of the Alps, and finds a way between the Alps and the Bohemian Forest Range. After penetrating some smaller ranges it turns south in several parallel channels across the flat plain of Hungary, which plain was probably once a great lake. It is joined by the Drave and Save from the Alps, and the Theiss from the Carpathians, as it crosses the nearly level plain. The narrow channel of the Iron Gate, between the opposed ranges of the Carpathians and Balkans, allows the Danube to enter the open plain, across which it flows to a delta on the Black Sea.

387. **Rivers of Eastern Europe.**—The long southern slope of Eastern Europe is traversed by the great rivers Dnieper and Don, flowing through gorges cut in the low plain to the Black Sea. The still greater Volga (§ 89) rising in the Valdai Hills winds eastward and southward, always encroaching on its right bank, which is high and steep, and always leaving successive alluvial terraces on its low left bank. The Oka is the most important of its many tributaries on the right, and on the left the Kama, flowing from the Ural Mountains, is the largest. When the Volga reaches sea-level its course is directed south-westward, parallel to that of the Don and very near that river, but the great stream turns sharply south-eastward, splitting into numerous channels, and finally enters the closed Caspian Sea (§ 335) by a great delta. The short northern slope of Eastern Europe is occupied by the basins of the Pechora flowing to the Arctic Sea, and the Northern Dwina to the White Sea.

388. Lake District of Northern Europe.—North of the Baltic the long slope of the peninsular mass of land, including Scandinavia and Finland, is toward the south. The great Lake Ladoga, which discharges its overflow by the short swift Neva into the Baltic, receives the drainage of a vast lake district—Lake Onega on the north, Lake Ilmen on the south, Lake Saima and innumerable connected lakelets on the west, all draining to it. At Imatra, on the river joining Lakes Saima and Ladoga, the most impressive cataract in Europe is formed in a nearly flat country by the water pouring through a narrow and steep bed of hardest granite, which converts the course for more than a mile into a thunderous mass of feathery foam and leaping yellow waves. All the lake-basins of this district are due to glacial action, and date from the same period as those of North America. They are, as a rule, shallow, some having been scooped out of a flat floor of crystalline rock, while others are formed by the irregular accumulation of glacial detritus (§ 332). About one-sixtieth of the area of Europe is covered with lakes, but in the district of Finland the proportion is one-tenth.

389. The British Islands.—An upheaval of 300 feet would convert the bed of the North Sea, south of a line drawn from St. Abb's Head to the Skaw, into a low plain continuous with that of England and of Northern Europe. During the evolution of Europe elevation and subsidence have repeatedly raised the whole region into land and again lowered it under water. Viewed as a whole, the island of Great Britain is higher toward the west than the east (see Plate XVI.) The watershed lies near the west coast, giving a long east slope traversed by the longest rivers. The east coast is comparatively smooth, with occasional wide funnel-like estuaries and scarcely any islands; while the west coast is very deeply indented by winding fjords or sea-lochs, and many groups of large and often lofty islands. No true mountain ranges can now be traced in the British Islands (§§ 303, 329). Glacial action has been traced over all the British Islands except the extreme south of England, and the existing configuration has thus been modified in most places. Mountains, below the height of 3000 feet at least, have acquired a more or less flowing outline through glacial grinding; and the low land has been largely enveloped in boulder clay and similar accumulations.

390. **Scotland.**—On the map of vertical relief (Plate XVI.), the northern part of Great Britain is seen to be divided into three natural regions stretching across the island from north-east to south-west. Most of the area north-west of a line drawn from the Firth of Clyde to near Aberdeen is occupied by the Highlands. This is an old plateau, largely composed of crystalline schistose rocks, and pierced by many granite-like masses. The heights, separated by deep valleys, are rugged and often precipitous, crowned by crests of splintered rock. The Highlands are divided into a northern and a southern group by the Great Glen which unites the Moray Firth with the Firth of Lorne, and contains Loch Ness and Loch Lochy, two long narrow fresh-water lakes. The highest point of Great Britain is the mass of Ben Nevis (4400 feet), near the south-western extremity of the Great Glen, but a greater area of scarcely lower elevation occurs round Ben Macdhui. South of the Highlands stretches a broad low plain—the Midland Valley—diversified by lines of hills like the Pentlands, Ochils, and Sidlaws, and isolated precipitous crags such as those occupied by the castles of Dumbarton, Stirling, and Edinburgh. These abrupt heights are due to masses of hard volcanic rocks formed in the Carboniferous period or later, and now exposed by the more rapid erosion of the softer strata which had buried them. Along the border of the Highlands there is a strip of Old Red Sandstone sharply separated from the crystalline schists, slates, etc., by the Great Fault which runs from the Firth of Clyde to near Aberdeen. Along the southern edge of the plain a similar strip of the same formation is terminated by a line of faults stretching from near Ayr to near Dunbar. Carboniferous strata—with the coal-measures cropping out in several places—occupy the centre of the Lowland Valley. The Southern Uplands, which form the third division, are a group of rounded grassy and often peat-topped hills, lower

than the Highlands, divided from each other by gently sloping valleys, and composed mainly of Silurian rocks, although the Cheviot Hills on its southern boundary are largely of igneous origin.

391. **England.**—The mountainous Lake District of north-western England has been carved by erosion from great masses of Silurian rock, but numerous outbursts of ancient volcanic material have given ruggedness and grandeur to many of the summits. The mountains of North Wales culminating in Snowdon (3570 feet) generally resemble those of the Lake District in their geological structure. They slope in steep terraces to the sea on the west, and dip down more gently to the low plain of England on the east. In South Wales the mountains of circumdenudation are lower, and the Silurian rocks give place to Old Red Sandstone. This is in turn covered by a great expanse of Carboniferous rocks in the south, where the coal-measures come to the surface. Ancient Primary rocks, especially lower Carboniferous and Devonian strata, build up the peninsula of Devon and Cornwall, but great intrusions of igneous rock form the hard framework which the sea has wrought into a coast-line vying in grandeur with that of the north-west of Scotland. The band of high land in the north of England known as the Pennine Chain slopes to the sea on the east, adjoins the Lake District on the west, and to the south-west and south gradually spreads out and sinks into the low plain. The hills and dales of this region are carved out of a great anticline of Carboniferous rocks, comprising limestones, coal-measures, and grits or coarse sandstones. The crest of the anticline has been denuded down to the grits, while the coal-measures and limestones crop out on the slopes, forming extensive coal-fields. All the rest of England to east and south is occupied by a great low plain built up of Secondary and Tertiary rocks, the elevation of which scarcely anywhere exceeds 600 feet. From this plain the rugged heights of Primary rocks in the west and north rise as from a sea, the whole character of their scenery contrasting with its gentle ridges and low undulations. An irregular line of heights forming a steep escarpment to the west and

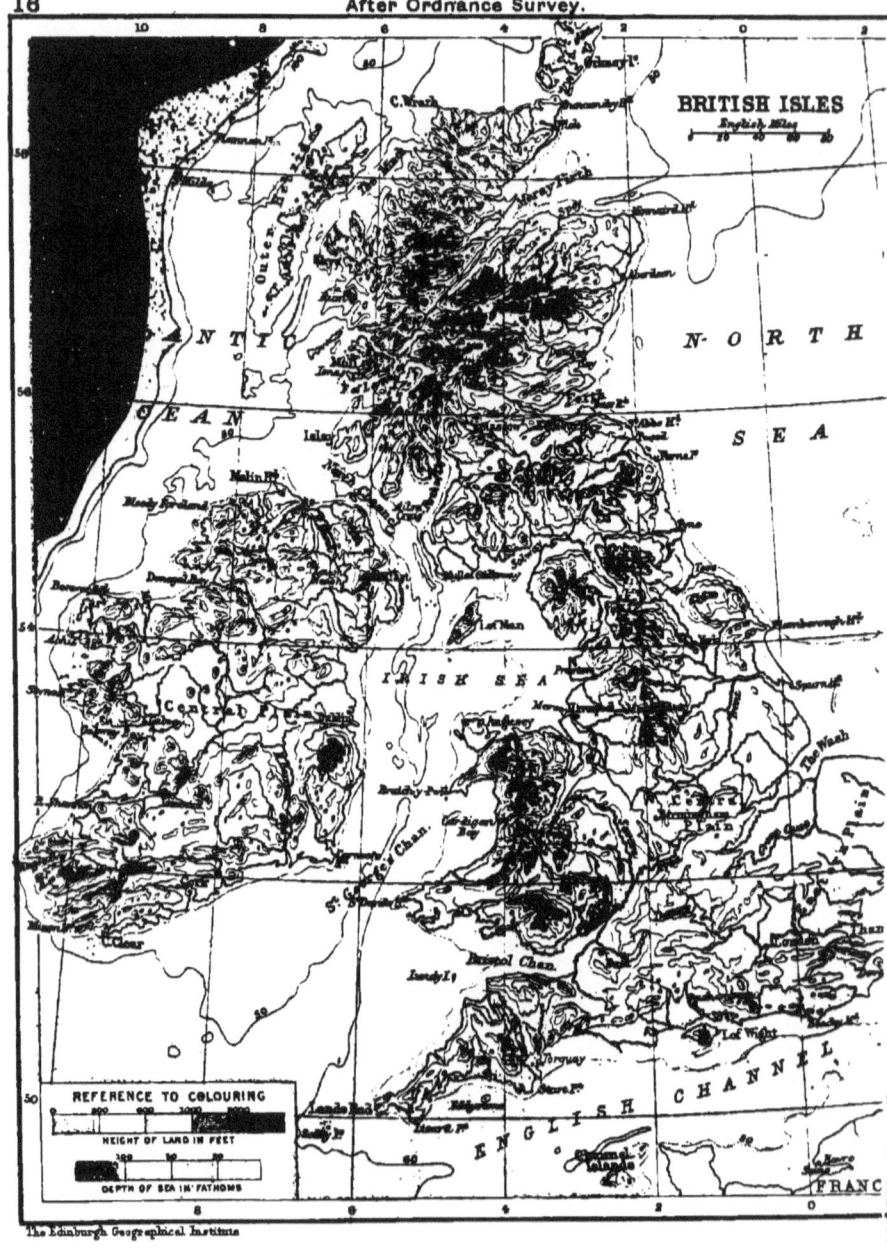

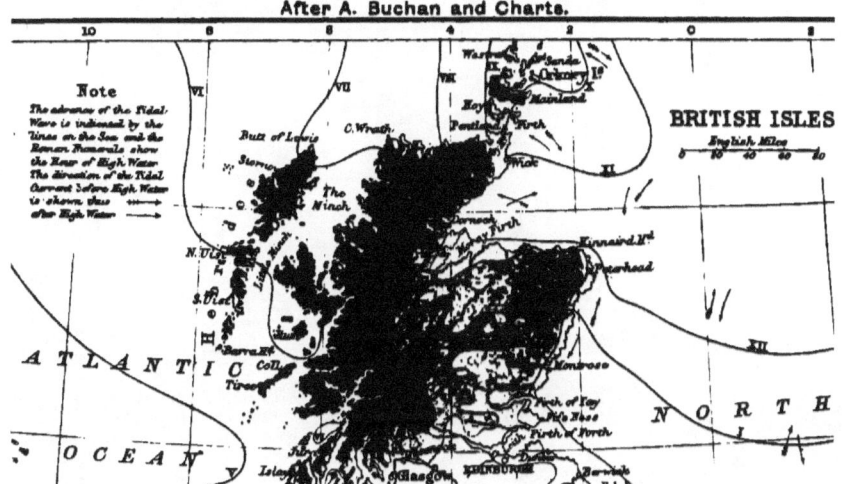

MEAN ANNUAL RAINFALL AND CO-TIDAL LINES.
After A. Buchan and Charts.

a gentle slope to the south-east overlooks the Severn Valley as the Cotswold Hills. It is continued north-eastward to the Humber and thence on the other side of the estuary northward, where it swells up into the Yorkshire Moors, and terminates in a line of cliffs along the coast. This edge is the outcrop of a great belt of relatively hard oolitic limestone (Jurassic period) which dips gently to the south-east, and is separated by a line of older but less durable Secondary rocks from the Primary system in the north and west. A similar but more broken escarpment is formed farther south by an outcrop of Cretaceous rocks, which also dip gently to the south-east. This Chalk ridge reaching its greatest height in Salisbury Plain, the Marlborough Downs, and the Chiltern Hills, is continued in the lower East Anglian heights running north-eastward through Norfolk. It appears north of the Wash as the Wolds of Norfolk, and north of the Humber as the Yorkshire Wolds, terminating in Flamborough Head. From Salisbury Plain two low chalk ridges diverge: one runs eastward as the North Downs, the other south-eastward as the South Downs, and both end in the Chalk Cliffs of Kent. The River Thames rising on the southern slope of the Oolitic ridge, flows through the Chalk ridge between the Marlborough Downs and Chiltern Hills, and turns eastward to the North Sea. Its triangular valley between the Chiltern Hills on the north and the North Downs on the south is occupied by tertiary rocks, consisting of clay, sands, and marls of Eocene age.

392. **Ireland.**—The east coast of Ireland is comparatively low and unindented, while the west coast is cut into many long inlets lined by lofty cliffs and fringed with islands. The configuration of Ireland is entirely different from that of Britain. A low plain occupies the whole interior, and its elevation is so slight that a subsidence of 250 feet would unite the Irish Sea and the Atlantic across the island. Isolated groups of lofty mountains rise at irregular intervals round the outer edge, the highest being Carn Tual (3400 feet) in the south-west. The Shannon, the largest river, flows southward along the centre of the plain, and turns westward into the Atlantic. Geologically the low plain of

Ireland is composed of a vast expanse of the Carboniferous formation, in which the coal-measures are only slightly developed. The mountains are islands of more ancient rock, Silurian and Old Red Sandstone, with metamorphic schist and gneiss, like those of the Highlands, in the north-west. Great masses of volcanic rock occur in the north-east, where the basaltic columns of the Giant's Causeway form one of the wonders of the world. These harder rocks are prominent on account of their resistance to the erosion which planed down the soft strata into a uniform surface. The centre of Ireland is full of shallow lakes surrounded by peat-bogs, formed by the decay of vegetation in the wet climate on ground too flat to allow of natural drainage.

REFERENCES

[1] J. Murray, "On the Height of the Land and the Depth of the Ocean," *Scot. Geog. Mag.* iv. 1 (1888).

[2] Rohrbach, "Continental Distances." See *Scot. Geog. Mag.* vii. 213 (1891).

[3] J. Murray, "Drainage Areas of the Continents," *Scot. Geog. Mag.* ii. 548 (1886).

BOOKS OF REFERENCE

Longmans' *New Atlas.*
J. G. Bartholomew, Macmillan's *School Atlas.*
H. R. Mill, *Elementary General Geography.* Macmillan and Co.
For explorations in little-known regions see *Proceedings of the Royal Geographical Society.*
For papers and references regarding Physical Geography see *Scottish Geographical Magazine.*

CHAPTER XVI

LIFE AND LIVING CREATURES

393. The World without Life.—The World as a whole may be compared to a great house. Geology describes its materials, records the process of building, and keeps account of the alterations which are always being carried out. Oceanography has to do with the currents of water interchanged between the tropical boilers fired by the central furnace of the Sun and the polar refrigerators. It explains the arrangements by which those rooms most exposed to the furnace are cooled down by iced water, whilst those more remote have their temperature raised by copious hot streams. Geology records many past contests between the furnace and icehouse in controlling the heating arrangements, and many changes in the direction of the hot and cold water-pipes. Meteorology discusses the still more complicated and variable methods of ventilation in use in various rooms, depending as they do on the circulation of water and on the structure of the buildings. Astronomy has something to say as to the arrangements for lighting the great house, explaining how each room is illuminated with a certain brilliancy for a special time. Astronomy also supplies reasons for the changes in the strength of furnace and refrigerators in the past. Geography concerns itself with the plan of the house so far as it is completed, showing the dominant style of architecture and tracing the modifications adopted in the several parts, and gives a general view of all the arrangements.

394. Life in the World.—Geology and Oceanography bear evidence of changes in structure which cannot be explained by the laws of matter and energy. These laws enable us to understand that water should in certain conditions dissolve carbonate of lime and silica. But they cannot account as yet for the opposite process which is at work in exactly the same physical conditions. Carbonate of lime and silica separate out from solution and assume the solid form, not with the uniform sharp angles and smooth faces of crystals, but with curved and varied outlines decorated with delicately-etched designs of infinite variety (§ 273). Fossils are evidently due to a similar temporary reversal of ordinary chemical and physical change. These reversed processes are recognised as the characteristic result of life. Geology may be said to present us with a view of the world as a vast cemetery full of monuments of past generations of living creatures. When we look around us in the open country our eye is not, as a rule, attracted by bare rocks or soil, but by a covering of grass, flowers, and trees, amongst which beasts and birds and insects are moving. These are the living inhabitants of the great World House. Between them and the rooms they inhabit there is a close and ever-varying relation, the comprehension and description of which is the central aim of Physiography.

395. Classification of Living Creatures.—Every one can tell at a glance that a bush and a cow belong to widely different classes; indeed a close observer might fail to find anything in common between them. It is easy and natural to class trees, bushes, herbs, grass, and even seaweeds, as essentially similar, and to recognise them all as *Plants*. Similarly, although four-footed beasts, birds, reptiles, insects, and fishes, differ a good deal amongst themselves, they are classed together, almost without a thought, as *Animals*. A great gulf seems to separate the Vegetable and Animal kingdoms, to use the names given by Linnæus who laid the foundations of the modern classification of creatures. Plants are rooted in the soil; animals are free to move over the land, through the water or air. When carefully studied both of the great kingdoms are found to fall into a number of natural groups, the members of which show a regular advance in complexity of structure. Between the simplest groups of each kingdom it is difficult and often impossible to trace any difference. All living creatures are termed organisms, and the science which takes account of them with special regard to their common characteristics is termed *Biology* (literally Life-lore). The classification and life-history of plants are the objects of the department of Biology known as *Botany*, while the department known as *Zoology* is similarly occupied with the study of animals.

396. **Classes of Plants.**—Botanists group plants into sub-kingdoms, classes, natural orders, genera, and species. A *species* includes all the individual plants, which are so much alike as to make it certain that they are descended from the same stock and which are mutually fertile. A *genus* includes a group of species closely related to each other. A group of related genera forms a *family*, a number of allied families forms an *order*, and the orders are themselves grouped in *classes*. Thus, for example, in the class of Dicotyledons there is an order called Ranunculaceæ, which includes several families and many genera, amongst others that of Ranunculus, which in turn includes many distinct species. Following the suggestion of Linnæus, each species, that is each separate kind of plant, is known to botanists by the name of its genus, followed by a specific name. One particular kind of buttercup is thus termed *Ranunculus acris*. The classes of plants, with a typical example of each, are as follows:—

    I. THALLOPHYTES (*no stem*).
        PROTOPHYTA—Bacteria.
        ZYGOSPOREÆ—Diatoms.
        OOSPOREÆ—Fucus.
        CARPOSPOREÆ—Most Seaweeds and Fungi.
    II. MUSCINEÆ.
        HEPATICÆ—Liverworts.
        MUSCI—Mosses.
    III. VASCULAR CRYPTOGAMS.
        EQUISETINEÆ—Horsetails.
        FILICINEÆ—Ferns.
        LYCOPODINEÆ—Club-mosses.

IV. PHANEROGAMS (*flowering plants*).
  GYMNOSPERMS—Pines and Firs.
  MONOCOTYLEDONS—Lilies.
  DICOTYLEDONS—Buttercups.

397. **Classes of Animals.**—Animals are more numerous and varied in their kinds than plants, and their classification, according to resemblances and differences, is in consequence more complex. Species, genera, families, and orders are distinguished much in the same way as with plants, and animals also are named after both genus and species. The great groups into which they are divided (and the classes of the last group), with typical examples, are as follows :—

  PROTOZOA—Radiolarian, Foraminifera, Amœba, etc.
  PORIFERA—Sponge.
  CŒLENTERATA—Jellyfish, Sea-anemone, Coral.
  ECHINODERMATA—Starfish, Crinoid, Sea-urchin.
  VERMES—Worms.
  ARTHROPODA—Lobster, Barnacle, Millipede, Spider, Insects.
  MOLLUSCA—Oyster, Snail, Pteropod, Cuttlefish.
  PRIMITIVE VERTEBRATES—Tunicate, Lancelet.
  VERTEBRATA—*Fishes*—Flounder, Salmon, Shark.
    Amphibians—Frog, Newt.
    Reptiles—Turtle, Serpent, Lizard.
    Birds—Eagle, Ostrich, Sea-gull, Sparrow.
    Mammals—Kangaroo, Lion, Ox, Whale, Ape, Man.

398. **Functions of Living Creatures.**—The simplest organism or the unit-mass of any living creature is merely a jelly-like speck made visible by means of the microscope. Part of the jelly-like substance may form a darker nucleus in the interior, and in some cases a tougher film is seen to surround and contain the whole. The organism is said to consist of a single cell. The jelly-like substance called *protoplasm* is a complex kind of matter, the precise nature of which is unknown, but it consists mainly of carbon, oxygen, and hydrogen, with minute quantities of nitrogen, sulphur, and phosphorus. Living protoplasm is continually undergoing two opposite sets of changes—building up or renewal, and breaking down or decay. The process of building up, which is distinctive of living creatures alone,

involves nutrition or the taking in of food-substance, digestion or the elaboration of food, and assimilation or absorption into protoplasm. While this process goes on the organism grows by the assimilation of unlike substances, which are transformed into protoplasm and added to the mass from within and throughout. The simultaneous breaking-down process, on the most commonly accepted theory, is brought about by respiration or the absorption of oxygen. Protoplasm is an extremely unstable compound, always ready to combine with oxygen and break up into carbonic acid, water, and a very small proportion of a few other stable compounds. The living protoplasm is purified by the process of excretion, which is simply the thrusting out of the burnt products (carbonic acid, water, etc.) and of those parts of the food which escape digestion. When life ceases, protoplasm ceases to grow, oxidation continues unchecked, and the organism breaks up and decays away by slow combustion. In the process of growth, matter which is not living may be built into the substance. For example diatoms and radiolarians, which are single-celled organisms, form coats or skeletons of silica, and foraminifera, also consisting of one cell, secrete hard shells of carbonate of lime (§ 273). All organisms, except the protozoa and the simplest plants, consist of many cells containing protoplasm, built up into organs set apart for special purposes. These cells are usually supported in a framework of matter such as wood or bone, elaborated by the living organism and sharing its life for a time, but becoming practically lifeless as they grow older. When a cell grows, it increases in size to a certain limit and then divides into two cells, the process being termed *reproduction*. In the protozoa the division of a cell is complete separation, producing two individuals; but in higher organisms a single cell, termed an *ovum* or egg-cell, is separated from the rest, and grows by subdivision into a separate many-celled organism similar to the parent form. Most often, both in plants and animals, this liberated cell is unable to develop until it unites with a cell of another kind (termed a male cell) from the same species. Thus the continuance of the species is secured in spite of the death of the individual.

399. Constructive Plant Life.—Plants alone are able to raise inorganic substances, such as water, oxygen, carbonic acid, into the sphere of life-wrought or organic material. They cause the elements to combine into *proteids*, the raw material of protoplasm. This power in its entirety is confined to those plants which possess green leaves, and is exercised by them only when the energy of sunlight falls on the green colouring matter known as chlorophyll. Then the leaf is able to break up carbonic acid derived from the atmosphere, to restore the oxygen to the air, and cause the carbon to combine with the elements of water, forming starch which is at first stored up amongst the cells of the leaf. Subsequently the starch is transformed into sugar, which dissolves in the sap and is carried through the whole plant. On meeting the nitrates, sulphates, phosphates, and other salts of lime or potash, absorbed from the soil by the roots, the sugar combines with them, producing proteids and various waste products in a manner not yet discovered. The influence of green leaves on the air in sunlight is to unburn or decompose (§ 44) the carbonic acid. The solar energy used up in this work is converted into potential energy of chemical separation, which is restored to the kinetic form when wood or coal (§ 347) unites with oxygen. The oxygen given out by the action of chlorophyll in the leaf laboratory is more than enough to supply the ceaseless respiration of the plant in daylight and darkness so that, on the whole, green plants diminish the proportion of carbonic acid and increase that of oxygen in the air.

400. Destructive Animal Life.—Contrasted with the constructive processes of plants, changing lifeless into living matter and kinetic into potential energy, animals are wholly destructive. They cannot utilise solar energy, but derive all their power of doing work from oxidation of their own substance. They cannot manufacture proteids, so that all their food has to be prepared for them by plants. Animal life would indeed be impossible if plant life did not precede it. In their respiration animals are always removing oxygen and increasing the amount of carbonic acid in the atmosphere (§ 154). Those plants which do not contain chlorophyll, such as the fungi, moulds, and bacteria, are as powerless as animals to manufacture food from carbonic acid and water. But unlike animals they have the power of manufacturing proteids if they obtain starch or sugar and the various salts amongst their food. Thus fungi—all the mushroom kind —grow abundantly only in decaying vegetable substance, which supplies plenty of starch. To sum up in a metaphor, the green plant, like a coal-laden steamer, conveys solar energy—using up some in the process—to the animal, which like a stationary steam-engine converts it into work.

401. **Micro-organisms.**—Many minute one-celled organisms, probably plants allied to the fungi and moulds, known as bacteria, bacilli, microbes, or classed together as micro-organisms, play a very important part in the course of their life-history. One of these, known as the *nitrifying ferment*, changes the salts of ammonia derived from the atmosphere, or from decomposing animal matter, into nitric acid in the soil, thereby greatly facilitating the growth of plants (§ 311). Another known as *yeast*, when cultivated in a weak solution of sugar, uses up some of the sugar, and changes the rest into carbonic acid and alcohol, hence it is extensively used in raising bread and in making wine and beer. A different micro-organism changes alcohol in the presence of air into vinegar, and is extensively cultivated for that purpose. The spores, or young undeveloped cells, of many kinds of micro-organisms form a considerable part of the dust in air (§ 161), and are present everywhere. When these find a suitable place to grow in—for example, the blood or the tissues of a person not in strong health—they develop and multiply, producing by their vital processes certain poisons, which give rise to disease. Different species of micro-organisms have been detected as the cause of cholera, consumption, diphtheria, and other maladies. The recognition of this cause of disease has led within the last few years to the greatest modern advances in medical treatment.

402. **Evolution.**—As the Earth, like other members of the solar system, is the result of a slowly unfolding series of

changes; as the continents have by long and gradual degrees come to their present form, and are still undergoing alteration,—so also living creatures display a progressive evolution. The classifications of plants (§ 396) and of animals (§ 397) are ascending scales, showing in each group a more complex structure and organs more distinctly set apart for special purposes. Amongst animals, for example, the protozoa have no organs at all; the single cell acts as a whole in every function. In the echinoderms, eyes and a separate stomach appear; in the arthropoda, limbs adapted for walking; and an internal skeleton connected to a backbone, and supporting the framework of the body, is only found in the vertebrata. Similar progressive advancement is to be found within each group, and even in the same species individuals vary so much that a regular gradation may often be traced into other species making it difficult to draw the dividing line. Transition types, such as Archæopteryx (§ 349), a bird partly resembling a reptile, and the Australian duck-bill, which although a mammal has a beak like a bird and lays eggs, connect the different classes of animals or of plants. When this regular order of succession from lower to higher forms in plants and animals became apparent to biologists they were convinced that different species had not been created separately in different places, but had gradually developed in the course of ages from a common parent form. The late Charles Darwin and Mr. A. R. Wallace almost simultaneously framed a theory to account for organic evolution—the gradual unfolding of the progressive design of plant and animal life; and the period of most rapid advance in modern biology dates from the publication of Darwin's *Origin of Species* in 1859. The original views of Darwin and Wallace are gradually being modified as new facts are encountered and the general principles of evolution stand out more clearly.

403. **Heredity and Environment.**—Darwin explained the origin of different species of living creatures by the two great influences of *heredity* or likeness to parents and *environment* or surrounding circumstances. As a rule the young of plants and animals resemble their parents, but no

two are precisely like each other. General similarity is associated with small variations of structure. Sometimes these variations produce no influence on the life of the organism, and may pass unnoticed. But when they happen to make one individual better fitted for obtaining food or escaping danger than the others, that one has a better chance of living, thriving, and handing on its fortunate peculiarities to its descendants. If the variation of structure throws an individual out of harmony with its environment, making it weakly or stupid, that individual has a smaller chance of surviving and leaving offspring. According to Darwin's view the constant struggle for life is always weeding out the weak and improving the position of the strong, leading by a process of natural selection to the survival of the fittest. But climate, and even the outline and configuration of the land, are not constant; hence organisms, hitherto victorious in the struggle for existence, have to contend with an altered environment, and their development, according to natural selection, must after a time take place in a new direction with great sacrifice of life, and possibly the extinction of some species. This subject is far from simple, many of the facts have still to be discovered, and none of the hypotheses as yet can compare for certainty with theories that admit of mathematical proof. An excellent idea of the difficulties and the fascinating interest of biological facts and theories will be obtained from Professor Geddes's *Modern Botany*, and Mr. J. Arthur Thomson's *Animal Life*, in this series.

404. **Conditions of Plant Distribution.**—Plant life, as a rule, is most luxuriant where there is abundant sunlight, high temperature, copious rainfall, and soil abounding in the soluble salts necessary for nutrition. In the course of the ages plants have gradually been modified, so as to adapt themselves to their environment. Thus not only the comparative luxuriance, but also the species of plants, depends to a large extent on the conditions of their growth. Where natural conditions change abruptly, as, for example, on the sea-coast, on the slopes of a snow-clad mountain, or the edge of a desert, the kinds of creatures inhabiting the two regions differ in a very marked way. If such barriers are

developed in a region formerly of uniform configuration and climate, similar plants may become separated by quite different species produced by the new conditions. While all animals are absolutely dependent on plants for food, some plants are in great part dependent on animals for their continued existence, and bright flowers, perfumes, and honey have an important office in attracting them. Insects especially carry pollen from one flower to another, and so secure cross fertilisation, which greatly improves the strength of the seed.

405. **Floral Zones.**—Speaking widely, the luxuriance and variety of vegetation decrease from the equator to the poles, and from sea-level toward the summit of mountains. Fig. 63, adapted from Smirnoff's Russian *Physical Geo-*

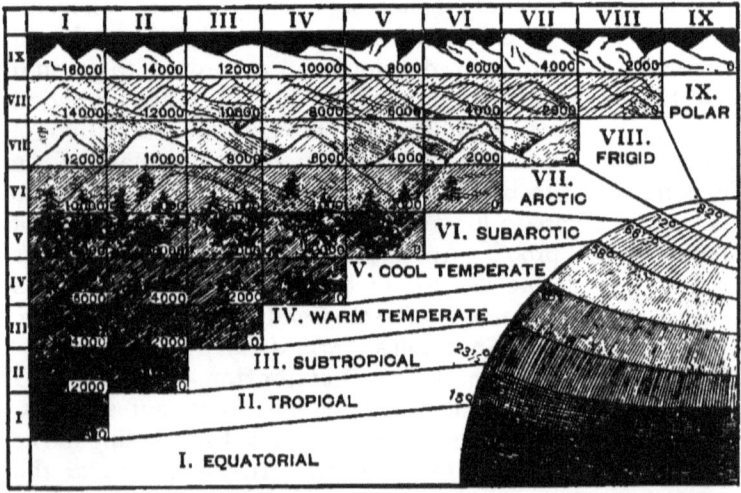

FIG. 63.—Zones of climate and vegetation in latitude and altitude (after Smirnoff).

graphy, represents a quadrant of the Earth's surface divided into climate zones at sea-level. The *Equatorial* zone corresponds to the region of maximum heat and rainfall; the *Tropical* to the region of maximum heat and small rainfall. The *Subtropical*, *Warm Temperate* and *Cool Temperate* zones show a gradual transition to the *Subarctic*, in which long cold winters produce a dwarfing effect on vegetation. The *Arctic* zone of stunted plants leads to the *Frigid*, and that to the unchanging ice-deserts of the *Polar* zone. The vertical columns represent slices of 2000 feet of mountain-sides from the region above the snow-line (shown at the top of each column) down to sea-level. The horizontal rows show by their connecting lines at what average height the climate and vegetation corresponding to each of the sea-level zones is attained. Dr. Oscar Drude divides the Earth according to the affinities of its vegetation into three great divisions—the *Boreal* or Northern, the *Tropical*, and the *Austral* or Southern. In each one of these the species of plants are closely allied to each other, but distinct from those inhabiting the other divisions. The **Austral Group** includes the parts of the three southern continents south of the tropic of Capricorn, and falls naturally into an *American*, *African*, and *Australian* division. The flora of Australia is unlike all the others; there are trees, such as the eucalyptus or gum-tree, which are evergreen but shed their bark yearly; the wattle (a kind of acacia) and the beef-tree, which bears long green branchlets instead of leaves. The **Tropical Group** extends from the tropic of Capricorn northward to the Tropic of Cancer in America, to the centre of the Sahara in Africa, and to the Himalaya in Asia. It also contains three main divisions. Cinchona, mahogany, and the cactus are characteristic of the *American section*; the oil-palm, baobab, and giant euphorbias of the *African*; and teak, banyan, and sandal-wood of the *Oriental*. The **Boreal Group** is remarkable for the wide range of plants of similar species, such as the pine, birch, and oak, over the *Northern division* in the three continents—in America north of the Great Divide, in Europe north of the southern peninsulas, in Asia the whole northern slopes. The other divisions of this group are the *Eastern Asiatic*; the *Central Asiatic*, comprising the vast plateau region; the *Mediterranean lands*, where the olive, mulberry, chestnut, orange, and cork-oak flourish; and the *Central North American*, the natural home of maize, tobacco, and the giant pines of California.

406. Deserts.—In many respects Plate XVIII. gives the most interesting division of the world according to its vegetation. It shows three great barren zones forming broken girdles round the Earth, and covering, according to Mr. Ravenstein's calculation, more than 4,000,000 square miles. *Ice deserts* surround the north pole, and are succeeded in the north of Europe, Asia, and America by a belt of frozen land called the *Tundra*, which thaws on the surface in summer and supports a thin growth of moss and stunted grass. *Arid deserts* occur in all areas of great heat and very small rainfall. A northern zone includes the vast Sahara, the interior of Arabia, and Central Asia, terminating in the dreary Gobi, and the Great Basin of North America. Horny cactuses, the saxaul with foliage like wire, and the dull-gray sage-bush, are characteristic of the scanty plant life. A similar set of smaller deserts appears in the southern hemisphere, near the cooler but drier western sides of the continents, the Kalahari in Africa, the great Victoria Desert in Australia, and the small salt desert of Atacama in South America, forming links in the chain. Solar energy here falls on barren land, and, not being absorbed by plants, does the work of heating air and maintaining the permanent winds of the globe, which carry rain to more favoured regions. Thus in a sense the existence of fertile lands is a consequence of deserts. Treeless plains are common in all regions of scanty rainfall and great range of temperature, such as the borders of deserts. They occupy about 14,000,000 square miles of surface, covered with rich grass during part of the year, transformed into deserts of driving dust in the dry season, and flooded or covered with snow, according to the climate, during the rainy season or winter. The fertile *prairies* of North America, the *llanos* and *pampas* of South America, and the *steppe-lands* of Russia and Central Asia, are examples of such semi-deserts.

407. Tropical Forests.—When the grassy plains surrounding the tropical deserts on the equatorial side begin to receive a larger rainfall, bushes first break their monotony, and then great forests are formed, the trees standing well apart, but growing closer as the heavy rains of the equatorial zone are approached. The densest forests naturally extend on both sides of the equator, where heat and rainfall unite to produce a paradise for plants. The *Selvas* of the Amazons, the darkest forests of the Congo and its tributaries, the forests of the Western Ghats of India, of the west coast of the Malay Peninsula, and of the islands of the Malay Archipelago, vie with each other as types of the utmost wealth of vegetation. Soft leafy canopies borne by lofty evergreen trees meet and intercept the light, so that no grass can grow in the dark depths of the woods, but climbing and twining plants innumerable, with stems like ropes or cables, force their way up on the trunks of their stouter rivals, and push on to expand their crown of leaves in the sunlight. The decaying vegetation below supplies abundant nourishment for pale-coloured parasitic plants, which, deprived of sunlight, have lost their chlorophyll and the power to manufacture food, and therefore live on their fellows.

408. **Temperate Forests.**—On the temperate side of the tropical deserts, the plains reaching into regions of moderate warmth and moderate rainfall become covered with less luxuriant but very extensive forests. These are most developed around the great lakes of North America, in Scandinavia, and as a broad belt from the Carpathians north-eastward to the Baltic, eastward to the Ural Mountains, and beyond them across Asia north of 50° to the Pacific Ocean. In Western Europe the ancient forests—which appear to have once formed an unbroken belt across all the northern continents—have been cut down and the land cultivated. The warm temperate forests are composed of deciduous trees, that is, trees whose leaves wither and drop each winter, the leaf laboratories being shut up in the comparatively sunless months. Oak, beech, elm, ash, lime, and many other kinds of forest tree, are found in their greatest luxuriance in this zone. Toward the pole, where the winters are longer and more severe, the deciduous trees vanish, the hardy birch, with its silvery bark, reaching farthest north. Pines and firs, clad in small, hard, needle-shaped leaves, can alone resist the climate, and vast forests

of these characterise the subarctic zone and the higher slopes of mountains.

409. **Animals and their Life Conditions.**—The life conditions of some of the marine animals of most importance from the physiographical standpoint, have already been touched on (§§ 273, 279). Amongst all animals the struggle for life is harder, or at least more apparent, than with plants, the stronger hunting down and devouring the weaker. Animals in their native haunts should therefore be inconspicuous if they are not to attract the attention of their enemies, or to arouse the suspicion of their prey. Almost all fishes, and many caterpillars, rapidly assume the colour of their surroundings. The hare and ptarmigan, living amongst the brown heather of northern hillsides in summer, are brown in fur or plumage, but in winter, when the land is white with snow, their colour also changes to white, and they remain inconspicuous in their new surroundings. This periodical adaptation to environment, which is common in Arctic animals, is one of the causes which has led to the preservation of the race. Some insects are so like withered leaves or twigs that even an experienced eye is often deceived by them. Strange resemblances have also been traced out between entirely different species of animals; and since the similarity is always brought about by the weaker or inferior type assuming the appearance of the stronger or superior, almost as if of purpose to impose on enemies, it is called *mimicry*.

410. **Faunal Realms.**—Animals exhibit more marked peculiarities of distribution than do plants. (Similar forms are usually, though by no means always, found in like conditions.) The *fauna*, or collection of animals, of each one of the northern continents bears a close resemblance to that of the others; while the *faunæ* of the three southern continents are similar in a much less degree, and, as a rule, totally unlike that of any of the northern. The most generally accepted division of the Earth into realms occupied by different faunæ is that suggested by Dr. Sclater, shown in Plate XIX. The names adopted for these divisions or realms are—the *Palæarctic* or Old Northern, the

Ethiopian, the *Oriental*, the *Nearctic* or New Northern, the *Neotropical* or New Tropical, and the *Australian*. Professor Heilprin, another eminent authority, prefers to class the Palæarctic and Nearctic realms together, on account of their general similarity, as the *Holarctic* or Entire Northern. He also recognises a region of transition to the Neotropical realm occupying the south of North America, and another of much greater extent forming a transition to the Ethiopian and Oriental realms and including the whole Mediterranean region.

411. **Northern Realms.**—In both the Old and the New Northern realms the white polar bear frequents the northern snow-deserts. Farther south occur reindeer and elks, bears—black, brown, or grizzly,—foxes, wolves, beavers, hares and squirrels, and the bison, now almost extinct in Europe and rapidly being exterminated in America. The representatives of various families become more unlike each other toward the southern border. Moles, rats and mice, badgers, sheep and goats, the camel and the yak, are confined naturally to the Palæarctic realm. On the other hand, the musk-ox, skunk, prairie dog, racoon, and jumping mouse, are exclusively restricted to the Nearctic realm. Compared with the southern realms, those of the north are remarkable for the high place in the scale of development occupied by their most common animals. But the very complete study of the fossil forms of life preserved in the rocks shows that in past ages the northern lands were inhabited by a gradually developing series of more primitive types, from which the existing creatures are evidently descended.

412. **Ethiopian Realm.**—Africa, south of the Sahara, and Arabia, contain few or none of the animals which make their home round the Mediterranean at the present time. There are no wolves, foxes, bears, or tigers, the flesh-eating animals being represented by the lion, "the king of beasts," the leopard, panther, hyæna, and jackal. This purely tropical realm is the exclusive home of the hippopotamus and the giraffe, tallest of living animals. The elephant and rhinoceros are common also to the Oriental realm. Swift-footed,

graceful, and fantastically striped zebras and quaggas frequent the grassy plains. Of all African animals the most widespread and characteristic are the antelopes, which gallop in vast herds over the plains, and, ranging in size from an ox to a rabbit, inhabit bush, forest, and desert as well. Apes—narrow-nosed, tailless creatures of the monkey kind—are very common in all parts of the continent. The forests are the chosen home of the most highly developed and fiercest, the gorillas and chimpanzees. The ostrich, the largest bird in the world, is typical of Africa, being found in all the open plains and deserts both in the north and south. The adjacent island of Madagascar contains very few of the animals common in the Ethiopian realm, but abounds in lemurs, a kind of half-monkey.

413. **Neotropical Realm.**—South America is richer in varieties of animal life than any other realm, and it is also peculiar for the very large number of species which are found nowhere else. The true monkeys are confined to the great forests, where they swarm in amazing numbers. They differ from the African and Oriental apes mainly in having a broad nose and a long prehensile tail, by which they swing from branch to branch. Vampires and others of the leaf-nosed bats, the rabbit-like chinchilla of the Andes slopes, the beaver-like coypu rat of the plains, and the little agouti, allied to the guinea-pig, are all exclusively South American. So are the more peculiar sloths which swing back downward from the trees, the great bushy-tailed ant-eaters with long slimy tongues specially modified to lick up ants, and the curious armour-clad armadilloes resembling in their habits the hedgehogs of Europe. Although no bears, foxes, or wolves penetrate south of the transition zone, the jaguar, resembling in many respects the tiger of the Oriental realm, ranges over the entire continent, and the puma or American lion even extends far into North America. The llama, alpaca, and vicuna, confined to the upper slopes of the Andes, are closely allied to the camel family, which inhabits only the Palæarctic realm. Neotropical birds are numerous and distinctive, ranging in size from the huge unsightly condor to gem-like humming-birds, which are

smaller than many insects. The rhea of the southern plains belongs to the ostrich family, but, as a whole, the bird-fauna of South America is more allied to the Oriental than to the Ethiopian.

414. **Oriental Realm.**—Animals common in the Palæarctic and the Ethiopian regions meet together in the Oriental realm, and give it a characteristically mixed fauna. Lions, leopards, rhinoceroses, and elephants, almost or quite identical with those of Africa, are found along with bears, wild dogs, foxes, and the true deer so distinctive of Northern Eurasia. Lemurs akin to those of Madagascar are abundant in the south, and the mixture is completed by tapirs and many birds with strong South American affinities. The tiger is peculiar to the Oriental realm, but ranges from Java northward within the borders of the Palæarctic as far as Sakhalin, and is curiously enough absent from Ceylon and Borneo. This realm abounds in squirrels, mice, and bats, and, together with some Ethiopian forms of apes, it affords a home in Borneo to the man-like oran-outan. Although to north and west the Oriental merges gradually into other realms it has a sharp boundary to the southeast, where Wallace in his exploration of the Malay Archipelago found the Oriental species, even of birds and insects, stop at a line drawn between the small islands of Bali and Lombok, and thence between Borneo and Celebes south of the Philippines. Celebes, however, seems to be occupied by a transition fauna.

415. **Australian Realm.**—So peculiar and distinctive is the fauna of Australia and the surrounding islands that many naturalists class it as a main division opposed to all the rest of the globe. Except the dingo or native dog, which may have been introduced by man, the flying foxes (of the bat family), and some birds, none of the animals of other realms occur in it. Their place is taken by the least developed of mammals, the monotremes, of which the duckbill is the type, and the marsupials, represented by the kangaroo. Opossums, living in trees, are the only Australian form of animals, and indeed the only marsupial, found in other continents, a few species occurring in America. The

emu and cassowary are allied to the ostrich family; the bower-bird, which delights in laying out the ground in front of its nest like a garden ornamented with pebbles and flowers, cockatoos, and the black swan, are characteristic birds. Australian animals are found in all the islands of the Archipelago northward and westward to Celebes and Lombok.

416. **Island Life.**—From Wallace's researches in the Malay Archipelago it appears that an entirely different fauna and a largely different flora live on adjacent islands in identical physical conditions. Hence he concludes that the islands on the Australian side of the dividing line have not been united with those on the Asiatic side since the fossil marsupials of the northern hemisphere were alive. It is equally evident that the islands of Lombok and Celebes have been connected with Australia, and that Bali and Borneo have been connected with Asia by land which has been submerged so recently that the organisms have not yet had time to be much modified from the type of their continental contemporaries. Similarity of faunæ between the Malay Archipelago and South America, and many resemblances in the flora of the three southern continents, indicate the probability of a former Antarctic land connection right round the world, which is not contradicted by the configuration of the bed of the Southern Ocean. Purely oceanic islands are usually inhabited only by species which might have been conveyed by sea from the nearest continent, and often contain very remarkably modified forms.

417. **Action of Living Creatures on the Earth.**—The processes of erosion by which the continents are carved into their present form are largely modified by the action of living creatures. Corals and other marine organisms are powerful agents in rock-making (§ 280). Forests, and the growth of vegetation generally, bind the soil together, preventing denudation on mountain slopes, reclaiming alluvial terraces in rivers, and often putting a stop to the drift of sand-dunes. Vegetation also affects climates, producing a uniform rainfall, checking evaporation, and regulating the flow of rivers by absorbing the water of heavy

rain, saving sudden floods, and by keeping up continuous oozing in dry weather, preventing the streams from dwindling away. Disintegrating action is on the whole more frequent. The roots of plants and the little root-like fibres of lichens serve as wedges, splitting up rocks and aiding the formation of soil. Earthworms, termites, and ants (§ 311) aid largely in mixing and pulverising the ingredients of the soil. Boring molluscs drive long narrow holes into the rocks below sea-level, and enable the breakers to produce a much more rapid disintegration of the cliffs than would be possible otherwise. Cray-fishes, burrowing under the banks of rivers, are important agents in causing changes in the direction of the stream and the position of its bed. Beavers have a strange instinct of felling trees and constructing dams across streams to provide an expanse of water in which to build their "lodges." These dams serve to accumulate a head of water, and when burst by a flood the destructive force of the current works great changes on surface scenery. There is no living creature, large or small, which does not leave some trace of its life-work impressed upon the solid globe, and although the individual result of the action of most creatures may be little, the sum of the life of the globe is a very potent factor in the evolution of the conditions which ultimately determine it.

BOOKS OF REFERENCE

(In addition to those mentioned in the text)

Charles Darwin, *Origin of Species, Insectivorous Plants, Formation of Vegetable Mould*, and other books.

A. R. Wallace, *The Malay Archipelago* and *Island Life*. Macmillan and Co.

A. Heilprin, *The Distribution of Animals.* International Science Series.

CHAPTER XVII

MAN IN NATURE

418. **Man as an Animal.**—One genus of the animal kingdom separates itself from the rest in a manner so complete as to require special consideration. It is the genus to which we ourselves belong, and it contains the one species—Mankind. Varieties of this species differ so much amongst themselves physically that there is nearly as great a gap between the most highly developed and the most degraded as between the latter and some of the most developed apes. Organic evolution seems liable to no exception at this point. Mankind is subject to the same natural conditions as other animals, being dependent on plant life for food, and always under the control of heredity and environment. In some respects the human species is inferior to the less developed animals, particularly in the possession of a thin skin without fur or feathers, and in the absence of claws, tusks, or any natural weapons of offence.

419. **Man as Man.**—The differences between Man and the lower animals are so numerous, definite, and distinctive that until within the present century they obscured, even in scientific minds, the full significance of the similarities. (There are no limits to the geographical distribution of the species.) Men live in all the continents, and from the equator to 84° of latitude, but there is no reason to believe it impossible to support life at the poles when they can be reached. Although the contrast between Man and other animals becomes more distinct amongst the higher mem-

bers of the human species, it may be traced in all. It is less of degree than of kind, and is rather intellectual and spiritual than physical. (The use of reason with the associated power of language, the recognition of a Creator, and as a necessary consequence the sense of religious duty, are distinctively human attributes. As these powers become developed, strengthened, and purified, Man advances in the scale of being, independently of his physical development. Heredity and environment acquire new importance, and indeed their existence and potency were first recognised by the way in which birth and education determine the higher powers of the mind. The intellectual as well as the physical unity of the human species is strikingly shown by the fact that even amongst the most advanced peoples there are individuals who exhibit the untamed instincts of the savage, while in the most degraded tribes individuals with some higher powers and finer feelings occasionally rise far above the level of the rest. (By the use of reason men are able to modify or choose their environment,) and thus, consciously or unconsciously, to direct the course of their own development toward advancement or degradation. This power gives to the individual man far greater influence and independence than is exercised by individuals of any other species.

420. **Civilisation** may be defined as the result of men using the power of changing their natural surroundings, and regulating their natural wishes or impulses in order to increase the wellbeing of the community to which they belong. Each variety of the human species appears to be capable of attaining a certain degree of mastery over themselves and their surroundings, this degree being much higher in the case of some varieties than in others. The position occupied by different groups of the human species with respect to civilisation is intimately connected with their conceptions of religion. Tribes of the lowest civilisation live, as a rule, in a state of vague fear of evil spirits and of the ghosts of their ancestors, which they try to appease by worship and sacrifices. They believe that the spirits dwell in rude idols or *fetishes*, to which they

accordingly pay great respect. More civilised peoples, reasoning on the appearances of Nature, are *Polytheists*, or believers in many separate gods, to whom the creation of different parts of Nature is ascribed. *Pantheism* (illustrated by Buddhism) is a development of Polytheism, from which it differs in conceiving God to be present everywhere, and all existing things, Man included, to form part of Him. The highest and most civilised races are *Monotheists*, recognising one God, who created the World and directs its processes of endless evolution. Three forms of Monotheism are prevalent—the Jewish, in which the Old Testament is held as a divine revelation ; the Christian, all sections of which accept also the teachings of the New Testament ; and the Mohammedan, following the Koran, a book compiled from the Jewish and Christian Scriptures by Mohammed.

421. **Environment and Man.**—External conditions do much to determine Man's position in the scale of civilisation. It is matter of dispute whether the different races of mankind result merely from the different conditions in which they have developed, or if changes consequent on moral advancement or degradation have had a large share in producing them. The races lowest in civilisation are most completely slaves to their environment, exercising only the purely animal powers. Where the climate makes clothing unnecessary, and abundant fruit-bearing plants supply the means of life without labour or forethought, as in tropical forests, mankind is found in the least developed or most degraded form. On the other hand, when natural conditions are very hard, the climate severe, and the means of life only to be obtained by chance success in hunting or fishing, the development of intelligence appears to stop short when the prime necessities—food, clothing, and shelter—are secured. The fur-clad Eskimo, feeding on blubber in his ingeniously-constructed house of ice, is certainly an advance on the naked homeless savage of the tropics, who satisfies his hunger with fruits and insects. But both are so exclusively fitted to their environment that the Eskimo pines by the Mediterranean, and the forest Pygmy sickens and dies in the sunlit grass-lands. Intel-

lectual development appears to be stimulated by conditions which make life neither too easy nor too hard. In temperate regions, necessitating shelter and warm clothing, where there is a regular succession of seasons, forethought and thrift are encouraged by the need of providing in summer for the coming winter. Ingenuity has to be exercised in evading the effects alike of heat and cold, and the skill thus acquired finds additional employment in providing ornaments and luxuries to gratify an awakened and cultivated taste. Strength and self-reliance come from the successful struggle with adverse conditions, and many of the characteristics of nations are due as much to the nature of the land they dwell in as to the inherent qualities of the race. Mountaineers of every race are hardier, more independent, and more attached to their native land than the dwellers on low plains, who, on the other hand, work more, excel in perseverance, and are as a rule more successful in obtaining a sufficiency of the means of life. Seafaring peoples, compelled to be continually watching for signs of change in weather, and often called upon to decide quickly and act promptly in circumstances of danger, acquire a distinctive steadiness of nerve and quickness of resource which lead to a general advance in civilisation. (Climate and scenery exercise a powerful influence on moral as well as on physical conditions.) By contrasting the stolid earnestness and ceaseless activity of the dwellers in Northern Europe with the passionate vivacity and general listlessness of Southerners, an ingenious author once went so far as to say that *Character is a function of latitude.* The poetry and the religious systems of all peoples are closely connected with the nature of their land. Patriotism also is a quality derived from the same source, and is shown most intensely by peoples long settled on small but clearly characterised natural regions. The tendency of civilisation is gradually to modify the influence of environment, widening the field of view from that of the family or tribe to that of all mankind, and merging love of country into cosmopolitanism.

422. **Races of Man.**—Certain distinct types of mankind

may be easily recognised, but the transition between them is so gradual that it is almost impossible to draw the dividing line. Students of Ethnology form classes of mankind partly by taking account of physical resemblance and difference, partly by considering the nature of the languages spoken. Following Professor Keane, we may group mankind around three main centres, corresponding respectively to the Black, Yellow, and White types of humanity. The table expresses some of the larger groups, with a selection of illustrative races :—

BLACK	YELLOW	WHITE
WESTERN { Negro, Bantu	MONGOL- { Kalmuck, Kirghiz	ARYAN { Kelt, Teuton, Slav, Hindu
NEGRITO	TIBETO-CHINESE	
EASTERN { Papuan, Australian	FINNO-UGRIAN { Eskimo, Lapp, Magyar	SEMITIC { Arab, Jew
	MALAYO-POLYNESIAN { Malay, Maori	HAMITIC { Berber, Somali
	AMERICAN	CAUCASIC

423. **Black Type.**—This represents the least civilised peoples, and around it is grouped about one-seventh of the World's population. As the name implies, the complexion is black or dark brown. The hair, also black, is woolly or frizzled, and each hair has an extremely characteristic form, resembling a minute flat ribbon. Most of the people of the Black type are tall and powerful, often with well-formed bodies, but with wide flat noses, thick lips, and projecting jaws. They are sensual and unintellectual; like children they are usually happy, light-hearted, and careless, but are subject to moods of depression and outbursts of appalling cruelty. They inhabit the tropics exclusively, except when removed as slaves to warm temperate regions. As a rule, in their own lands they go nearly unclothed, living by hunting or by cattle-rearing, and, in rare cases, following a primitive agriculture. The religion professed is usually a low form of Nature-worship, characterised by fetishism and the practice of witchcraft. Moham-

medanism makes rapid headway amongst some of the tribes, but Christianity seems less adapted to the nature of the Black type. The *Negro* tribes occupying the Sudan region of Africa are the most typical examples. The brown-skinned *Bantus* inhabiting the whole of the Great African Plateau are best known as the Zulu nation of the South. The eastern division of the Black type includes the frizzly-headed *Papuans*, or natives of New Guinea, and the *Australian Aborigines*, who, while probably the lowest race in point of civilisation, differ from the typical Black and approach the White in possessing abundant wavy hair and a full beard. The *Negritoes*, or " Little Negroes," are difficult to classify. They are usually small of stature and of slight mental power. The best representatives are the Pygmies of the Central African forests, the Bushmen of South Africa, and the diminutive natives of the Andaman Islands.

424. **Yellow Type.**—People grouped around the Yellow type make up considerably more than one-third of the World's inhabitants. Their complexion varies from clear yellow to coppery brown, and typically they have a small nose, frequently upturned, and narrow slit-like eyes. Their hair is black, coarse, and straight, and each hair forms a minute circular tube. They are usually under the middle height, and although of slight physical strength they have great powers of endurance, and are as a rule very laborious workers. Intellectually they show a fair degree of civilisation, and in many instances have attained considerable success in science and in art. Conceit and apathy are characteristic mental qualities. They are usually Polytheists and worshippers of ancestors; many are Buddhists, and a considerable number Mohammedans. Finns and Magyars, inhabiting Finland and Hungary, are included under the Yellow type only on account of the nature of their language; physically and intellectually they are indistinguishable from the highest members of the White type. The *Tibeto-Chinese* are possessed of an ancient civilisation, and the Japanese, a race of this group, show great aptitude in following modern western ways of life and

thought. The greater part of Asia is peopled by tribes of the Yellow type of a relatively high civilisation. Except where seafaring has called forth their powers, the people inhabiting the tropical Malay Archipelago are as a rule ignorant and uncivilised, although far above the level of the degraded peoples of the Black type. The Maoris of New Zealand, belonging to the *Malayo-Polynesian* section, contrast strongly with the Australian blacks. The *American* section shows some well-marked differences from the other representatives of the Yellow type. Their coppery complexion won for them the name of Red Indians in the days when the first Europeans reached America and believed it to be part of Asia. From the Arctic Circle to Cape Horn the race is, in its essential features, the same, although the degree of civilisation attained varies. In the hot forests of the Amazon they range as tribes of naked savages, as low in the scale as the African blacks. On the northern prairies they form nations of hardy warriors, brave in battle and inconceivably cruel to their captured foes, living by hunting, but scorning work, and rapidly dwindling away before the white settlers. The highest native American civilisation had its seat on the plateau of Mexico and in the Andes valleys; and although the strongly organised native empires of the Aztecs and the Incas were destroyed by the Spaniards in the sixteenth century the " Indian " element has always remained of importance, and appears now to be rapidly gaining ground in the countries of that region.

425. **White Type.**—The leading physical peculiarities of this type are a prominent and highly arched forehead, and abundant wavy hair, the cross-section of which is oval. Dark skins, almost approaching those of the Black type, occur in the Hamitic section, but the complexion of the White races is usually fair and ruddy. The White type is the centre of a more numerous group of mankind than either of the others, and intellectually it is the most advanced. Religion has its fullest and purest forms of expression, science has been studied to best purpose, the fine arts have been raised to the highest perfection amongst them. Enterprise in commerce and valour in war

are equally pronounced, and at the present time the White type, particularly the Aryan races spreading from Western Eurasia, dominate the whole world. No other peoples have ever succeeded in establishing democratic government. The classification given in the table is founded mainly on affinities of language. The *Aryan* group, for example, includes the speakers of the Aryan or Indo-Germanic languages, all of which contain many words of common derivation, notwithstanding the differences between English, German, Danish, Spanish, French, Italian, Latin, Greek, Russian, Persian, and Sanscrit. Consequently it is assumed by Professor Max Müller and other philologists that the races using these languages are also descended from a common ancestry, and much ingenuity has been applied to the discovery of the original seat of this primitive people—the Aryans.

426. **People of Europe.**—Professor Huxley has recently shown that, so far as the peoples of Europe are concerned, it is impossible to reconcile the linguistic with the physical classification. He points out that the difference between the Teutonic-speaking nations of Britain, Germany, Holland, and Scandinavia; the Romanic-speaking people of Spain, Portugal, Italy, France, and Rumania; the Slavonic-speaking people of Servia, Bulgaria, and Russia; and even the Magyar and Finnish-speaking people of Hungary, Finland, and Lapland, do not warrant a scientific classification by language. He recognises two extreme types of Europeans, which are rarely found pure, and occur mixed together in varying proportions in all parts of the continent. The first type is that of tall men, averaging about 5 feet 8 inches in height, with long heads, fair complexions, yellow or light brown hair, and blue eyes. Such people are most abundant in the north, round the coasts of the Baltic, and their character is typically solid, trustworthy, persevering, and deliberate. The second type is that of a shorter race, averaging about 5 feet 5 inches in height, with rounded heads, swarthy complexions, black or dark brown hair and eyes. They are most numerous in the south bordering on the Mediterranean. Their prevailing character is impulsive

and enthusiastic; they are passionate, inconstant, and fond of ease. These two types evidently represent different races; but they have mingled so thoroughly that any attempt at exact classification is now impossible, although some indefinite but very interesting subdivisions have been made out.[1]

427. **Distribution of the Human Race.**—The estimated population of the world is 1,470,000,000 people. These are all dependent for their means of life on the land, and the densest population, that is the greatest number living on a given area, is necessarily found where the land is richest in useful productions. Deserts are practically unpeopled; the few inhabitants live on the produce of the date-trees of the oases and on the aid given them by passing caravans, to which the oases afford invaluable halting-places. Steppe-lands can carry more inhabitants, who as a rule are wandering shepherds feeding their flocks on the best grass they can find, and moving on to "pastures new" when the ground is cropped bare. Well-watered lands, when naturally treeless, or after the trees have been cleared, yield to agriculture abundance of food and material for clothing, hence such countries can support many inhabitants. The crowded Nile delta, the river-plains of China, and the valley of the Ganges are the most densely peopled parts of the Earth, on account of the fertility of the ever-renewed soil allowing large crops of food-plants to be raised at moderate expense. The question of the production of food is the most important in order to find how many people a given country can support. Mr. Ravenstein calculates that with proper treatment of the land about 6,000,000,000 inhabitants should be comfortably provided for on the Earth, a number which, if the present rate of increase continues, will be attained in less than 200 years.[2] There are other wants besides food, and by the division of labour made possible by the organisation of civilised life, a large population may be engaged in working mines or carrying on manufactures in regions where sufficient food for them cannot be grown. The supply of bread and meat is kept up by trade with their fellow-workers on lands

yielding a superabundance. Western Europe has a dense population on this account. Traffic, or carrying commodities to and fro, gives rise, at points where a change of routes or means of conveyance occurs, to a local concentration of population, and thus trading towns arise at harbours, fords, and the intersections of roads or railways.

428. **Centrifugal Migrations.**—In a primitive state of society the migrations of tribes are not unlike the migration of the lower animals, being directed from regions in which the means of life no longer suffice for the inhabitants. They are of the nature of evictions. A much larger population formerly resided in Central Asia, the margin of the Gobi (§ 381) being lined with remains of ruined cities; but the desiccation of the continent drove the people outward into whatever lands afforded food for their cattle or plunder on which to live. The people against whom the hordes of wanderers were driven were in turn dispersed in all directions, and the disturbance spread throughout every part of Eurasia. Overcrowding in countries of dense population also necessitates migration to more thinly peopled regions; but here as a rule the human power of discrimination and choice regulates the resulting movement. Lands are sought out which afford similar natural conditions to those in which the emigrants have formerly lived, and promise an easier or more prosperous life than the overcrowded country could offer. Thus the people of North-western Europe, and particularly of the British Islands, have thronged in millions to North America, South Africa, and Australia; while numbers of the people of Southern Europe have migrated to South America and Northern Africa. Another form of Centrifugal Migration is the voluntary exile of people persecuted for holding particular religious or political opinions. The settlement of New England by the Puritans, of Maryland by Irish Catholics, and of Utah by the Mormons, illustrates the action of this principle.

429. **Centripetal Migrations** have exercised an extraordinary influence in modern times. They are the result

of attraction rather than repulsion, and take place toward, and not from, a special region. The most potent magnet is gold. This led the Spaniards to Mexico and South America on the discovery of the new continent. In 1849 the discovery of gold in California caused a rush of fortune-seekers from all parts of the world, and led to the very rapid settlement of the Pacific coast of North America. Victoria was the scene of a similar rush in 1850, and tropical South Africa presents the same phenomenon, though in a less intense form, at the present time. Diamonds have had a like effect in attracting a large population to Kimberley, in Cape Colony. In each case many of the people attracted by the abundance of precious and portable products remained after these ceased to be readily available, in order to develop the agricultural resources of the land. Coal-fields and regions where petroleum or natural gas abound now rapidly attract a large population, on account of the facilities afforded for carrying on manufactures of every kind. Rich agricultural lands such as those of Dakota and Manitoba also give rise to concentration of population from all sides, when means are provided by railways or rivers to carry the wheat or other farm products to a profitable market.

430. **Geography** takes account of the relations between regions and races. Physiography is concerned with the study of Man in relation to the Earth, while Geography treats of the Earth in its relation to Man. The branch of geography dealing with the useful or desirable things which occur in or on the Earth's crust, and the effects which the discovery, production, transport, and exchange of these have on mankind, is known as Commercial Geography. Communities of civilised people associated together under one government form nations, and the definite region of the Earth's surface occupied by a nation is called a country. Countries have sometimes arisen from the centrifugal or centripetal migration of peoples under natural influences; but more commonly their limits have acquired their present position by the conquest or loss of territory in struggles against neighbouring nations. Wars carried on by kings

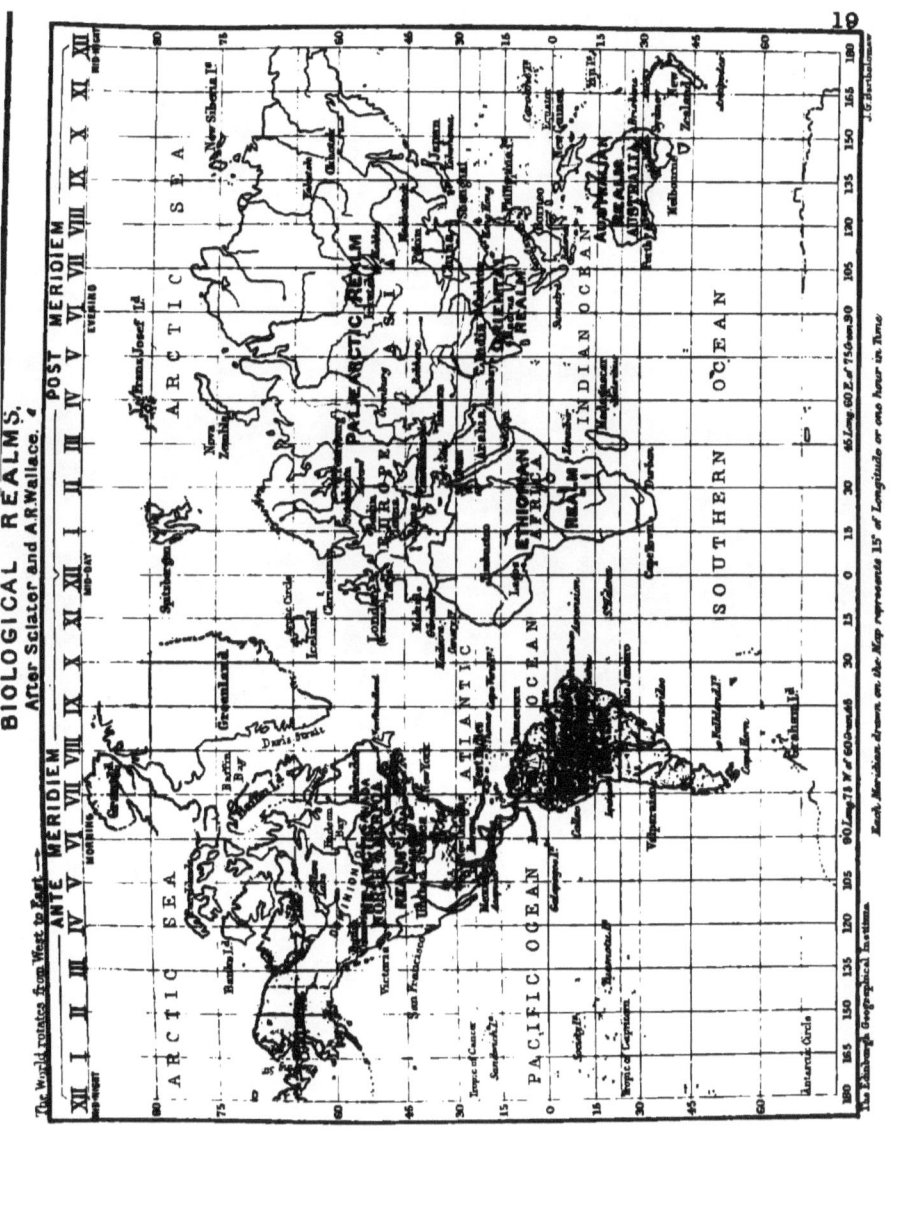

or governments, usually without the consent of the people concerned, have drawn most of the boundary lines on "political" maps. Historical Geography concerns itself with tracing out the changes in the extent of territory exclusively occupied or controlled by each nation at different times.

431. **Man's Power in Nature.**—Man more than any other animal leads a destructive life. The use of wood in construction and for fuel enables him to destroy forests so rapidly that in comparison the depredations of beavers and all other animals are insignificant. The need for communication between distant parts of the Earth has produced considerable changes in the configuration of coasts and in the distribution of land and water. Plants and animals also have been modified by cultivation, and their natural limits of distribution entirely altered. Much of Man's power in Nature is evasive. It consists in devising methods of utilising natural phenomena for the purpose of escaping uncomfortable consequences. Thus the invention of the umbrella and of the sun-helmet give a certain amount of independence of the weather; still more the methods of heating, cooling, and lighting houses. Lightning conductors reduce the risk to which life and property are exposed in a thunderstorm; knowledge of the laws of cyclone-motion often enables sailors to escape the fury of a storm. Steam-engines on land and sea, and above all the electric telegraph, deprive wide tracts of the Earth's surface of their natural influence as barriers. But in every case natural powers are not overcome; they are merely utilised.[3]

432. **Geographical Changes.**—When land becomes valuable it is often profitable to reclaim ground from the sea. This is done along the flat coasts of most civilised countries, and to an unequalled extent in Holland, where most of the people actually live below sea-level. The sea is kept out by a grand system of artificial dykes and regulated sand-dunes, while continual pumping by steam or wind power keeps the water-tight compartments of the reclaimed land dry. On the other hand, there are many projects for

flooding the sunk plains of arid regions, so as to provide new sea-routes, or modify desert climate by the presence of a sheet of water. Examples of these are the proposal to admit the Mediterranean and Red Sea to the great Jordan Valley (§ 335), in order to open a new sea-route to India; and the suggestion of admitting the Mediterranean water to some of the shotts of the northern Sahara (§ 377). Land-masses necessitating long sea-routes have frequently been severed by artificial channels, of which the Suez Canal is the most remarkable example. A German canal for large vessels across Jutland, from the North Sea to the Baltic, is in progress; a French canal to admit war-vessels is about to join the Bay of Biscay and the Mediterranean north of the Pyrenees; a Greek canal has severed the isthmus of Corinth. Several attempts have been made to cut the isthmuses of Central America and have hitherto failed, not because the task is impossible, but on account of financial or political bungling. Rivers are continually being interfered with, their mouths deepened into harbours, their course levelled into canals, the current split up into irrigation channels, or diverted bodily to prevent floods, or to furnish a route for railways. The greatest project of river diversion ever proposed is that of a Russian engineer to restore the Oxus to its ancient bed (§ 382) and bring it into the Caspian once more, thus affording a water-way from Europe into Central Asia. Tunnels such as those through the Alps, through the Khojak Hills in North-western India, and under the Andes in South America, are other examples of geographical changes wrought by human power. So too are the subsidences which follow mining operations, and sometimes alter the direction of streams.

433. **Biological Changes.**—By diligent cultivation and careful selection the food-grains of the modern farmer have been produced from various species of wild grasses, which naturally had small and innutritious seeds. In like manner many varieties of animals have been obtained by careful breeding, which are specially fitted for the use of man. Without his interference they would never have existed, and in many cases, if left to their own devices, they would

be unable to make a living. Savage or useless creatures have been exterminated over wide areas, and useful forms of life introduced in their place. Sheep are now far more numerous in Australia and temperate South America than any indigenous species of mammal ever was. Human interference can never overcome, but only take advantage of, natural conditions; and the rabbits accidentally introduced to Australia happened to be so much in harmony with their new surroundings that they have thriven and multiplied, so as to be an intolerable plague in some districts. By human agencies the horse, dog, sheep, and cow are no longer confined to any faunal realm, and the useful plants of each of the continents have been transplanted wherever suitable conditions are found in all the others. Maize and tobacco brighten the fields of Southern Europe, while wheat, sugar-cane, and coffee spread over vast expanses of America. The American cinchona and the Australian eucalyptus are now invaluable to the fever-haunted lands of India, and the latter tree flourishes in the swampy lowlands of the Mediterranean, while the vine and olive gladden the heart of the Australians.

434. **Meteorological Changes.**—The regulating effect of vegetation on rivers (§ 417) is accompanied by an actual increase in the rainfall of wooded as compared with barren regions. This is so clearly recognised that in many of the treeless plains of North America and Australia tree-planting is encouraged by the institution of an annual holiday called Arbour Day, on which each citizen is expected to plant a tree. In Russia the cutting of trees is prohibited in the whole belt of forests which covers the Ural-Carpathian ridge, whence all the rivers of Eastern Europe flow to north and south. Palestine presents a very striking example of climate altered by human action. In the days of the Israelites the steep mountain slopes were terraced artificially by walls supporting a narrow strip of soil, on which grain, vines, olives, and fruit-trees of many kinds were grown. The rainfall was regular and gentle; and after percolating through the terraces, formed perennial springs at the foot of the slopes, feeding the brooks which rippled

through the valleys. Now by neglect the terraces have been broken down, and the soil is all swept into the valleys. The mountain-sides, being bare and rocky, allow the occasional heavy showers to dash down in impetuous torrents to flood temporary streams, which, when the rain passes, give place to channels of dry stones. The land becomes baked in the fierce rays of the sun by day, and chilled by intense radiation through the clear dry air at night, the range of temperature having increased as the rainfall diminished.

435. **Man and the Degradation of Energy.**—Men are continually at work altering the distribution of matter and energy on the Earth. Gold is sought for in all lands, and accumulated in enormous quantities in London, Paris, Berlin, and other towns. Diamonds are more numerous in Amsterdam than in Africa, India, or Brazil; and so with other mineral commodities. The salts of the soil on which its fertility depends are being removed by every crop of wheat, to be ultimately cast as useless sewage into the sea. Land deprived of its salts ceases to yield crops; the natural process of restoration by weathering (§ 310) is too slow, and manures, which every year are becoming scarcer, must be sought far and near to replace them. No animal but man is so improvident. All others restore the mineral constituents to the land from which they gathered their food, and so insure a continuous supply. The potential energy laboriously stored in growing trees is destroyed by reckless timber-cutting, and the use of wood as fuel. The accumulated savings of energy stored up in coal are being expended in every industrial occupation, and coal is rapidly becoming scarcer. Every consumption of energy, except that of the regular income of solar radiation (§ 119), is impoverishing the Earth, and accelerating the natural process of the degradation of Energy (§ 75). The great steamer, driving its giant bulk across the ocean at 20 miles an hour, consumes as much potential energy in every revolution of the propeller as served in former days for the stately clipper, rising and dipping over the crests of the sea under the impulse of the sun-driven winds, to make the whole journey. Tidal

power, already utilised to some extent, and likely to be made use of increasingly, simply does work off the energy of the Earth's rotation (§ 103), and, although in a very minute degree, its employment hastens the time when Earth and Moon will have the same period of rotation. Similarly, all processes now proudly being increased in power and speed dissipate ever faster the wealth of potential energy that Nature lays up at an ever diminishing rate. Wind and water power and the Earth's store of internal heat are the only non-wasteful sources of work. Nothing is given for nothing, and even the knowledge revealed by the scientific study of Nature, that the power for effecting these processes will not last for ever, has been dearly bought. Since the true part played by energy has been understood in fact, though possibly not in name, the governments of all civilised nations have exerted themselves to encourage the most economical processes of manufacture, the most satisfactory systems of agriculture, the most intelligent methods of sewage disposal, and particularly to ensure the continuance, and if possible the increase, of the forests of the world, on which its prosperity, and even its habitability, largely depend.

436. **Man's Place in Nature.**—The grand distinction between Man and other creatures is that he can take advantage of his environment, so as to modify his development in any desired direction. He need not, except wilfully, drift before the wind of natural changes, but can sail close up to it like a well-handled ship. Man's higher nature can, and in many cases does, completely control his lower or animal existence. The sense of moral duty overcomes even the first law of animal nature—the preservation of life; it reverses the struggle for existence by substituting the principle of self-sacrifice, on which the stronger protects, instead of destroys, the weaker. Man, when most truly human, or in the highest attained stage of the evolution of civilisation, ceases to be in harmony with the system of Nature in the sense true of the lower animals—

"Know, man hath all which Nature hath, but more,
And in that *more* lie all his hopes of good.
Nature is cruel, man is sick of blood;
Nature is stubborn, man would fain adore;

"Nature is fickle, man hath need of rest;
Nature forgives no debt, and fears no grave;
Man would be mild, and with safe conscience blest.

"Man must begin, know this, where Nature ends;
Nature and Man can never be fast friends,
Fool, if thou canst not pass her, rest her slave!"

REFERENCES

[1] T. H. Huxley, "On the Origin of the Aryans," *Nineteenth Century*, November 1890.

[2] E. G. Ravenstein, "Lands of the Globe still available for European Settlement," *Proc. Roy. Geog. Soc.* xiii. 27 (1891).

[3] H. R. Mill, "Scientific Earth Knowledge as an Aid to Commerce," *Scot. Geog. Mag.* v. 302 (1889). "The Influence of Man on Nature," *Madras Christian College Magazine*, August 1888.

BOOKS OF REFERENCE

E. B. Tylor, *Anthropology*.

H. T. Buckle, *History of Civilisation*, vol. i. Longmans.

Keith Johnstone, *Physical, Historical, Political, and Descriptive Geography*, revised by E. G. Ravenstein. Stanford.

G. P. Marsh, *The Earth as modified by Human Action*. New York: Scribners.

G. G. Chisholm, *Manual of Commercial Geography*. Longmans.

H. R. Mill, *Elementary Commercial Geography*. Cambridge: Pitt Press Series.

APPENDIX I

SOME IMPORTANT INSTRUMENTS

437. Weights and Measures.—Standard masses called "weights" are used in a balance in order to find the mass of any body of convenient size by weighing it, that is by finding how many of the standard masses are attracted by the Earth with the same force as the body of unknown mass is attracted. The standard masses may be of any size or form, provided they can be easily obtained, and new ones exactly equal to them made if the originals be lost. Grains of seed were once used for this purpose, but now the standards are always made of dense metal of a kind which does not alter in the air. When a standard is once accepted it does not matter how it originated, as copies are always made by actual weighing. The British unit mass or pound avoirdupois is divided into 7000 grains or 16 ounces, and 2240 pounds are called a ton. In the United States the same unit pound is used, but 2000 of them are called a ton. In English-speaking countries the way in which masses are calculated is very contradictory and puzzling; but almost all other civilised nations employ a uniform system called the metric, the unit mass of which is the kilogramme (equal to about $2\frac{1}{4}$ lbs.) divided into 1000 grammes, and the gramme is similarly divided into 10 decigrammes, or 100 centigrammes, or 1000 milligrammes. These standards of mass are used by scientific men in every country, although the results have often to be translated into pounds and grains to make them popularly intelligible. The unit of length amongst English-speaking people is the yard, divided into 3 feet of 12 inches each, and 1760 yards are called a mile, although the sea-mile or mean minute of latitude contains rather more than 2000 yards. Measures on the *metric system* are like the weights subdivided decimally. The unit is the metre (about 39½ inches), divided into 10 decimetres or 100 centimetres or 1000 millimetres; and for measuring long distances 1000 metres are called a kilometre. It is convenient to remember that 25 millimetres are nearly equal to 1 inch, or, more exactly, that 33 centimetres are equal to 13 inches, and that 8 kilometres are equal to 5 miles. The measures of volume fluid ounces, pints, gallons, bushels, cubic inches, cubic feet, used in English-speaking countries are as confused as the other standards, while the unit volume of 1 litre (about 1¾ pints) divided into 1000 cubic centimetres is as convenient as the other parts of the metric system. The only connection between the British systems of weights and measures is that the gallon is fixed as the volume of 10 lbs. of pure water at 60° F. Relations of a much more intimate kind pervade the metric system. It is true that the metre is not quite the length originally intended, which was $\frac{1}{10000000}$ of a quadrant of the Earth's meridian, but the litre is a cube 1 decimetre in the side, and the kilogramme is the mass of 1 litre of pure water at 4° C., the gramme being similarly equal to the mass of 1 cubic centimetre of water at the maximum density point. Notwithstanding the simplicity and convenience of the metric system, it was considered advisable in this book to make use of the familiar British units in order to present the facts of science in the manner most easily grasped by English-speaking people.

438. **The Mariner's Compass** consists of a magnetised steel needle, or a series of such needles fixed parallel to each other, delicately pivoted in a box, which is loaded with lead and hung so as to remain horizontal in spite of the tossing of a ship. A light circular card is fixed above the needles and moves with them. The point over the north-seeking end of the needle is marked as the North, the opposite point is marked South, and the ends of the diameter at right angles East and West. The edge of the card is divided into 360 degrees, there being 90 in each quadrant, *i.e.* from N. to E. or from E. to S. The exact direction or bearing of a distant object may be stated as N. 45° E. if it appears midway between the north and east points of the horizon as estimated from the card. Sailors have another way of expressing direction. They divide the edge of the card into thirty-two "points," each containing 11¼ degrees, but divided into halves and quarters. For each point they have a special name; thus the quadrant from north to east is divided into *North, North by East, North-North-East, North-East by North, North-East, North-East by East, East-North-East, East by North, East*; and so on

round the card. (See compass in Plate I., where each alternate point is named.) The indications of the compass require to be corrected for variation (§ 98), and also for the local attraction of the vessel, in order to be as free as possible from which the standard compass is usually carried on the top of a high pole rising above the highest part of the deck.

439. **Barometers and Barographs.**—The simple mercury-tube (§ 146) mounted in a metallic case is the most accurate form of barometer. The height of the mercury in the tube is measured either to the fiftieth of a millimetre or to the thousandth of an inch by means of an arrangement called a vernier, due allowance being made for the change of level in the cistern as well as in the tube of mercury. In comparing atmospheric pressure at different stations it is necessary to correct the reading to some standard temperature (always 32° F. or 0° C.), because when mercury is heated it expands, its density becomes less, and a slightly higher column would be supported by the same atmospheric pressure. A correction for gravity, or rather for gravity and centrifugal force combined (§§ 38, 93), must also be made, as a column of mercury weighs less at the equator than near the poles. For popular purposes a barometer is sometimes made to show its rise or fall by the movement of a pointer round a dial, the change of quarter of an inch in level of the mercury being thus magnified on the dial to an inch or so. Glycerine barometers are in use in some places, and as the liquid is only about one-twelfth of the density of mercury, the tube has to be over 30 feet in length, and the fluctuations are shown in feet instead of in inches. The readings of a glycerine barometer are recorded daily on a diagram in the *Times*. Self-recording barometers are used in observatories. The simplest in principle (Fig. 64) produces a photographic record by a beam of parallel light from a lamp passing through the upper part of the tube ac, and falling on a cylinder $a'c'$ covered with photographic paper. and revolving once in twenty-four hours by means of clock-work. The paper opposite the clear space is blackened by the light, and Fig. 64 shows the sort of record left by a barometer rising irregularly, the height of which at any given moment can be estimated by seeing how much of the paper $b'c'$

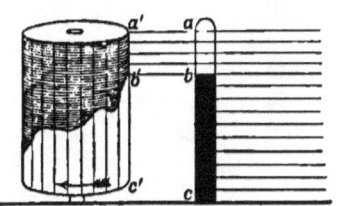

FIG. 64.—Photographic Barograph. (Diagrammatic View.)

was shielded from light by the mercury *bc* in the tube. The portable Aneroid barometer consists essentially of a metal box with an elastic top and exhausted of air. When the atmospheric pressure increases the top is forced in, when it diminishes the top curves out, and this movement is transmitted by suitable mechanism to a hand moving round a dial, or to a lever carrying a pen which records the fluctuations of pressure in a curve drawn in ink on a rotating cylinder. "Inches" and fractions are marked round the dial by comparison of the aneroid with a mercurial barometer, and a scale of heights is usually added, for aneroids are of most value in hill-climbing.

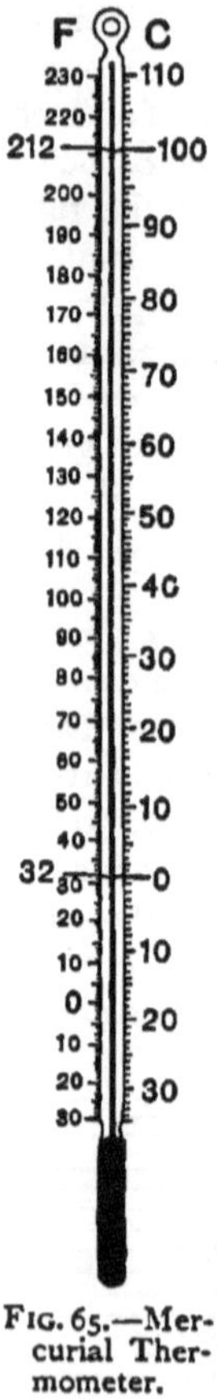

440. **Thermometers** are instruments for measuring temperature by means of the difference of expansion of a gas or liquid and the glass containing vessel. Mercury is usually employed as the liquid, because it has a low specific heat, great conducting power, expands considerably when heated, has a low melting-point and a high boiling-point. A mercurial thermometer consists of a globular or cylindrical bulb (Fig. 65), and a long tube of extremely small bore, which has been sealed while filled with boiling mercury, so that, after cooling, the bulb and part of the tube contain mercury and the remainder is a vacuum. The freezing and boiling points of any liquid depend only on the pressure, and if the pressure remains unchanged the liquid always freezes at one definite temperature, and always boils at one definite temperature. Thermometers are graduated by plunging them bodily into melting ice and after the mercury has contracted to the full, marking its position by a scratch on the glass; then by hanging them in the vapour of boiling water at ordinary atmospheric pressure, and when the mercury has expanded to the full, marking its new position by a scratch. Between the two fixed points any kind of subdivision might be made, but only three ways of dividing the space into "degrees" or steps are in use.

FIG. 65.—Mercurial Thermometer.

On the *centigrade* scale (often erroneously named after Celsius) the freezing-point is marked 0, the boiling-point 100, and the space between is divided into 100 equal degrees, which are continued above 100 and below 0 as far as may be necessary (C, Fig. 65). On *Fahrenheit's* scale, used popularly in English-speaking countries, the freezing-point is called 32, the boiling-point 212, the space between being divided into 180 equal degrees, which are continued downward and upward (F, Fig. 65). On the *Reaumur* scale, used popularly in Germany and Russia, the space between freezing and boiling point is divided into 80 degrees. The centigrade scale is used in scientific work all over the world, except for meteorological observations in English-speaking countries, for which the Fahrenheit scale presents too many advantages to be discarded. It is convenient to remember a quick way of translating centigrade into Fahrenheit degrees. *Multiply by 2, subtract one-tenth of the result, and add 32.* For example, to translate $15°$ C., $15 \times 2 = 30$, subtracting one-tenth $30 - 3 = 27$, adding, $27 + 32 = 59°$ F. Since mercury freezes at -40 (a temperature which happens to be expressed by the same figure on both centigrade and Fahrenheit scales), alcohol thermometers are used for measuring lower temperatures, such as those of the winter at Verkhoyansk. No two common thermometers read exactly alike, and those employed for accurate observations are always compared with standard instruments (those of Kew Observatory for the United Kingdom), and have their error ascertained and allowed for. **Thermographs** are constructed on the principle of the barograph, to furnish a continuous record of changes of temperature. **Deep-sea thermometers** require to be protected against the pressure at great depths by surrounding the bulb by a glass sheath partly filled with mercury or other liquid. They are constructed either to leave an index sticking in the tube at the points of highest and lowest temperature encountered while submerged, or to be inverted by appropriate mechanism, and so caused to register the temperature at any given point. (See article "Thermometer" in *Encyclopædia Britannica*, 9th edition.)

441. **Hygrometers** measure the amount of water-vapour in the atmosphere by finding either at what rate the air is taking up vapour by evaporation at its actual temperature, or how far the air must be cooled in order that its vapour may be saturated. The commonest form consists of two thermometers placed side by side, the bulb of one being left dry, while that of the other is covered by a piece of fine muslin, and kept wet by a thread dipping into a vessel of water. The farther the vapour of the air is from saturation the more rapid is the evaporation from the wet bulb, and since evaporation withdraws heat (§§ 71, 157), the temperature shown by the wet-bulb thermometer is lower than that shown by the dry. The greater the difference between the readings of the two, the smaller is

the relative humidity of the atmosphere, the exact value of which for each difference of temperature has been calculated and recorded in tables by Glaisher. Dew-point hygrometers, in various forms, invented by Regnault, Daniel, Dynes, and others, consist of a polished surface, the temperature of which can be lowered by evaporating a liquid, or by a current of iced water, until a film of moisture is condensed from the air. The temperature at which condensation takes place is that of the dew-point, at which the vapour of the air becomes saturated, and a table of the vapour-pressure of saturated vapour at different temperatures gives the absolute humidity (§ 158).

442. **Anemometers**, or instruments for measuring the force of the wind, are constructed either to record velocity or pressure. To show velocity a series of hollow metal cups, mounted on a light pivoted frame, are caused to revolve by the wind, and each revolution is registered by an arrangement like that of a gas-meter. Experiment shows what ratio the speed of the revolving cups bears to that of the wind. In pressure anemometers the wind blows against a large flat surface, the pressure exerted on which is indicated by the tension of spiral springs. These instruments, like all others for measuring phenomena subject to constant variation, can be made to write a continuous record on a revolving cylinder, from which the exact direction, force, and velocity of the wind may be ascertained at any moment.

443. **Deep-sea Soundings.**—The depth of calm water, when less than 200 fathoms, can easily be found by letting down a lead weighing 7 lbs. by a line marked at regular intervals. The impact of the lead on the bottom may usually be felt, and the line ceases to run out, or at any rate, if too much line is let out, a sudden increase in weight is felt when, on hauling it in, the lead is lifted off the bottom. At great depths a very heavy sinker must be used: its impact on the bottom cannot be felt, and the line runs out steadily. In making a deep sounding, the line—usually a fine steel wire—is marked at every 100 or 50 fathoms, and the intervals of time at which each mark disappears in the water are carefully noted. On account of the increasing resistance of the water on the lengthening line the time interval lengthens gradually and uniformly; but when the sinker reaches the bottom there is an abrupt increase in the time taken for the next 50 fathoms to run out, which is sufficient to assure the officer in charge that bottom is reached. From depths of 3000 or 4000 fathoms no ordinary line or wire is strong enough to haul up the heavy sinkers, which accordingly are so constructed as

to detach themselves after driving the brass "sounding tube" to which they were attached deep into the floor of the ocean, where it is filled with mud, and whence it can readily be raised to the surface. The process of making deep-sea researches of every kind is full of interest, and the student should, if possible, read the descriptions in the works referred to at the end of Chapter XI.

APPENDIX II

CURVES AND MAPS

444. Graphic Representations.—Self-recording instruments, like the barograph and thermograph, write their changes as continuous curves, which present to the eye a vivid picture of the nature and extent of these changes. The daily and annual changes of temperature and pressure are represented in the form of curves in Figs. 23, 24, and 28. When any one of the conditions under consideration varies uniformly, the curve form of expression can be used; thus Fig. 27 shows temperature at different latitudes, where position on the Earth varies uniformly, and Fig. 33 shows temperature at various depths in the sea, where depth varies uniformly. The highest point of a curve or any convex bend is called a maximum the lowest point, or any concave bend, a minimum; and a line drawn horizontally, so that the curve cuts off an equal area above and below, is called its mean. It is simply a matter of convenience that the space representing a degree of temperature, and that representing an hour, a day, a fathom, or a degree of latitude, should have the same length in a diagram. In the sections of oceans and continents there is a natural relation between heights and lengths; but if on a section of Asia 100 miles of length were represented by an inch, the greatest height of the continent would be shown by one-twentieth of an inch, and would scarcely be visible. Accordingly heights are drawn on a much larger scale, and the steepness of the slope is exaggerated in the same proportion, while the positions and relative amounts of change of level are brought vividly before the eye. It would be an excellent exercise for the student to reduce these sections to a true scale, either by reducing the heights on the paper to one-three-hundredth of their height (but this is scarcely possible), or by keeping

Appendix II

them unchanged, and lengthening the whole section, or a part of it, three hundred times. This would give the true average slopes of the continents and oceans.

445. Maps.—The plan or map of a room is simply an exact drawing of the outline of the floor, and the spaces occupied by each article of furniture, drawn so that one inch or any other definite length on the paper corresponds to one foot on the floor. The ratio of the lengths is called the scale of the map; thus the scale of a map in which one inch represents one foot is 1 : 12; the maps of counties on the Ordnance Survey of the United Kingdom are drawn on the scale of six inches to one mile, or 1 : 10,560; those of the country generally, in which one inch stands for one mile, are on the scale of 1 : 63,360; Plates IX. and X. represent the British Islands on the scale of 1 : 7,500,000; and Plates III.-VIII., etc., show the Earth on the scale of 1 : 200,000,000 along the equator. In the case of the plan of a room, the map, if increased twelve times in length and breadth, would make a carpet accurately fitting the floor, with spaces marked for the furniture to rest on; but if the map of the British Islands were magnified 7,500,000 times each way, it would not fit the country exactly, because the Earth's surface is curved, and a flat sheet cannot lie smoothly on a curved surface without being folded or stretched. In the case of the Earth as a whole, this difficulty of representing the whole surface in its true form and proportions is much greater. The surface of the sphere cannot be spread out flat, and many devices—termed projections—are adopted to represent it with as little distortion as possible. On *Mercator's projection*, shown in Plates I. and II., the parallels of latitude are shown as straight lines, the equator being unbent from a ring into a rod, so that we can see all round it at one glance. The other parallels are not only unbent, but stretched to the same length as the equator, so that the meridians become parallel straight lines, and, in latitude 60°, are just twice as far apart as they should be. In order to preserve the correct outline of the land, and to make the *directions* measured on the map correct, the parallels are not placed equidistant, but stretched out toward the poles, the degrees of latitude increasing in length in the same proportion as the degrees of longitude. Thus different parts of the map are on different scales; one square inch including Greenland, for example, represents only one-tenth of the area which one square inch including India comprises. It resembles a cylindrical projection, which may be supposed to be drawn on a great sheet wrapped round the globe, as shown in Fig. 66. Mercator's projection, although much less distorted than the true cylindrical projection of Fig. 66, is useless for comparing areas. But it is of unique value,

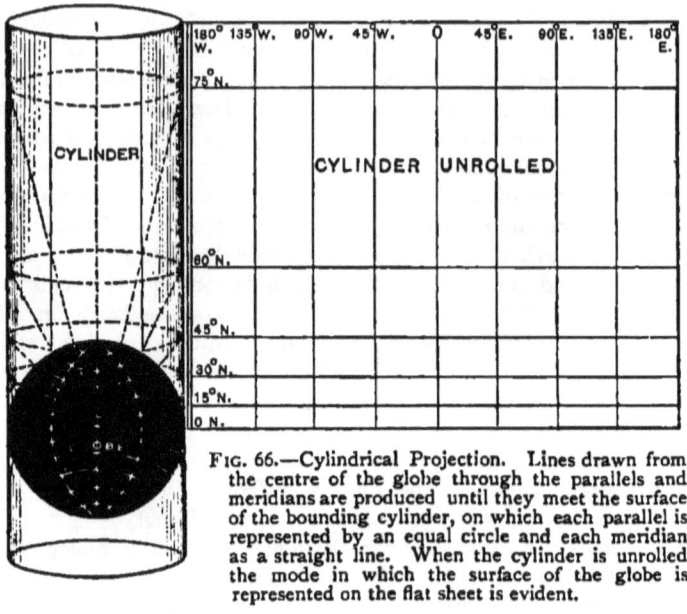

FIG. 66.—Cylindrical Projection. Lines drawn from the centre of the globe through the parallels and meridians are produced until they meet the surface of the bounding cylinder, on which each parallel is represented by an equal circle and each meridian as a straight line. When the cylinder is unrolled the mode in which the surface of the globe is represented on the flat sheet is evident.

because a line drawn between any two points cuts all the meridians at the true angle, and it is therefore much used in navigation. Plate III. and most of the other maps of the world shown are drawn on *Gall's stereographic projection*, which does not distort the areas so much, and does distort the angles considerably. Plate XII. shows beautifully the amount of distortion of area in this projection, the 250 mile coast belt appearing nearly three times as broad round Greenland as round Africa where the distortion is least. In maps of the world in hemispheres the meridians are shown converging to the poles, and there is an infinite number of projections employed for special purposes. *Lambert's equivalent area projection* (Plate XIV.) is valuable because, although it distorts angles greatly, it preserves the equality of areas; a square inch measured on any part of the map represents exactly the same number of square miles. The calculations of Dr. John Murray, referred to in previous chapters, were made by measuring areas on large-scale maps constructed on this projection. Maps of a small area can be more accurately shown

Appendix II 353

on a *conical projection*. Those of the British Islands (Plates IX. X. etc.), for example, are on a conical projection; the meridians converging to the proper degree and the parallels being arcs of circles. If a cone of transparent paper were placed over an artificial globe (Fig. 67) and the lines traced through, a map of this kind would result; the distortion being greatest at the greatest distances from the

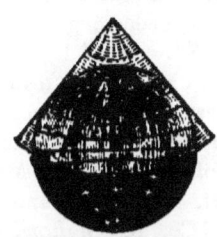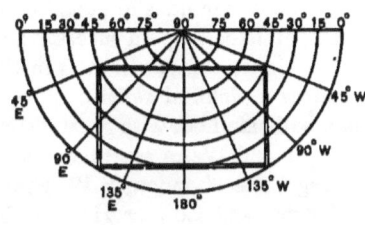

FIG. 67.—Conical Projection. The left hand figure represents a cone placed on the globe, the surface features of which are projected, as in Fig. 66, by lines drawn from the centre. The right hand figure shows the cone unrolled showing the parallels as semicircles and meridians radiating from a centre. The double lines show a map cut from the developed cone.

parallel along which the cone touched. When the cone is supposed to cut the globe along two parallels, the resulting map is much more accurate. In actual map-making the distance and curvature of the parallels and meridians for each projection are ascertained by mathematical calculations.

446. **Contour-lines** are drawn on maps to express differences of level in an exact manner. They express the height of the land in the same way as isotherms express the temperature. Each contour-line represents a string of figures of elevation having the same value. The sea-coast is a natural contour-line, and raised beaches are natural contour-lines etched on the hill-sides. Every contour-line represents the coast-line that would result, if the sea rose to that level. When contour-lines are far apart the gradient or slope is gentle; for example, along AB (Fig. 68) we could advance nine divisions of the

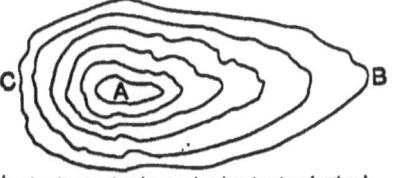

FIG. 68.—Contour-lines. The Line BC represents sea-level, each of the inner lines represents a level 100 feet higher than that next to it on the outside, the line round A being 500 feet above the sea. The scale below refers to horizontal distance.

2 A

scale before the elevation became 500 feet lower, but along AC this difference of height is reached in three divisions of the scale, or the slope is three times as steep and the contour-lines are much closer. The student, if residing in the United Kingdom, should procure and carefully study the Ordnance Survey maps (contoured) on the one-inch and six-inch scales for his own locality. He might advantageously follow the lines in pencil to make them more prominent, and then paint the map in successive washes, deepening the colour within the higher contour-lines as in Plates XI. and XVI. He will thus produce a pictorial relief map, on which all the features of hill and dale will stand out with great distinctness. The Ordnance maps for England and Wales are to be had from Mr. Edward Stanford, 55 Charing Cross, London, S.W.; those for Scotland from Messrs. John Menzies and Co., 12 Hanover Street, Edinburgh; and those for Ireland from Messrs. Hodges, Figgis, and Co., 104 Grafton Street, Dublin. Mountains and watersheds are frequently represented on maps by shading in certain conventional ways, so as to bring out the general appearance of the surface. One of these systems combined with contour-lines is shown in the map of a glacier in Fig. 54.

APPENDIX III

DERIVATIONS OF SCIENTIFIC TERMS

ABERRATION, L. *ab*, from ; *erro*, to wander
Absorption, L. *ab*, from ; *sorbeo*, to suck in
Agglomeration, L. *ad*, to ; *glomus*, a ball
Agonic, Gr. *a*, not ; *gonia*, a corner or angle
Amorphous, Gr. *a*, not ; *morphe*, form
Amplitude, L. *amplitudo*, large
Analysis, Gr. *ana*, up ; *luo*, to loosen
Anemometer, Gr. *anemos*, wind ; *metron*, measure
Aneroid, Gr. *a*, not ; *neros*, liquid
Annular, L. *annulus*, a ring
Anticline, Gr. *anti*, against ; *klino*, to lean or incline
Anticyclone, Gr. *anti*, opposite to, and CYCLONE
Aphelion, Gr. *apo*, from ; *helios*, the sun
Approximation, L. *ad*, to ; *proximus*, nearest
Aqueous, L. *aqua*, water
Arc, L. *arcus*, a bow
Archæan, Gr. *archaios*, ancient
Archæopteryx, Gr. *archaios*, ancient ; *pteryx*, wing
Arthropoda, Gr. *arthros*, a joint ; *pous*, foot
Asteroid, Gr. *aster*, star ; *eidos*, form
Atmosphere, Gr. *atmos*, air ; *sphaira*, a sphere
Aurora, L., the goddess of dawn
Austral, L. *auster*, the south wind, southern
Axis (*pl.* axes), L., an axle
Azote, Gr. *a*, not ; *zao*, to live

BAROGRAPH, Gr. *baros*, weight ; *grapho*, to write
Barometer, Gr. *baros*, weight : *metron*, measure

Biology, Gr. *bios*, life ; *logos*, a discourse
Bisect, L. *bis*, twice ; *seco*, to cut (to divide into two equal parts)
Boreal, L. *boreas*, the north wind, northern
Botany, Gr. *botane*, herb or plant

CAINOZOIC, Gr. *kainos*, recent ; *zoe*, life
Calcareous, L. *calcarius*, chalky
Capillarity, L. *capillus*, hair
Carposporeæ, Gr. *karpos*, fruit ; *sporos*, seed
Centrifugal, L. *centrum*, centre ; *fugio*, to flee from
Centripetal, L. *centrum*, centre ; *peto*, to seek
Chlorophyll, Gr. *chloros*, pale green ; *phullon*, leaf
Chromosphere, Gr. *chroma*, colour ; *sphaira*, a sphere
Chronometer, Gr. *chronos*, time ; *metron*, measure
Cirrus, L. *cirrus*, a curl
Cœlenterata, Gr. *koilos*, hollow ; *enteron*, bowel
Cohesion, L. *co*, together ; *hæreo*, to stick
Comet, Gr. *kometes*, long-haired
Complement, L. *complementum*, that which fills up
Concentric, L. *con*, with ; *centrum*, centre (having the same centre)
Conduction, L. *con*, together ; *duco*, to lead
Constellation, L. *con*, together ; *stella*, a star
Convection, L. *con*, together ; *veho*, to carry
Cretaceous, L. *creta*, chalk
Cryptogam, Gr. *kruptos*, concealed ; *gamos*, marriage
Cumulus, L. *cumulus*, a heap
Cyclone, Gr. *kuklos*, a circle

DATUM (*pl.* data), L. *datum*, given
Deciduous, L. *deciduus*, falling off
Desiccation, L. *desicco*, to dry up
Detritus, L. *de*, off ; *tero, tritus*, to rub
Devitrification, L. *de*, from ; *vitrum*, glass ; *facio*, to make
Diameter, Gr. *dia*, through ; *metrein*, to measure
Dicotyledon, Gr. *dis*, two ; *kotuledon*, a cup(-shaped leaf)
Discrete, L. *discretus*, separate

ECHINODERMATA, Gr. *echinos*, hedgehog (spiny) ; *derma*, skin
Elasticity, Gr. *elaso*, to drive
Electricity, Gr. *elektron*, amber (by rubbing which electric phenomena were first observed)
Ellipsoid, Gr. *en*, in ; *leipo*, to leave ; *eidos*, form

Appendix III

Eocene, Gr. *eos*, dawn ; *kainos*, recent
Equator, L. *æquus*, equal
Equisitineæ, L. *equus*, horse ; *seta*, bristle
Erosion, L. *e*, away ; *rodo*, to gnaw
Escarpment, Fr. *escarper*, to cut down steeply
Estuary, L. *æstuare*, to boil up, *i.e.* tumultuous tides
Ethnology, Gr. *ethnos*, a nation ; *logos*, a discourse
Evolution, L. *e*, out ; *volvo*, to roll
Experiment, L. *experior*, to try thoroughly

FAUNA, native animals supposed to be protected by the *Fauns,* or rural gods
Filicineæ, L. *filicis*, a fern
Flora, L. *flos*, a flower
Foraminifera, L. *foramina*, openings ; *fero*, to carry

GENUS (*pl.* genera), L. *genus*, birth (related by birth, of one kin)
Geography, Gr. *ge*, the earth ; *grapho*, to describe
Geoid, Gr. *ge*, the earth ; *eidos*, form
Geology, Gr. *ge*, the earth ; *logos*, a discourse
Glacier, Fr. *glace*, ice
Glauconite, Gr. *glaukos*, bluish gray
Gravitation, L. *gravis*, heavy
Gymnosperm, Gr. *gumnos*, naked ; *sperma*, seed

HEMISPHERE, Gr. *hemi*, half ; *sphaira*, a sphere
Hepaticæ, Gr. *hepatos*, the liver
Homogeneous, Gr. *homos*, one ; *genos*, kind
Horizon, Gr. *horizo*, to bound
Humidity, L. *humidus*, moist
Hydrosphere, Gr. *hudor*, water ; *sphaira*, a sphere
Hygrometer, Gr. *hugros*, wet ; *metron*, measure
Hyperbola, Gr. *huper*, beyond ; *ballo*, to throw

ICHTHYOSAURUS, Gr. *ichthos*, fish ; *saura*, lizard
Igneous, L. *ignis*, fire
Indigenous, L. *indu*, in ; *geneo*, to produce
Inverse, L. *inverto*, to turn round
Isobaric, Gr. *isos*, equal ; *baros*, weight
Isothermal, Gr. *isos*, equal ; *therme*, heat

LATERAL, L. *latus*, a side

Latitude, L. *latitudo*, breadth
Lithosphere, Gr. *lithos*, stone ; *sphaira*, a sphere
Littoral, L. *littus*, the shore
Longitude, L. *longitudo*, length

MEDIUM, L. *medius*, middle (anything coming between)
Meridian, L. *meridies*, mid-day
Mesozoic, Gr. *mesos*, middle ; *zoe*, life
Meteorite, Gr. *meteoron*, suspended beyond ; *lithos*, a stone
Meteorology, Gr. *meteoron*, suspended beyond ; *logos*, a discourse
Miocene, Gr. *meion*, less ; *kainos*, recent
Mollusca, L. *mollis*, soft
Monocotyledon, Gr. *monos*, alone ; *kotuledon*, a cup(-shaped leaf)
Monsoon, Malay *musim*, a season
Musci, L. *muscus*, moss

NEBULA, L. *nebula*, a little cloud
Nimbus, L. *nimbus*, a rain-cloud
Nitrogen, Gr. *nitron*, nitre ; *gennao*, to produce
Node, L. *nodus*, a knot
Normal, L. *norma*, a rule

OBLATE, L. *oblatus*, carried forward
Oligocene, Gr. *oligos*, few ; *kainos*, recent
Oolite, Gr. *oon*, an egg ; *lithos*, a stone
Oosporeæ, Gr. *oon*, an egg ; *sporos*, seed
Orbit, L. *orbis*, a ring
Oriental, L. *orior*, to rise ; hence the east
Orographical, Gr. *oros*, a mountain ; *grapho*, to describe
Oxygen, Gr. *oxus*, acid ; *gennao*, to produce
Ozone, Gr. *ozo*, to smell

PALÆOCRYSTIC, Gr. *palaios*, ancient ; *krustallos*, ice
Palæozoic, Gr. *palaios*, ancient ; *zoe*, life
Parabola, Gr. *para*, beside ; *ballo*, to throw
Parallax, Gr. *para*, beside ; *alasso*, to change
Parallel, Gr. *para*, beside ; *allelon*, one another
Pelagic, Gr. *pelagos*, the sea
Perihelion, Gr. *peri*, near ; *helios*, the sun
Perturbation, L. *per*, thoroughly ; *turbo*, to disturb
Phanerogam, Gr. *phaino*, to bring to light ; *gamos*, marriage
Phenomenon, Gr. *phainomenon*, an appearance

Appendix III

Philology, Gr. *philos,* loving ; *logos,* word (the study of language)
Photosphere, Gr. *phos,* light ; *sphaira,* a sphere
Physiography, Gr. *phusis,* nature ; *grapho,* to describe
Plane, L. *planus,* even, smooth
Planet, Gr. *planetes,* a wanderer
Pleistocene, Gr. *pleistos,* most ; *kainos,* recent
Plesiosaurus, Gr. *plesios,* near to ; *saura,* a lizard
Pliocene, Gr. *pleion,* more ; *kainos,* recent
Porifera, L. *porus,* a pore ; *fero,* to carry
Potential, L. *potens,* being able
Proteid, Gr. *protos,* first
Protophyta, Gr. *protos,* first ; *phuton,* plant
Protoplasm, Gr. *protos,* first ; *plasma,* form
Protozoa, Gr. *protos,* first ; *zoon,* animal
Pterodactyl, Gr. *pteron,* wing ; *daktulos,* finger
Pteropod, Gr. *pteron,* wing ; *podes,* feet

RADIATION, L. *radio,* to radiate
Radius, L. *radius,* a rod, ray
Rarefaction, L. *rarus,* rare ; *facio,* to make
Reflection, L. *re,* back ; *flecto,* to bend
Refraction, L. *re,* back ; *frango,* to break
Rotation, L. *roto,* to turn

SATELLITE, L. *satelles,* an attendant
Saurian, Gr. *saura,* a lizard
Secretion, L. *secretus,* from *se,* apart ; *cerno,* to separate
Sedimentary, L. *sedimentum,* from *sedeo,* to sit, to settle
Sequence, L. *sequor,* to follow
Sidereal, L. *sidus,* a star
Solstice, L. *sol,* the sun ; *sto,* to stand
Species, } L. *specio,* to look (that which is seen)
Spectrum,
Spicule, L. *spiculum,* a point
Stalactite, Gr. *stalaktos,* dropping
Stalagmite, Gr. *stalagmos,* a dropping
Stratum, } L. *sterno* (*stratum*), to spread out
Stratus (*pl.* strata),
Subtend, L. *sub,* under ; *tendo,* to stretch
Syncline, Gr. *sun,* together ; *klino,* to lean or incline
Synoptic, Gr. *sun,* with ; *opsis,* a view

TALUS, Fr. *talus,* a slope
Tangent, L. *tango,* to touch
Terrigenous, L. *terra,* the earth ; *geneo,* to produce
Thallophyte, Gr. *thallos,* a twig ; *phuton,* a plant
Thermometer, Gr. *therme,* heat ; *metron,* a measure
Transit, L. *trans,* across ; *eo,* to go
Trias, Gr. *trias,* union of three
Trigonometry, Gr. *trigonon,* triangle ; *metron,* a measure
Tropic, Gr. *tropos,* a turning

UNIVERSE, L. *unus,* one ; *verto,* to turn

VACUUM, L. *vacuum,* empty
Vermes, L. *vermis,* a worm
Vernal, L. *ver,* spring
Vertebrata, L. *vertebra,* a joint
Vertical, L. *vertex,* the top
Vibration, L. *vibro,* to quiver
Vortex, L. *vorto,* to turn or whirl

ZENITH, Arabic, *semt-ur-ras,* way of the head
Zero, Arabic, *sifr,* nothing (a starting-point)
Zone, Gr. *zone,* a girdle
Zoology, Gr. *zoon,* an animal ; *logos,* a discourse

INDEX

The Figures refer to the sections.

ABERRATION of light, 108
Absolute Zero, 68
Absorption and radiation, 63
Absorptive power of air, 160
Abysmal area, 255, 257, 277
Accuracy, 10
Acids, 44
Adelsberg caves, 317
Adriatic Sea, 216, 325
Africa, 373-377, 412
Agassiz, Lake, 368
Agonic lines, 98
Agulhas current, 248
Air, 151; in sea-water, 225; temperature of, 187, 189, 190
Albert, Lake, 375
Aletsch glacier, 337
Alluvial deposits, 322
Alps, 303, 385
Altai Mountains, 381
Altitude of the Sun, 124
Amazon, 219, 230, 269, 318, 319, 361
America, see *North*, and *South America*
American race, 424
Amu Daria (Oxus), 382, 432
Amur River, 319, 381
Analysis, 40; spectrum, 63
Andes, 201, 353, 358
Anemometer, 442
Aneroid barometer, 439
Angles, 31 *et seq.*

Angular measurement, 31-33
Animals, 397, 400, 409-417
Antarctic circle, 122
Antarctic continent, 276, 340
Anthracite, 347
Anticlines, 302, 303
Anticyclones, 205
Anti-trade winds, 181
Apennines, 385
Appalachian Mountains, 366
Approximation, 10
Arabia, 379, 412
Aral, Lake, 333, 382
Arbour Day, 434
Archæan rocks, 346
Arctic circle, 122; sea, 216, 234
Aryans, 425
Ash, volcanic, 294
Asia, 379-383
Asteroids, 129
Atlantic Ocean, 216, 243-246, 258
Atmosphere, 84, 145-213
Atmospheric electricity, 172
Atolls, 280
Atoms, 47, 48
Aurora, 99, 174
Austral group of plants, 405
Australia, 353, 370-372
Australian people, 423; realm, 415
Autumnal equinox, 123
Avalanches, 336

The Figures refer to the sections.

Axis of continents, 356; of the Earth, 90
Azote, see *Nitrogen*

BAIKAL, Lake, 333, 382
Balkan Range, 385
Balkash, Lake, 382
Baltic Sea, 216, 238, 325, 388
Bangweola, Lake, 376
Banks, submarine, 325
Bantu race, 423
Barograph, 439
Barometer, 146, 439
Barrier reefs, 280
Bars of rivers, 325
Basalt, 290, 294
Bases, 44
Basin, Australian, 371, 372
Basin, the great, 365
Basins, river, 319; see also *Ocean*
Bayous, 324
Beaches, formation, 265; raised, 284
Benguela current, 243, 245
Bermuda Islands, 279, 307
Biela's comet, 135
Black Forest Mountains, 385
Black Sea, circulation, 238
Black type of mankind, 423
Blow-holes, 266.
Blue mud, 270
Bode's Law, 129
Bohemian Forest Range, 385
Boiling, 70, 72
Bonneville, Lake, 365
Bore, the tidal, 219
Boreal group of plants, 405
Boulder clay, 338, 368, 389
Boyle's Law, 148, 163
Brahmaputra River, 380, 383
Brazil, High Plain, 359
British Islands, climate, 202-204; surface, 389-392
Buys Ballot's Law, 192

CAINOZOIC rocks, 350
Calcareous organisms, 273
Cambrian rocks, 346
Canals, 432
Cañons, 328; submarine, 326
Capacity for heat, see *Specific heat*
Capillarity, 39, 310, 314
Carbonic acid, 154, 294, 317, 399, 400
Carboniferous rocks, 347, 366, 371, 390, 391
Carpathian Mountains, 385
Cascade Range, 364
Casiquiare River, 360
Caspian Sea, 333, 335, 382, 387
Caucasus Mountains, 379
Cause and effect, 16
Cavendish experiment, 85
Caverns, 317
Cells, 398
Centigrade scale, 440
Centrifugal and centripetal migrations, 428, 429
Centrifugal force, 51
Chalk Ridge, 391
Challenger expedition, 183, 188, 251, 268
Charlestown earthquake, 301
Chlorophyll, 399
Circulation, atmospheric, 176, 177; of deep lakes, 228; of enclosed seas, 237, 238; of water by wind, 240-242
Cirrus cloud, 168
Civilisation, 420
Classification, 4, 13; of animals, 395, 397; of elements, 47; of plants, 395, 396; of stars, 138
Clay, 311
Cleavage of rocks, 290
Climate, 125; of British Islands, 202-204; of Earth, 186-201
Cloud-bursts, 210
Clouds, 167, 168
Coal, 29, 347
Coast Range, 364
Cohesion, 39
Colorado River, 327, 329, 364

Index

The Figures refer to the sections.

Colour, 64
Columbia River, 364
Comets, 132, 133
Common-sense, 9
Comparison, 4
Compass, mariner's, 438
Compounds, 42
Compressibility, 35
Condensation, 70-73, 159, 166
Conduction, 59
Conductors of electricity, 77
Configuration and climate, 186
Congo, 269, 319, 326, 376
Conical projection, 445
Continental Area, 255, 354-392
Continental Shelf, 263, 264, 267
Continents, evolution of, 353; statistics of, 355
Contour-lines, 446
Convection, 68
Coral Islands, 280-282
Coral mud and sand, 272
Corals, 279
Corona, solar, 116
Cosmic dust, 161, 277
Cotswold Hills, 391
Counter equatorial currents, 243, 247, 248
Cretaceous rocks, 349, 391
Crevasses, 337
Crystals, 30
Cumulus cloud, 168
Currents of the ocean, 242-249
Curvature of the Earth, 81
Curves, use of, 444
Cyclones, 206-208

DAILY range of temperature, 182; of pressure, 183
Dalton's Law, 155
Danube River, 331, 386
Day, longest, 124; period of, 94
Dead Sea, 333, 335
Declination, magnetic, 98
Deductive reasoning, 17
Deep-sea soundings, 443

Definiteness, 11
Degradation of energy, 75, 435
Degree, angular, 31; of latitude, 92; of longitude, 97; of temperature, 440
Deltas, 325
Density, 29; of air, 148; of the Earth, 85; of sea-water, 223
Denudation, 305
Deposits in the ocean, 268-277
Deserts, 406
Devonian rocks, 346
Dew, 165
Diatom ooze, 273, 276
Differential attraction, 103
Dip of horizon, 81; of a magnet, 98; of rocks, 290
Direction on the Earth, 91
Disruptive discharge, 78
Distance of stars, 137
Distribution of animals, 409; of mankind, 427; of plants, 410
Doldrums, 179
Dolphin Ridge, 258
Drainage areas of ocean, 356
Dust, 134, 151, 161, 162, 297
Dykes, volcanic, 295

EARTH, the, 81 et seq.; and Moon, 104; orbit, 109
Earthquakes, 299-301
Eclipses, 113
Ecliptic, 112
Elasticity, 35; of Earth's crust, 299
Elbe, river, 386
Elburz Mountains, 379
Electrical energy, 76
Electrification of the atmosphere, 172
Electro-magnetism, 80
Elements, 45
Elevation and subsidence of land, 284
Energy, 25, 49, 53-56, 60, 163, 250, 283, 304, 305, 399, 435
England, 391

The Figures refer to the sections.

Environment, 403, 421
Eocene rocks, 350
Equator, 91
Equinoxes, 121, 123; precession of, 115
Erie, Lake, 330, 369
Erratics, 338
Eruptions, volcanic, 296, 297
Erzgebirge, 385
Estuaries, 231
Ether, the, 60
Ethiopian Realm, 412
Etive, Loch, 339
Eurasia, 378
Europe, 384-392; people of, 426
Evaporation, 70, 159
Evolution, organic, 402; of continents, 353
Exclusiveness, 34
Expansion by heat, 67, 68
Experiments, 18
Eyre, Lake, 372

FAHRENHEIT scale, 440
Faults, 290
Faunal Realms, 410
Felspar, 41, 286, 310
Ferrel's Law, 89
Figure of the Earth, 82
Fingal's Cave, 266
Firths, 231
Fjords, 229
Flinders Range, 372
"Floating" of dust in air, 161
Floes, ice, 234
Floods in rivers, 324
Floral Zones, 405
Fog, 167
Foraminifera, 273
Forests, 407, 408, 417
Form, 30
Fossils, 343-345
Foucault's pendulum, 87
Frigid Zones, 125
Fringing reefs, 280
Function of lakes, 334; of living creatures, 398; of the sea, 250

GALL'S Projection, 445
Ganges River, 269, 318, 331, 383
Genus, 396
Geography, 430
Geoid, 83
Geological theories, 341
Geysers, 316
Ghats, Eastern and Western, 383
Giant's Causeway, 392
Glacial action, 351, 352
Glaciers, 336-338
Globigerina ooze, 273, 275
Gobi, desert, 381, 406, 428
Gradient (barometric), 175
Granite, 29, 41, 43, 310
Graphic representations, 444
Gravitation, 19, 36-38, 52
Great Basin, 365, 406
Great Divide, 369
Great Dividing Range, 371
Great Fault of Scotland, 390
Great Lakes, 333, 369
Great Plateau of Africa, 374
Great Salt Lake, 365
Greenland, 340, 363
Green mud, 271
Guiana, High Plain, 359
Guinea, current, 243
Gulf Stream, 244, 279
Gyroscope, 51

HAIL, 171
Halley's comet, 132
Heat, 65-74; in air, 163, 164; in rocks, 306; of the Sun, 118; in water, 227
Height of land, 369
Heredity, 403
Highlands of Scotland, 390
Himalaya Mountains, 303, 380
Hindu Kush, 379
Hoang Ho, see *Yellow River*
Hoar-frost, 165
Humidity, 158
Huron, Lake, 369
Hurricanes, 208

Index

The Figures refer to the sections.

Hydrosphere, 84, 214-282
Hygrometer, 441
Hypothesis, 18

ICE, 69, 336-340
Ice Age, 351, 352
Icebergs, 234
Ice-caps, 340
Ice deserts, 406
Igneous rocks, 287
Impenetrability, 34
Indian Ocean, 216, 248, 261
Indian peninsula, 383
Inductive reasoning, 17
Indus River, 380
Inertia, 50
International Deep, 258
Inverse squares, 36
Ireland, 392
Islands, 262; life on, 416
Isobars, 192
Isotherms, 188

JENOLAN Caves, 317
Joints in rocks, 290
Jupiter, 127, 130
Jura Mountains, 385
Jurassic rocks, 349

KALAHARI desert, 406
Karakorum Mountains, 380
Kepler's second Law, 109
Krakatoa, eruption of, 297
Kuen Lun Mountains, 380
Kuro Siwo, 247

LADOGA, Lake, 388
Lahontan, Lake, 365
Lake district of Europe, 388; of North America, 368
Lakes, 332-335, 339; circulation of water in, 228
Lambert's equal-area projection, 445
Land and sea breezes, 184
Land and sea climates, 191
Land and water, 214

Land sculpture, 305
La Plata River System, 362
Latent heat, 69
Latitude, 92
Lava, 294
Laws of Nature, 14, 20
Lena River, 382
Life in the world, 394
Light, 62, 64
Lightning, 173
Lignite, 347
Limit of saturation in rocks, 314
Lisbon earthquake, 301
Lithosphere, 84, 251 et seq.; interior of, 292
Llanos, 360, 406
Loam, 311
Loess, 308
Longitude, 97
Low plains, 264
Lukuga River, 376

MAGNETISM, 79; terrestrial, 98
Malay Archipelago, 407, 414-416
Malayo-Polynesian race, 424
Mammoth Cave, 317
Man, 418-436
Maps, 445
Mars, 127, 128
Mass, 28
Mathematics, 8
Matter, 24, 27, 48
Mean sphere level, 254
Measures, 437
Mechanical equivalent of heat, 74
Mekong River, 380
Mercator's projection, 445
Mercury (metal), 66, 146, 440
Mercury (planet), 127
Mer de Glace, 337
Meridians, 91
Mesozoic rocks, 349
Metamorphic rocks, 289
Meteorites, 135
Meteoritic hypothesis, 144
Meteors, 134

The Figures refer to the sections.

Mica, 41, 286
Michigan, Lake, 369
Micro-organisms, 401
Midnight Sun, 122, 150
Migration of peoples, 428, 429
Milky Way, 140
Mimicry, 409
Minerals, 285, 286
Miocene rocks, 350
Mirage, 150
Mississippi, 319, 322, 324, 331, 367
Mist, 167
Mixture, 41; of gases, 155; of rivers and sea, 230
Molecular vibrations, 59
Momentum, 50
Monsoons, 185, 195, 198
Moon, 100 et seq.
Moraines, 337
Morar, Loch, 339
Motion, first law of, 50; energy of, 53
Mountains, of accumulation, 295; of circumdenudation, 329; of elevation, 303.
Muds, oceanic, 269
Murray River, 372

NATURAL law, 14
Nature, 2, 13, 21, 23
Nautical Almanac, 20, 92
Neap tides, 114
Nearctic Realm, 411
Nebulæ, 141, 142
Nebular hypothesis, 143
Negrito race, 423
Negro race, 423
Neotropical realm, 413
Neptune, 127, 131
Nevis, Ben, 163, 390; Loch, 339
Niagara Falls, 330
Nile, 318, 325, 375
Nimbus cloud, 168
Nitrifying ferment, 401
Nitrogen, 151, 152
North America, 363-369, 411

North equatorial currents, 243, 247, 248

OB River, 382
Objective things, 5
Oceanic currents, 242-249
Oceans, 215, 216
Old Red Sandstone, 346, 390
Oligocene rocks, 350
Onega, Lake, 388
On-shore and off-shore winds, 241
Ontario, Lake, 369
Oolitic limestone, 349
Oolitic Ridge, 391
Oozes, oceanic, 273
Orange River, 374
Ordnance Survey maps, 446
Organic evolution, 402
Oriental Realm, 414
Orinoco basin, 360
Oxus, see *Amu Daria*
Oxygen, 151, 153
Ozone, 153

PACIFIC Ocean, 216, 247, 259
Palæarctic Realm, 411
Palæocrystic sea, 234
Palæozoic rocks, 346
Palestine, 434
Pamir, the, 379, 380
Pampas, 406
Paraguay River, 362
Parallax, 97
Parallels of latitude, 91
Parana River, 362
Peat, 347
Pendulum, 54
Pennine Chain, 391
People, see *Races*; of Europe, 426
Periodic law, 47
Permian rocks, 348
Perpetual motion, 55
Peru current, 247
Phases of the Moon, 101
Philosopher's Stone, 46
Photosphere, 116
Physiography, 1, 6, 22, 23

Index

The Figures refer to the sections.

Planets, 126 et seq.
Plants, 396, 399, 404, 417
Pleistocene rocks, 352
Pliocene rocks, 350
Polar currents, 245; seas, 234
Polarity, 79, 88
Pole star, 90, 137
Poles, of the Earth, 88; magnetic, 98; of a magnet, 79
Post Tertiary, see *Quaternary*
Prairies, 367, 406
Precession of the equinoxes, 115
Pressure, and change of state, 72; of atmosphere, 147; and sea-water, 226
Primary rocks, 346-348
Probability, 15
Prominences (solar), 116
Proteids, 399
Protoplasm, 398
Pteropod ooze, 274
Pumice, 294
Pyrenees, 385

QUARTZ, 30, 41
Quaternary rocks, 351

RACES of man, 422-427
Radiant energy, 60
Radiation and absorption, 63
Radiolarian ooze, 273, 276
Rain, 169, 312
Rain-band, 160
Rainfall, 200, 318; of world, 201; of British Islands, 204
Raised beaches, 284, 352
Reason, 8
Reaumur scale, 440
Red Clay, 277, 278
Red Sea, 279, 373; circulation, 237; temperature, 233, 236
Reflection and refraction, 61
Refraction, atmospheric, 150
Religion, 420
Rhine River, 386
Rhone River, 386
Rivers, 318-331; water, 221

"Roaring Forties," 180
Roches Moutonnées, 338, 351
Rock-basins, 332, 338, 339
Rocks, 285, 287-290
Rocky Mountains, 364
Roraima, Mount, 312
Rotation of the Earth, 87, 93, 94; of the Moon, 102

SAHARA desert, 377, 406
Saima, Lake, 388
St. Anthony Falls, 330
St. Elmo's fire, 172
St. Lawrence River System, 369
Salinity of the ocean, 223, 224
Salt Lakes, 335
Salts, 44; of river-water, 221; of sea-water, 222
Sargasso Sea, 246
Saturn, 127, 130
Scale of maps, 445
Science, 3
Sciences, scope of, 21, 22
Scientific method, 9, 18
Scoriæ, 294
Scotland, 390
Seas, classes of, 215; level, 252; water, 222-227
Seasons, 120-123
Secondary rocks, 349
Sedimentary rocks, 288, 304, 345-351
Seiche, 239
Seismometers, 301
Selvas, 361, 407
Senses and their use, 7
Shoals, 262
Shooting-stars, 134
Sidereal time, 94, 111
Sierra Madre, 364; Nevada, 364
Siliceous organisms, 273
Silurian rocks, 346, 390, 391
Simoom, 209
Slopes of continental edges, 263; of land, 356
Snow, 170

The Figures refer to the sections.

Snow line, 163
Sogne Fjord, 339
Soil, 311
Solar spectrum, 117; system, 126 et seq.; time, 111; tides, 114
Solstices, 122, 123
Sound, 58
Soundings, deep-sea, 443
South America, 357-362, 413
South equatorial currents, 243, 247, 248
Southern Ocean, 216, 249
Southern Uplands of Scotland, 390
Species, 396
Specific gravity, 29
Specific heat, 66, 227, 306
Spectrum, 62; analysis, 63; of comets, 133; of stars, 138; of Sun, 117
Springs, 314, 315
Spring tides, 114
Stars, 136 et seq.
States of matter, 68
Steam, 71
Steppe-lands, 406
Storm-warnings, 213
Strain, 35
Stratus cloud, 168
Stress, 35
Subjective things, 5
Subsidence and elevation of land, 284
Summer solstice, 122
Sun, 105 et seq.
Sunspots, 116
Superior, Lake, 333, 369
Synclines, 302, 303
Synthesis, 40
Syr Daria (Jaxartes), 382

TANGANYIKA, Lake, 376
Tarim basin, 381
Temperate forests, 408
Temperate zones, 125
Temperature, 65; of air, 187-191; of British Islands, 203; of Earth's crust, 291; of lakes, 228; of ocean, 233, 235; of river entrances, 232; of seas, 236
Tension, surface, 39
Terms, 12
Terraces, pink and white, 316; river, 321
Terrigenous deposits, 269
Tertiary rocks, 350
Theory, 18, 19
Thermograph, 440
Thermometers, 440
Thrust-planes, 302
Thunder, 173
Tian Shan Mountains, 381
Tibet Plateau, 380
Tibeto-Chinese race, 424
Tides, 103, 114; in bays, etc., 219; currents, 218; oceanic, 217
Time, 95, 96, 111
Tornado, 209
Torrens, Lake, 372
Torrents, 320
Torrid zone, 125
Trade winds, 179
Transitional area, 255, 263
Trias rocks, 349
Trigonometry, 33
Tropical, group of plants, 405; forests, 407
Tropics, 122
Tsad basin, 377
Tuff, volcanic, 295
Tundras, 406
Tunnels, transalpine, 432
Tuscarora Deep, 260
Typhoons, 208

UNCONFORMABILITY, 342
Underground water, 313; temperature, 291
Ural Mountains, 384
Uranus, 127, 131

VALLEYS, 321, 327, 328

Index

The Figures refer to the sections.

Vapour pressure, 157, 158
Variation, magnetic, 98; biological, 403
Venus, 127
Vernal equinox, 121
Victoria desert, 406
Victoria Nyanza, 375
Volcanic action, 293; eruptions, 296; materials, 272, 294
Volcanoes, 295, 298
Volga River, 387
Volume, 29

WATER, 66, 69-71, 220; work of, in Nature, 293, 309 *et seq.*; 313, 318
Watershed, 319
Waterspouts, 210
Water-vapour, 71, 156, 157
Wave-length, 62; motion, 57, 61; sea waves, 239, 265
Weather, 207; charts, 211; forecasts, 212
Weathering of rocks, 310
Weight, 38
Weights, 437
Wells, Artesian, 314
Whirlpools, 219
Whirlwinds, 209

White type of mankind, 425
Wind, 175; in British Islands, 202; and currents, 242; prevailing, 193-198; work of, 307, 308
Windings of rivers, 323
Winnipeg, Lake, 368
Winter solstice, 123
Work, 49, 52; of rivers, 327-331; of wind, 307, 308
World ridges, 256
Wrinkling of Earth's crust, 302
Wyville-Thomson ridge, 246

YANG-TSE-KIANG, 219, 319, 380
Year, 110
Yeast, 401
Yellow River, 324, 331, 380
Yellowstone Park, 316, 364
Yellow type of mankind, 424
Yenisei, river, 382
Yukon, river, 364

ZAMBESI, river, 374
Zodiac, 112
Zones, of climate, 125; of vegetation, 405; of winds and calms, 178

THE END

Printed by R. & R. CLARK, *Edinburgh*

2 B

.

www.ingramcontent.com/pod-product-compliance
Lightning Source LLC
Chambersburg PA
CBHW051723300426
44115CB00007B/433